METHODOLOGY
IN HUMAN GENETICS

Sponsored by

THE UNIVERSITY OF UTAH

THE GENETICS STUDY SECTION
DIVISION OF RESEARCH GRANTS AND
DIVISION OF GENERAL MEDICAL SCIENCES
NATIONAL INSTITUTES OF HEALTH *

* Grant RG 7097

METHODOLOGY *in* HUMAN GENETICS

Edited by

WALTER J. BURDETTE, A.B., A.M., Ph.D., M.D.

Professor and Head of the Department of Surgery and
Director of the Laboratory of Clinical Biology,
University of Utah College of Medicine;
Surgeon-in-Chief, Salt Lake County Hospital;
Chief Surgical Consultant, Veterans Administration Hospitals;
Salt Lake City, Utah

 HOLDEN–DAY, INC., *San Francisco*
1962

PREFACE

In the annals of science, the passage of half a century is temporally but a small fragment; many periods of this magnitude have witnessed an unimpressive list of scientific achievements. However, the last sixty years has encompassed the rebirth, development, and emergence of genetics as one of the most enlightening fields of biologic endeavor. Some of the oldest of man's questions have been answered and new ones posed; the vanguard of medical contributions is at hand; and the hope that some global, socio-economic problems may have a biologic solution is a possibility. With this turn of events has come a Babelic diversity of disciplines within the field, offering investigative opportunities exceeding the supply of available, trained intellect to utilize them effectively. At least temporarily, traditional methods of instruction must be supplemented in order to place the means in the hands of the able. Therefore, there is some justification in publishing a treatise on methodology which of necessity must be presumptuous in conception if modest in attainment. Not quite a complete compendium for experts nor a treatise for beginners, this book is hopefully offered as a beginning for those who seek to broaden the scope of their scientific inquiry. No doubt time will bring obsolescence to some of the contents and enhanced value to others. If additional progress in the field of genetics is the net result, the many who have made the offering possible will consider it worthwhile.

WALTER J. BURDETTE
Chairman,
Genetics Study Section

Salt Lake City, Utah
November, 1961

PARTICIPANTS

V. Elving Anderson, Ph.D.

> Perinatal Research Branch, Collaborative and Field Research, National Institute of Neurological Diseases and Blindness, National Institutes of Health, Bethesda, Maryland

Louis Baron, Ph.D.*

> Division of Immunology, Walter Reed Army Institute of Research, Walter Reed Army Medical Center, Washington, D.C.

W. C. Boyd, Ph.D.

> Department of Immunochemistry, Boston University, Boston, Massachusetts

J. A. Buckwalter, Jr., M.D.

> Department of Surgery, State University of Iowa, College of Medicine, Iowa City, Iowa

Walter J. Burdette, Ph.D., M.D.*

> Department of Surgery, University of Utah, College of Medicine, Salt Lake City, Utah

Barton Childs, M.D.

> Department of Pediatrics, Johns Hopkins University, School of Medicine, Baltimore, Maryland

Rody P. Cox, M.D.

> Department of Research Medicine, Hospital of the University of Pennsylvania, University of Pennsylvania, Philadelphia, Pennsylvania

James Crow, Ph.D.*

> Department of Medical Genetics, University of Wisconsin, Madison, Wisconsin

Virginia J. Evans, Sc.D.

> Tissue Culture Section, Laboratory of Biology, National Cancer Institute, National Institutes of Health, Bethesda, Maryland

C. E. Ford, D.Sc.

> Medical Research Council, M. R. C. Radiobiological Research Unit, Atomic Energy Establishment, Harwell, England

Eldon Gardner, Ph.D.

> Department of Zoology, Utah State University, Logan, Utah

Francis B. Gordon, Ph.D.*

> Naval Medical Research Institute, National Naval Medical Center, Bethesda, Maryland

John W. Gowen, Ph.D.

> Department of Genetics, Iowa State University, Ames, Iowa

Earl L. Green, Ph.D.*

> Roscoe B. Jackson Memorial Laboratory, Bar Harbor, Maine

T. S. Hauschka, Ph.D.

> Roswell Park Memorial Institute, Buffalo, New York

C. Nash Herndon, M.D.

> Department of Preventive Medicine and Genetics, Bowman-Gray School of Medicine, Wake Forest College, Winston-Salem, North Carolina

Leonard Herzenberg, Ph.D.

Department of Genetics, School of Medicine, Stanford University, Palo Alto, California

Walter E. Heston, Ph.D.*

National Cancer Institute, National Institutes of Health, Bethesda, Maryland

R. L. Hill, Ph.D.

Department of Biochemistry, Duke University, Durham, North Carolina

Richard T. Jones, Ph.D.

Division of Experimental Medicine, University of Oregon Medical School, Portland, Oregon

H. Warren Kloepfer, Ph.D.

Department of Anatomy (Genetics), School of Medicine, Tulane University, New Orleans, Louisiana

Joshua Lederberg, Ph.D.*

Department of Genetics, School of Medicine, Stanford University Medical Center, Palo Alto, California

A. M. Lilienfeld, M.D.

Division of Chronic Diseases, Johns Hopkins School of Hygiene and Public Health, Baltimore, Maryland

Colin M. MacLeod, M.D.

Department of Research Medicine, New York University, New York, New York

N. E. Morton, Ph.D.

Department of Medical Genetics, University of Wisconsin, Madison, Wisconsin

Arno G. Motulsky, M.D.

   Department of Medicine, University of Washington, Seattle, Washington

J. V. Neel, M.D., Ph.D.

   Department of Human Genetics, University of Michigan, School of Medicine, Ann Arbor, Michigan

H. B. Newcombe, Ph.D.

   Atomic Energy of Canada, Limited, Chalk River, Ontario, Canada

Edward Novitski, Ph.D.

   Department of Biology, University of Oregon, Eugene, Oregon

Ray D. Owen, Ph.D.*

   Division of Biological Sciences, California Institute of Technology, Pasadena, California

Klaus Patau, Ph.D.

   Department of Pathology, University of Wisconsin, Madison, Wisconsin

Theodore T. Puck, Ph.D.

   Department of Biophysics, University of Colorado, Denver, Colorado

Charles M. Rick, Jr., Ph.D.*

   Department of Vegetable Crops, University of California, College of Agriculture, Agricultural Experiment Station, Davis, California

J. A. Fraser Roberts, M.D., D.Sc.

   Clinical Genetics Research Unit, M. R. C. Institute for Child Health, Hospital for Sick Children, London, England

Katherine K. Sanford, Ph.D.

   Tissue Culture Section, Laboratory of Biology, National Cancer Institute, National Institutes of Health, Bethesda, Maryland

Jane Schultz, A.B.

Department of Medicine, Johns Hopkins University, School of Medicine, Baltimore, Maryland

Arthur G. Steinberg, Ph.D.

Department of Human Genetics, Western Reserve University, School of Medicine, Cleveland, Ohio

Gunther Stent, Ph.D.*

Virus Laboratory, The University of California, Berkeley, California

Curt Stern, Ph.D.

Department of Zoology, University of California, Berkeley, California

Wilson S. Stone, Ph.D.*

Genetics Foundation, The University of Texas, Austin, Texas

H. Eldon Sutton, Ph.D.

Genetics Foundation, The University of Texas, Austin, Texas

J. H. Tjio, Ph.D.

National Institute of Arthritis and Metabolic Diseases, National Institutes of Health, Bethesda, Maryland

Katherine S. Wilson, Ph.D.*

Genetics Study Section, Division of Research Grants, National Institutes of Health, Bethesda, Maryland

Charles M. Woolf, Ph.D.

Department of Zoology, Arizona State University, Tempe, Arizona

George Yerganian, Ph.D.

Children's Cancer Research Foundation, Boston, Massachusetts

* Member, Genetics Study Section.

CONTENTS

ANALYSIS OF HUMAN HEREDITY

A. M. Lilienfeld, M.D.

SAMPLING TECHNIQUES *and*
SIGNIFICANCE TESTS

Two specific topics of importance to genetic investigations of the human population which do not generally receive much, if any, attention have been chosen for discussion. The first concerns certain methods and problems involved in sampling human population groups. In studies of human heredity, it is often desirable to obtain samples of either individuals or families for estimating the frequency of a condition or to serve as a control group for comparative purposes with groups of people with a disease. Despite the clear need for population sampling, it is quite surprising how often data are collected from highly selected or biased groups of the population and treated as if they were probability samples. Since investigators often think that adequate sampling is difficult or very costly, a brief review of some relatively simple procedures for drawing samples of human populations is in order. Within the past few years, several excellent texts on this subject have appeared and they should be consulted for a detailed discussion.[145, 950]

SAMPLING TECHNIQUES

"Probability sampling" is a term applied to samples selected in such a way that every member of the population has a known probability of being chosen which is not zero for any member of the population. Sampling is preferred since sampling theory for estimating standard errors, confidence limits, and so on, can be utilized only for this type of sample.

The probability sample most familiar is the simple, random type. To draw a 10 per cent sample of persons from a community with a total population of

3

1,000, a list of names of persons living in this community is taken, numbered from 0 to 999, referred to a table of random numbers, and 100 numbered and associated names are selected in a familiar, prescribed procedure.

The major practical problem facing the investigator is that lists of names of a population in the community are often not available. Consequently, in most human-population surveys, use is made of "cluster sampling." In this method the individual is not the sampling unit, but a cluster or aggregate of persons compose the unit that is sampled. This type of sampling procedure takes advantage of the fact that people usually live in various types of social groups. Also, it would seem to be more appropriate in genetic studies of man, since interest in families is greater than in individuals.

The form of cluster sampling used most frequently is "area sampling," in which the clusters consist of individuals or families living in small geographic areas. For this type of sampling, the only requisite is a map to provide the list of areas. For example, an urban map showing blocks may be used. A certain number of blocks is selected at random (the number depending upon the size of sample desired, et cetera). Then all the households on the designated blocks are listed, and a sample of these households selected. Since the probability by which each block was selected can be determined along with the probability by which each household was chosen within the selected blocks, the probability of each household being drawn into the sample can be given. To illustrate, suppose 100 city blocks were selected from a list of 1,000 blocks and every household on the selected blocks was studied. Clearly each block had 100 out of 1,000 chances or 1 in 10 of being included in the sample. Once a block was chosen, each household was studied. Therefore, each household in the population had a probability of 1 in 10 of being included in the sample. If it were decided for various reasons not to interview all the households in the blocks and only 10 per cent of households in each selected block were interviewed, the probability of any household in the city being included in the sample would be $\frac{1}{10}$ (probability of selecting a block) multiplied by $\frac{1}{10}$ (probability of selecting a household); thus, each household would have a probability of 1 in 100 of being included in the sample.

Many variations of cluster sampling are possible, particularly with respect to the ways in which clusters are formed and how individuals within clusters are selected. Obtaining a list of clusters can be simplified if a recent city directory is available. The use of such a directory was found quite satisfactory by Woolsey in a recent study in Hagerstown.[941] We also used directory sampling in a survey carried out in Buffalo in 1956.[508] Most objections to the adequacy of a directory have been concerned with using the directory as a list of names of individuals and households, for which it is inadequate. However, the directory is excellent as a list of addresses in the city, each address representing a primary sampling unit. Either all households at each address may be included in the sample or a subsample of households at an address may be obtained. If sampling is being done in a city undergoing a large amount of new construction, it is well to supplement the directory listing with the addresses of the construction, obtained from the Bureau of Buildings issuing building permits. Since there is still a possibility that some addresses may be omitted from the directory list, Yates has suggested the use of the "half-open interval" method to take into account those omissions.[950]

In our Buffalo survey the accuracy of the directory list of addresses was investigated. All addresses in a sample of city blocks were compared with the directory listings. Approximately three per cent of the addresses had been omitted from the directory. However, use of the "half-open interval" provided half these omissions, so that the method omitted 1.5 per cent of addresses from the sample. In view of sampling variation and the errors that result from listing operations, this was considered an acceptable degree of error. Another possible source of addresses is the listing for property taxes, which consists of all structures for tax purposes. No doubt, such lists are kept up to date and are complete.

The important point to be emphasized in this brief discussion on techniques of sampling is that there are methods available for obtaining adequate samples of human populations and that these methods are not difficult or too costly. Therefore, no reason exists for failing to obtain an adequate population sample when it is clearly required in a particular study.

SIGNIFICANCE TESTS

The second topic of importance relates to statistical methods developed for testing conformity of observations with a given genetic hypothesis. Our attention was drawn to related problems as a result of some work with a specific type of statistical test. This test was used by Steinberg and Wilder [832] in analyzing the inheritance of diabetes mellitus, which was similar to one used by Allan in 1933 [4] to study the same disease entity, although it has not been used to any great extent during the intervening years. Dahlberg also has discussed this particular approach in some detail.[832] Steinberg and Wilder collected data from a clinic group of diabetic patients who supplied information on the age and history of diabetes in their parents and siblings. The data are used to test the hypothesis that diabetes is due to a single recessive gene under the following assumptions: (1) random mating, (2) equal fertility of all types of matings, (3) equal viability of all offspring, and (4) independence of age of onset of diabetes in the affected offspring from the type of parental mating.

Assuming that the clinic population is equivalent to a random sample of diabetic offspring from a general population, the following test for the simple, recessive gene is made:

1. The number of each of the three types of parental mating is determined: i.e., (a) the number of patients, neither of whose parents had diabetes, (b) the

number of patients, one of whose parents had diabetes, and (c) the number, both of whose parents had diabetes. These represent the observations.

2. The gene frequency, denoted by p, is determined from the frequency of diabetes among the parents of the diabetic patients by standard methods.

3. The relative distribution of each of the mating types that is expected on the basis of the recessive gene hypothesis, should be q^2, $2pq$, and p^2 for the respective matings: (a) neither-parent-affected, (b) one-parent-affected, and (c) both-parents-affected matings, where $q = 1 - p$.

4. The observed distribution of parental mating types is tested by means of a chi-square test to see if it agrees with the expected distribution.

The derivation of the test, which is designated the "binomial test," is presented in table 1. It is based on the Hardy-Weinberg equilibrium, from which

Table 1

DERIVATION OF BINOMIAL MODEL FOR TESTING RECESSIVE GENE HYPOTHESIS

Number of affected parents	Parental genotypes	Population frequency of parental matings by genotypes (a)	Frequency of parental matings in total population who are capable of producing affected offspring (b)	Proportion of recessive offspring in each mating class (c)	Frequency of affected offspring derived from parental mating class in total population of offspring $(d) = (c) \times (b)$	Relative distribution of parental mating classes among affected offspring $(e) = \dfrac{(d)}{p^2}$
0	$AA \times AA$	q^4				
	$AA \times Aa$	$4q^3p$				
	$Aa \times Aa$	$4p^2q^2$	$4p^2q^2$	$\frac{1}{4}$	p^2q^2	q^2
1	$AA \times aa$	$2p^2q^2$				
	$Aa \times aa$	$4p^3q$	$4p^3q$	$\frac{1}{2}$	$2p^3q$	$2pq$
2	$aa \times aa$	p^4	p^4	1	p^4	p^2

the frequencies of those parental matings capable of producing affected offspring are determined. These frequencies are then reduced by the ratios, $\frac{1}{4}$, $\frac{1}{2}$, and 1, to obtain the frequency of affected offspring in each parental mating class in the total population of affected offspring. The relative distribution of each parental mating class among the affected offspring is then determined by division of the frequencies previously obtained by p^2. The resultant frequencies are distributed binomially. Essentially, this model determines the relative distribution of the three different types of parental mating classes among a group of affected individuals, provided the disease is due to a recessive gene.

Steinberg and Wilder applied this test to other published data on diabetes. Table 2 is reproduced from their paper containing the results of this application.

Table 2

COMPARISON OF EXPECTED FREQUENCIES OF MATINGS YIELDING DIABETIC
OFFSPRING WITH OBSERVED FREQUENCIES FOR FOUR SETS OF DATA *

Parental mating class	Steinberg and Wilder		Pincus and White		Allan		Harris	
	Ex-pected	Ob-served	Ex-pected	Ob-served	Ex-pected	Ob-served	Ex-pected	Ob-served
Both parents diabetic	21.6	22	3.6	3	0.8	2	3.1	8
One parent diabetic	370.8	370	78.8	80	19.4	17	118.8	109
Neither parent diabetic	1,588.6	1,589	440.6	440	122.8	124	1,119.1	1,124
χ^2(1 d.f.)	0.009		0.119		2.109		8.573	
P	>0.90		>0.70		>0.10		<0.01	

* From Steinberg and Wilder. [832]

In four sets of data, three fit the model for a recessive gene, as indicated by P values of more than .05; whereas one set does not fit the model, the P value being less than .01.

We were interested in the adequacy of this particular test in discriminating between genetic and nongenetic factors, since we thought that it might be possible to obtain a similar kind of model on nongenetic assumptions. Dr. Paul Meier and I examined this possibility on theoretical grounds, and it became apparent that some nongenetic causes of familial aggregation may be represented by a binomial model. However, the most reasonable way of studying this was to obtain similar data on an individual characteristic that occurred with a certain degree of familial aggregation primarily due to nongenetic social or environmental conditions. These data could then be analyzed as if one were attempting to determine whether this characteristic were inherited as a recessive gene by means of the binomial test. Two rather similar characteristics were investigated in this manner.

The first characteristic studied was attendance at medical school. By means of a questionnaire, all students attending the University of Buffalo Medical School were asked whether or not their parents were physicians. The questionnaire was completed by 261 students; less than 20 students did not return the questionnaire. Thus, the characteristic was considered as attending or having attended a medical school some time during an individual's span of life. Table 3 contains the distribution of the students by the following three parental classes: (1) neither parent had attended a medical school, (2) one parent had attended a medical school, and (3) both parents had attended a medical school. From these data, the equivalent of the gene frequency, p, was estimated as .0613. Expected numbers based on a binomial distribution were then obtained, and chi-square test of goodness of

Table 3

COMPARISON OF OBSERVED AND EXPECTED NUMBERS OF PARENTAL MATING
CLASSES AMONG OFFSPRING ATTENDING MEDICAL SCHOOL

Parental mating class	Number of medical students in mating class	
	Observed	Expected
Both parents physicians	0	1.0
One parent physician	32	30.0
Neither parent physician	229	230.0
Total	261	261.0
p = "gene frequency"	0.0613	
Chi-square (1 d.f.)	1.14	
P	$0.30 > P > .20$	

fit of the observed numbers with expectations resulted in a probability level of between 0.2 and 0.3. From a purely formalistic viewpoint, the inference could be made from these data that attendance at medical school was consistent with a recessive-gene hypothesis. It is quite apparent that the test of the genetic hypothesis was not able, in this instance, to distinguish a type of familial aggregation that is principally determined by socio-cultural factors.

Because the number of individuals in the sample was small and the parental mating class with two affected parents was absent, another characteristic was selected for a similar analysis. This characteristic was defined as attendance at the University of Buffalo, and all students attending the University were considered as being affected. At the time of registration at the University all students were asked to indicate on the registration card whether or not any of the members of their family previously attended the University of Buffalo. On the basis of this information the following three types of parental matings may be defined: (1) neither parent had previously attended the University, (2) one parent had attended the University, and (3) both parents had attended the University.

The distribution of all 3,886 students by type of parental mating is presented in table 4. The gene-frequency, p, was estimated, binomial expectations were computed, and a chi-square test applied to the data. Chi-square was quite high at a probability of less than .001, indicating a lack of fit between expectancy and observation. On the basis of this lack of fit, the inference is made that attendance at the University is not consistent with a recessive gene hypothesis. However, since investigations of human diseases rarely deal with a group as large as 3,886 affected individuals, it was considered desirable to see what effect smaller samples would have on probability. This was studied by sampling the population of the University students and determining the proportion of samples that have a good fit with binomial expectations. Such a sampling experiment was carried out with random number cards and an electronic computer.

Table 4

COMPARISON OF OBSERVED AND EXPECTED NUMBERS OF PARENTAL MATING
CLASSES AMONG OFFSPRING ATTENDING THE UNIVERSITY OF BUFFALO

Parental mating class	Number of students in each mating class	
	Observed	Expected
Both parents attended University of Buffalo	14	4.4
One parent attended University of Buffalo	232	251.6
Neither parent attended University of Buffalo	3,640	3,630.0
Total	3,886	3,886.0

p = "gene frequency"	0.0335
Chi-square (1 d.f.)	22.51
P	<0.001

Two sizes of samples were investigated: a sample of 900 representing approximately 25 per cent of the population and one of 1,900 representing approximately 50 per cent of the population. For each size, the computer generated 100 random samples. For each sample of each size, a p value was computed and binomial expectations obtained. Also chi-squares for each sample were determined. Table 5 contains the frequency of samples by various chi-square values, together with their probability.

Table 5

DISTRIBUTION OF SAMPLES DERIVED BY EXPERIMENTAL SAMPLING BY CHI-SQUARE
AND PROBABILITY VALUES FOR SAMPLE SIZES OF 900 AND 1,900

Chi-square values (1 d.f.)	P	Per cent distribution of samples Sample size	
		900	1,900
<3.841	>0.05	54	27
3.842–6.635	0.05<P<0.01	13	11
6.636–10.827	0.01<P<0.001	12	14
>10.827	0.001	21	48

In 54 per cent of samples of size 900 (table 5), chi-square was less than 3.841, with probability values greater than .05. Expressing this in terms of the genetic hypothesis, this indicates that, in 54 per cent of instances, a sample of 900 would have a result consistent with a recessive gene, even though the population data, from which this sample was drawn, was not consistent with a recessive gene. When the sample size increased to 1,900, the proportion of samples that fit expectations decreased to 27 per cent at levels of 5 per cent probability. In other words, given an individual sample of size 900 in which a good fit with expectation is obtained, a genetic inference would be wrong about 54 per cent of the time, and, with a sample size of 1,900, the chance of error would be about 27 per cent. The lack of discriminatory ability of the chi-square test in this situation probably results from the small numbers expected in the category of "both parents affected" and this in turn is a function of the estimated gene frequency, p.

The importance of these data lies in the results obtained by the sampling experiment. They are a specific example of what is termed the "power" of a test in statistical theory. This concept will be reviewed briefly and in a general manner. Details can be found in most texts on statistics. In testing hypotheses, one is liable to make two types of errors, as follows:

1. Type I is the error made if one rejects the null hypothesis when the null hypothesis is true.

2. Type II is the error incurred if one accepts the null hypothesis when the null hypothesis is false.

The usual significance test which is familiar to all provides us with the probability of making an error of type I. This probability is usually represented as *alpha* (α). Thus, when one fits expectancies to observations and uses a chi-square test, the resulting probability that is determined is that of making an error of type I. The probability of an error of type II is usually expressed as *beta* (β), and $1 - \beta$ is known as the power of the test. These two types of errors are related to each other in that, for a fixed number of observations, *beta* will be automatically determined if *alpha* is determined. Also, for a fixed number of observations, *beta* will be increased if *alpha* is decreased. To decrease both *alpha* and *beta,* the number of observations must be increased.

The above terminology may be somewhat confusing when translated into terms concerning the testing of genetic hypotheses. In this situation, acceptance of the null hypothesis really represents acceptance of a genetic hypothesis, and we should substitute the term, "genetic," for "null" in the above definitions. It is clear that the results of the sampling experiment indicate in broad and general terms that the error of type II in this particular test is high. As was indicated earlier, this may be partially a result of the low gene frequency. It would be interesting to see the results of similar sampling experiments with different frequencies of genes and with other sizes of samples. Such results would be of assistance in determining the size that should be used in a particular investigation.

However, these results have a broader significance. When various statistical methods available to the human geneticist today are reviewed, it is apparent that the power of most of these tests is not known. In view of the results obtained with respect to one particular test, it would seem highly desirable to determine the discriminating sensitivity of other testing procedures. This could be done either by mathematical analytical methods or by experimental sampling, using an electronic computer.

This attempt to outline procedures for sampling the human population indicates the need for additional statistical research into methods for genetic investigation in man. Perhaps this discussion will stimulate further studies of these problems in methodology.

DISCUSSION

DR. BURDETTE: Dr. Morton will open the discussion of Dr. Lilienfeld's paper.

DR. MORTON: Dr. Lilienfeld is to be congratulated for presenting a very clear discussion of the epidemiologic problems in human genetics. The conjunction of these problems and genetics is quite a recent development, and interest in them as well as financial support for studying them are even more recent. It is clear that the synthesis of these two disciplines is only beginning. There are problems in genetics which can be handled only by epidemiologic methods, and there are other problems for which the classical epidemiologic methods are grossly unsatisfactory.

Dr. Lilienfeld has urged the use of good sampling techniques. No one should question this, but I will argue that this is not an acute problem in human genetics and that in many cases the genetic method offers techniques more precise and valuable than epidemiologic methods, even for such standard epidemiologic problems as determining prevalence and incidence. This could easily become a rather empty, invidious discussion, which it should not be, since each discipline provides methods that can be used to advantage. This disagreement has arisen, I think, because of failure to specify the genetic problem more closely. For example, when we are dealing with a frequency of the order of $\frac{1}{15,000}$, in the case of a rare, genetic condition such as albinism, determination of the incidence by taking a random sample of the population is out of the question because of the enormous sample required. Besides, questions about the randomness of the sample, accuracy of diagnosis, and prior death of affected individuals would inevitably arise. In such a case, I defy any epidemiologist to demonstrate that the great cost of taking a population sample would give more precise information on incidence than the much more efficient techniques which the geneticist has developed to cope with problems of incomplete ascertainment. Furthermore, most geneticists would probably argue that precise knowledge of population incidence is far from indispensable for a genetic study and not worth the labor of collecting a random sample of the size required.

Now let us turn to the other extreme of a relatively common, highly heterogeneous condition such as hypertension. Here the geneticist must yield completely to the epidemiologist, at least on the problem of determining prevalence and incidence. It is easy to cite examples in which the objectives of the geneticist and epidemiologist are opposed. One is the National Health Survey,[892] which satisfies all the strictures Dr. Lilienfeld has laid down. It is a completely representative sample of the American population. However, once the sample is chosen, the survey consists of sending rather untrained interviewers to these carefully selected houses and inquiring of their occupants whether they are ill and what they think is wrong with them. It will take many years of careful study and enormous expenditure to determine just how unreliable these data are. This is a clear case of sacrificing diagnostic accuracy to the statistical fetish of a random sample. It is difficult to imagine how any information of biologic value can be extracted from such a source. The study of cerebral palsy also has all the disadvantages of a rather large and ill-conceived program.

I agree with Dr. Lilienfeld that the studies of diabetes have lacked power. However, it should be said in their defense that they provide other tests of the mode of inheritance based on predicted risks for each mating type. Both the statistical methods of human genetics and the biochemical tests to detect carriers have improved very much in the last few years. To study the genetics of diabetes today, one would use different procedures.

Again, I want to raise the question of how the areas are defined where epidemiologic methods are uniquely satisfactory, genetic methods are superior, or both may be used to advantage. The epidemiologist usually regards an accurate estimate of the population incidence as indispensable and precise knowledge of familial incidence as less important. To him complete ascertainment is almost essential. He is usually concerned with heterogeneous disease categories tabulated in morbidity and mortality lists. Rare diseases not recognized in these lists are much less interesting to him. Because his categories are heterogeneous, he can afford to sacrifice diagnostic precision for representativeness of the sample. The geneticist finds the population incidence of secondary interest, whereas precise knowledge of the familial incidence is usually essential. He is seldom interested in common disease categories and often works with conditions that present unusual diagnostic problems such as galactosemia, primaquine sensitivity, or hemoglobin G. Geneticists and epidemiologists must be perpared to exchange techniques, but a real synthesis of methods and interests has not yet been achieved.

DR. LILIENFELD: Dr. Morton's comments on the United States Health Survey are rather broad,[892] and some epidemiologists have also been critical of this survey. As he pointed out, most of the survey data is obtained by interview. Evaluation of morbidity surveys conducted in Baltimore and Hunterdon County, New Jersey, have indicated that information from interviews on morbidity is inadequate from the viewpoint of certain epidemiologic studies.[132] The individuals responsible for the National Health Survey are very much aware of these problems, and they are working out methods and procedures for conducting physical examinations on a subsample of the interviewed group to validate information obtained by interview. They have already done this on a trial basis in three areas of the country.

Admittedly the National Health Survey may be inadequate for the kinds of data epidemiologists and geneticists find of interest, but the Survey has a broader function. For example, it provides information of value in planning medical-care facilities, such as hospitals, and is not designed for epidemiologic or genetic research alone.

With respect to the cerebral palsy study that was mentioned, I do not know how many are familiar with this study that is being conducted in approximately twenty medical centers in this country and that will continue for the next fifteen or twenty years. Dr. Morton said it was ill conceived; I think the basic idea is a good one. Despite a number of difficult administrative problems, a fair amount of useful information should be derived from this particular study with respect to malformations as well as other characteristics and diseases.

Dr. Morton also commented on the use of population sampling for rare conditions. No doubt, this is a problem. Taking a random sample to determine the incidence of albinism is not feasible and does not make sense to me. In studying a rare condition other methods must be used. On the other hand, I think the same principles apply, though, even to those rare conditions in the sense that no matter how these cases are ascertained, they should be related to the population from which they come. The analyses of much of the original data on albinism were based on the collection of reports from the literature. One may excuse this approach on the basis of the rarity of the condition. On the other hand, the population from which the cases were obtained and the bias of the selection remains unknown. Inferences made from compilation of reports in the literature are questionable, and this method is rarely used in genetic studies today.

Dr. STEINBERG: Dr. Lilienfeld has pointed out, as noted when Dr. Morton talked about diabetes, the question concerns not diabetes but the method. I agree that the binomial test is weak, but I want to mention two omissions in Dr. Lilienfeld's analysis which he and I have been discussing since 1952 when my study was published.[832] One is that the geneticist does not ask the proband simply how many parents are affected with the disease or how many parents attended the University of Buffalo and then proceed to do the analysis. He asks another very important question: "How many siblings are affected?" Unless there is a suggestive ratio, namely, approximately the ratio of one to two to four, with neither parent, one parent, or both parents affected, respectively, he does not go further with the above technique.

Dr. Lilienfeld pointed out to me that, if the children are in the ratio of one to two to four, then of necessity the distribution of the parental matings will be in accordance with the binomial. Hence this is not an independent test of recessive inheritance, and I agree that the method is weak for that reason; but I think he has left out the basic point that leads a geneticist to look into subsequent relations, namely, the frequency in the family, not the number of people affected.

Dr. Gordon Allen of the National Institute of Mental Health has called attention to the fact that the method of analysis used by Steinberg and Wilder [832] is fundamentally a test for random mating (i.e., without regard to the presence of diabetes) among the parents of the diabetic probands and not purely a test for the frequency of affected offspring among the different matings as Dahlberg and Hultkranz, who originally derived the method, and as we, who derived it independently, believed.

Dr. LILIENFELD: I am not specifically interested in the ratio. In most of the statistical methods available in the literature for testing genetic hypotheses, such as those developed by Fisher, Haldane, and others, there has been no regard for their power in terms of their ability to discriminate genetic from nongenetic hypotheses.[233, 315] The errors of type II in these tests have not been examined, and I believe that many of the procedures used in human genetic studies have very low power, although some may have very high power. Here is a type of error that has not been investigated statistically, and it seems to me that it should be.

One particular method comes to mind, which you, Arthur, pointed out as having very low power.[828] This is a method suggested by Haldane for use in cases of incomplete multiple selection. He suggested that p (frequency of affected offspring) be estimated twice: once on the assumption of complete ascertainment and once on the assumption of single ascertainment. These two estimates provide confidence limits, and, if these limits include 0.25, the recessive-gene hypothesis is accepted. As you pointed out, these confidence limits are very broad. This broad range suggests that this approach has very low power and the method would not have a very good chance of excluding nongenetic hypotheses or genetic hypotheses other than the recessive one. This is the important point I would like to make. I do not think it really is a dead issue, generally speaking. The dead issue is the binomial distribution, since I am not particularly interested in this binomial test. I am using this because it is the only data we have where we are able to show that the power of a particular genetic test is somewhat low at the moment. This is also a general criticism of most of the tests available to the human geneticist today.

DR. KLOEPFER: The participants are to be commended for directing the discussion to a genetic problem that is as common as diabetes; the mathematical analysis of a rare genetic entity from a population isolate would be less open to criticism. In the discussion on diabetes, the assumption has been made that all cases are caused by the same mutation. The question that one should consider, it seems to me, is whether it is feasible to attempt to analyze, in one batch, data that probably represent effects of a number of different genotypes at a number of different chromosomal loci. The study under discussion has assumed that all genes causing diabetes follow the same pattern of inheritance. Actually, some genes for diabetes may be dominant and others recessive, and there is an additional problem of penetrance for each major gene involved.

This problem in analysis is mentioned in order to suggest a procedure for securing data which could be analyzed by the mathematical methods being discussed. This consists of finding population isolates in which diabetes is a rare disease and describing the various specific major genes as separate entities in one isolate at a time. Unfortunately, the location of ideal population isolates for the study of diabetes is unknown. One possible outcome of the cerebral-palsy study may be a method to locate population groups having advantages for use in special studies. Perhaps a method can then be found to locate population isolates in which only one major gene is contributing to the occurrence of diabetes. A mathematical analysis of the type being discussed would be valid when applied to data collected from such a population isolate.

DR. STEINBERG: I would like to ask Dr. Lilienfeld whether the power of a test is increased if a relative and not very powerful weakness is repeated on a series of samples and the fit is as good as $p = 0.5$. What is the significance of the result when this test, which is known to be weak, is used repeatedly on independent samples?

DR. LILIENFELD: I do not think it does anything to the power of the specific, statistical test. Clearly, one can be more confident in the inferences made from a series of studies with similar results than from a single study.

DR. STEINBERG: Then the geneticists need not discard all the data they have collected even if the tests used were poor.

DR. MORTON: I would like to go back to the Haldane method [315] to which Dr. Lilienfeld referred. It was introduced at a time when human geneticists depended a great deal on case histories from medical literature, where the degree of ascertainment is usually unspecified. The disadvantages in this source of material were common knowledge, but medical coöperation and financial support for better studies were often not available. Now geneticists have almost abandoned this method because they seldom cull families from the literature, and methods have been devised to extract much more information when the degree of ascertainment is specified. Therefore this particular method is quite obsolete. Haldane would be among the first to support this view, since he was among the first to draw attention to the limitations of unsystematic collections of pedigrees.

Although the National Health Survey [892] may not be the best example of an epidemiologic study, it still illustrates very well the points I have in mind. After all, if one simply wanted to get a rough notion of various administrative and possible scientific problems involved in ill health, the logical way of doing it would be through a morbidity register, supplemented by intensive follow-up studies. This is the pattern that birth and death registries have followed, and, thanks largely to legal sanction, they have come to satisfy the standards of even the most fanatical statistician with regard to completeness of the sample. Because of the size of such registries, they have a value for the ascertainment of rare diseases that no small, random sample could match. Of course, intensive follow-up is essential for most conditions, for either genetic or epidemiologic studies. By too parochial an interest in special surveys, by too great an emphasis on the initial incompleteness of a morbidity register, the epidemiologist has sacrificed the advantages of a morbidity register, immediately for ascertainment of rare diseases and ultimately, perhaps, as a complete sample of them. Only by separating the two functions of ascertainment and intensive study can reasonable diagnostic standards be maintained in collecting and studying a large sample of a rare disease.

DR. LILIENFELD: It is very nice to have disease registers. However, I have worked in health departments in areas where communicable diseases and selected chronic diseases are reportable by law. The results of reporting have, on the whole, been poor. For example, studies have shown that the percentage of cases of measles, chicken pox, or whooping cough reported by physicians have been of the order of 30 to 40 per cent.[860, 861] For some diseases, such as diphtheria, reporting is better, of the order of 80 per cent. In upstate New York where cancer is a legally reportable disease, about 80 per cent of cases are reported. This includes reporting by pathology laboratories and hospitals, and therefore reporting by physicians is somewhat less than this. Also, the reporting is improved by the fact that whenever a death is certified as being a result of cancer, it is added in the register when it has not previously been reported as a case of cancer. In this situation, one has difficulty in visualizing the value of establishing a register for reporting all kinds of diseases on a routine basis.

This statement should be amended for diseases customarily requiring hospital care, since a case register based on reports of hospitalized cases could be set up. Of course, the possibility that all diagnosed cases of a disease are not hospitalized remains. It is often difficult to obtain the coöperation of a large number of physicians to report cases with a disease. Sometimes, if patients with certain diseases are usually referred to a group of specialists and this is a small group in a particular community, then it may be easy to set up a reporting system with a relatively small group of physicians. For example, in Baltimore where we are conducting a study on leukemia, it is our impression at the moment that practically all diagnosed cases of leukemia are referred to the hematologists in the city. This permits us to set up a system of reporting with the coöperation of the hematologists.

Now that we have mentioned leukemia, we can illustrate how a community study can be conducted with respect to a rare disease, since leukemia is uncommon. In the kind of study with which we are usually concerned, we are interested in making comparisons between cases with the disease with a group of controls. We establish our reporting system which provides us with the diagnosed cases in the community. We then take a probability sample of the general population which provides us with our control group. This will permit us to make the best types of inferences from our data. What happens too frequently in these types of studies is that the type of control that is selected is usually not a sample of any known population. Usually one obtains the most accessible group such as medical students, the investigator's friends, and so on. One is never certain whether this type of control leads to biased inferences. It seems to me that the logic of most hypotheses in epidemiologic and human genetic studies almost always require community studies with probability samples at some point in a series of investigations. Admittedly, as I have pointed out elsewhere, nonprobability samples, such as hospital populations, are important in the initial phase of the investigation of an hypothesis,[509] but, after this exploration, confidence in results can be immeasurably increased by carrying out adequately controlled, community studies.

N. E. Morton, Ph.D.

SEGREGATION *and* LINKAGE

The study of human biology has many resources, including the methods and results of medicine, biochemistry, genetics, anthropology, immunology, cytology, and tissue culture. The assets are commensurate with the difficulties, for which a characteristic statistical methodology has been developed. At the risk of oversimplification, these methods may be divided into three types, which for brevity will be termed experimental, epidemiologic, and genetic.

The experimental method achieves validity by eliminating many sources of variation deemed irrelevant and randomizing the remainder. The epidemiologic method, lacking randomization and experimental control, seeks to achieve group comparability (1) by assuming that uncontrolled factors are irrelevant or fortuitously randomized (the least conservative approach); (2) by using stratified sampling, matching, and covariance adjustment to force the groups to be as nearly alike as possible with respect to variables not under study; or finally (3) by adding to these statistical controls tests of internal consistency, prospective studies, independent replications, and examination of concomitant variables that might obscure or simulate treatment effects. In this most conservative form, epidemiology approaches the precision of the genetic method, which consists of comparison of sample statistics with population parameters or with a functional relation among parameters specified by genetic theory.

Each of these methods has its own peculiar advantages and disadvantages. The experimental approach is the most powerful but the least frequently applicable. The epidemiologic method is the most complex and dangerous, since it encompasses no rigorous safeguards against error and is prone to accept ostensible freedom from bias as an adequate substitute for precise observation. The genetic method has the narrowest scope, since it requires specification of population parameters; but, properly applied, it has the high degree of reliability which complete formulation of an hypothesis entails.

According to this classification, many of the methods used in human genetics are either experimental or epidemiologic. Twin studies, empiric risks, and most estimates of heritabilities, incidences, gene frequencies, selection coefficients, and mutation rates are aspects of epidemiology. Only segregation and linkage analysis satisfy the definition of the genetic method as "comparisons of sample statistics with population parameters or with a functional relation among parameters specified by genetic theory." It is the purpose of this paper to summarize the uses and restrictions of this part of the armamentarium of human biology.

EVIDENCE FOR A GENETIC ETIOLOGY

Many criteria have been used to suggest the significance of genetic factors in a trait of unknown etiology, including:

(1) an elevated risk in relatives compared with the general population,
(2) greater concordance of the trait in identical than in fraternal twins,
(3) an excess of consanguineous parentage for a presumed recessive trait,
(4) failure of the trait to "spread" to nonrelated individuals,
(5) onset at a characteristic age without a known precipitating event,
(6) variation among populations and behavior on outcrossing in accordance with genetic theory,
(7) occurrence of an homologous, inherited condition in other mammals.

Consideration of these criteria is sufficient to show that, individually and conjointly, they cannot be regarded as rigorous evidence of a genetic etiology. Environmental factors common to relatives can simulate genetic determinants, and infectious diseases like syphilis were long thought to be hereditary. Identical twins on the average differ from fraternals in placentation, arrangement of the chorion, birth hazards, and postnatal environment. Social and biologic factors related to the trait may lead to consanguineous marriage, or there may be a fortuitously high frequency of the trait in a kindred with high consanguinity. The other criteria depend on exclusion of alternative environmental hypotheses, which cannot be absolutely exhaustive. It must be concluded that, while all these points are suggestive of a genetic etiology, none provides a proof that is beyond cavil. Excellent discussions of this problem will be found in David and Snyder [170] and Neel and Schull.[611]

One paradox of human genetics is that it is not possible to prove a trait is genetic except by showing it could be due to one of several simple modes of inheritance. The genetic method consists in attempting to extract from a possibly heterogeneous trait one or more genetic entities due, in increasing order of refinement, to a single inheritance pattern, locus, or allele. The recognizable modes of inheritance and criteria for their recognition include:

1. Autosomal dominant.
   a) Transmission continues from generation to generation, without skipping.
   b) Except for mutants, every affected child has an affected parent.
   c) In marriages of an affected heterozygote to a normal homozygote, the segregation frequency is $\frac{1}{2}$.
   d) The two sexes are affected in equal numbers.
2. Autosomal recessive.
   a) If the trait is rare, parents and relatives except siblings are usually normal.
   b) If the recessive genes are alleles, all children of two affected parents are affected.
   c) In marriages of two normal heterozygotes, the segregation frequency is $\frac{1}{4}$.
   d) The two sexes are affected in equal numbers.
   e) If the trait is rare, parental consanguinity is elevated.
3. Sex-linked dominant.
   a) Heterozygous mothers transmit to both sexes in equal frequency with a segregation frequency of $\frac{1}{2}$.
   b) Hemizygous, affected males transmit the trait only to their daughters, the segregation frequency being 1 in daughters and 0 in sons.
   c) Except for mutants, every affected child has an affected parent.
   d) If the trait is rare, its frequency in females is approximately twice as great as in males.
4. Sex-linked recessive.
   a) If the trait is rare, parents and relatives, except maternal uncles and other male relatives in the female line, are usually normal.
   b) Hemizygous, affected males do not transmit the trait to children of either sex, but all their daughters are heterozygous carriers.
   c) Heterozygous, carrier women are normal but transmit the trait to their sons with a segregation frequency of $\frac{1}{2}$, and $\frac{1}{2}$ of the daughters are normal carriers.
   d) Excluding *XO* daughters of carrier mothers, affected females come only from matings of carrier females and affected males, and their frequency in the population is approximately the square of the frequency of affected males.
   e) Except for mutants, every affected male comes from a carrier female.

Other modes of inheritance have been claimed in man, including Y-linkage, partial sex linkage, attached-*X* chromosomes, and cytoplasmic inheritance, but the evidence for them is inadequate.[571, 654, 838]

Some traits are limited in expression to one sex, although members of the opposite sex may transmit a genetic factor for them. Such characteristics are called sex-limited and include premature pattern baldness, imperforate vagina, and other traits associated with reproductive physiology or sex differentiation. If the gene is recessive, all the criteria of this mode of inheritance hold, except that only one sex is at risk. If the gene is an autosomal dominant and limited to males, there is a possibility of confusion with recessive sex-linkage. In both cases, normal carrier females have a segregation frequency of $\frac{1}{2}$ among sons and 0 among daughters. However, the following differences are apparent:

1. An affected, heterozygous male has a segregation frequency of $\frac{1}{2}$ in his sons under autosomal sex-limitation, but 0 under sex linkage.

2. A sex-linked gene may show reduced recombination with other sex-linked genes, like color-blindness and primaquine sensitivity, while an autosomal gene can show reduced recombination only with other autosomal genes.

3. Through nondisjunction, a carrier for a sex-linked recessive may produce an affected XO daughter, recognizable both clinically and cytologically.

4. If the gene is rare, the frequency of sporadic mutants among all affected cases is $m/(m+1)$ for an autosomal sex-limited gene and $mu/(2u+v)$ for a sex-linked recessive, where m is the selection coefficient against affected males and u and v are the mutation rates in egg and sperm, respectively. For a nearly lethal gene, m approaches 1 and $m/(m+1)$ approaches $\frac{1}{2}$. Consequently, a frequency of sporadic cases significantly less than $\frac{1}{2}$ is evidence for sex-linkage.[573]

Sex-limited manifestation is not the only complication in expression of genetic traits. There may be a quantitative difference between the sexes in manifestation rate. For example, hemochromatosis tends to be most severe in males and postmenopausal females who do not have the safeguard of menstruation to reduce the deposition of iron pigments in tissues. Other factors, either genetic or environmental, may modify the expression of a genetic trait (variable *expressivity*) or may even prevent the characteristic symptoms (incomplete *penetrance*). Variable expressivity presents no difficulties for genetic analysis, providing the alternative types of manifestation (*phenotypes*) are recognized, but incomplete penetrance modifies genetic expectations. Segregation frequencies are reduced, and phenotypically normal individuals may be either genetically normal, preclinical cases, or carriers who may transmit but will not themselves express the trait. Rigorous genetic analysis is impossible if penetrance is low. In general, analysis of incomplete penetrance is facilitated by rarity of the trait, since some of the possible genotypes for normal people will be negligibly rare. A satisfactory genetic study of common factors with low penetrance, such as diabetes, must wait until methods are found to increase penetrance by detection of genetic carriers, such as more sensitive glucose-tolerance tests.

Sometimes genetic traits depend on the presence of several genes (*complementary epistasis*). If the effect of each gene is recognizable, as for the Lewis and *ABO* secretor components of the Le^b trait, each component may be studied separately. Otherwise, the analysis is beyond the limits of available methods in human genetics, even though a segregation frequency near $\frac{1}{16}$, or some other epistatic proportion, might be suggestive. Again, the only reliable approach in a nonexperimental system is to defer genetic analysis until the responsible genes are individually recognizable.

THE SEGREGATION FREQUENCY (p)

The segregation frequency is of critical importance, not only for establishing simple modes of inheritance, but also for linkage analysis, empiric risks, concordance in twins, and other problems. The basic data of formal genetics consist of the frequencies with which a trait appears in different types of matings. We shall refer to persons with the trait as affected and to persons without the trait as normal. In simple situations, for any phenotypic class of mating (such as normal × normal, normal × affected, or affected × affected), we may conceive of every mating that has produced an affected child as being of either of two types: a low-risk type, in which the probability of an affected child is so small that the risk for two or more affected children is negligible, and a high-risk type, in which the chance for other cases after the first is appreciable. The risk for high-risk families is called the *segregation frequency (p)*.

The most familiar values of the segregation frequency are $\frac{1}{2}$ and $\frac{1}{4}$, but proportions modified by differential mortality, partial manifestation, and conceivably by meiotic drive [746] or gametic selection are also included. For example, if $\frac{1}{4}$ of the children from a particular mating type are expected to be of a certain genotype, but only 80 per cent of them develop a characteristic abnormal phenotype, we shall say that the segregation frequency is $(.80)(\frac{1}{4}) = .20$. This concept of incomplete penetrance is liable to abuse by uncritical workers, but under certain conditions it is a valid and useful concept. These conditions include dominant genes whose characteristic manifestations occasionally skip a generation and rare diseases with delayed onset for which the penetrance may be calculated from the distribution of age at onset and the ages of family members, without any genetic assumptions. One way of doing this is as follows: [570] Let $f(z)$ be the frequency of age z at death or last examination among normal and affected siblings, $f_1(z)$ be this frequency with the index cases excluded, and $G(z)$ be the cumulative frequency of onset at age z among affected cases. Then if incomplete penetrance is entirely due to delayed onset, the estimate of the average penetrance in the sample lies between

$$y = \int f(z) \ G(z) \ dz \quad \text{and} \quad y = \int f_1(z) \ G(z) \ dz$$

depending on the method by which the families were ascertained. The effective segregation frequency will then be $p = yp_0$, where p_0 is the theoretical value (say $\frac{1}{4}$ or $\frac{1}{2}$) specified by genetic hypothesis.

For the case of a rare dominant gene which sometimes skips a generation, penetrance may be estimated from the number, n, of unaffected carriers in the direct line of descent from an affected ancestor to the first affected descendant in that line. The probability of any value of n is the geometric distribution $(1 - y)^n y$, where y is the penetrance. The mean value of n is $(1 - y)/y$, with variance $(1 - y)/y^2$. As the penetrance approaches 1, the mean value of n approaches the frequency of skips. As the penetrance approaches 0, this method tends to overestimate penetrance by failure to recognize large values of n.

Not only incomplete penetrance but also differential mortality can be incorporated into the segregation analysis. If the viability up to the age of diagnosis relative to normal is f, the effective segregation frequency is

$$p = f p_0 / (1 - p_0 + p_0 f).$$

Segregation analysis provides one of the most powerful methods for studying selective forces in human population.[138, 574] It is especially valuable for early prenatal mortality and mortality before the age of diagnosis, which are difficult to detect or estimate in any other way. Absence of adequate data for segregation analysis of major genetic polymorphisms, such as the abnormal hemoglobins, serum proteins, and many blood groups, especially from populations under rigorous selection, is the most serious limitation on our knowledge of the dynamics of balanced polymorphism in man.

Traits determined by many genes, not individually recognizable, are unfavorable for genetic analysis, but the medical importance of such conditions as mental retardation, cleft palate, breast cancer, and schizophrenia has led to preliminary investigation of the empiric risks to which relatives of probands are subject. Traditionally, it has been assumed that the risk is uniform but unknown among all families within a category. On this assumption, the epidemiologic method of comparing two groups of relatives and controls, ostensibly matched for attributes other than the familial risk, is appropriate. Because of sporadic cases, this usually leads to estimation of a segregation frequency too low either to suggest a simple genetic hypothesis or to be of much medical interest. An alternative approach, with greater heuristic value, is to use segregation analysis to test whether the data are in principle divisible into high-risk and low-risk families and, if so, to estimate the segregation frequency in high-risk families and the frequency of the low-risk (sporadic) types.[570]

This has several advantages. By introducing an additional parameter, empiric risks should be more accurately specified. The assumption of sporadic cases may lead to recognition of heterogeneity in data hitherto regarded as homogeneous and may even show that the risk in high-risk families is great enough to suggest a simple genetic hypothesis. The concept of sporadic cases is forced on us by mutations, instances of heterozygous expression of recessive genes, and causes of acquired cases that must usually be nonrecurrent among siblings, such as intrauterine infection and birth trauma in mental deficiency, infections of infancy in deaf mutism, and carcinogenic radiation in cancer. Finally, empiric risks estimated by simple group-comparison tend to overestimate the risks in families with only one affected and to underestimate them in families with two or more affected.

If the force of these considerations be acknowledged, then empiric risks must be reassessed before genetic counseling can have a sound basis. Not only do we not know what proportion of cases of cleft palate, congenital heart disease, mental defect, or breast cancer is sporadic, but the question has not even been clearly formulated in the empiric risk studies that have been carried out. Fortunately, this error can be avoided by using the general form of the segregation analysis, with no assumed restriction on the frequency of sporadic cases, and it will be interesting to see how materially this contributes to the specification of empiric risks in the future.

An epidemiologist might distrust the genetic method as biased because it demands no control sample. However, traits of genetic interest are determined objectively, by physical or laboratory examination, so the possibility of bias is removed. Furthermore, risks in high-risk families are usually so much above the incidence in a control sample that bias is negligible even in anamnestic data. For example, the risk for albinism in a sibling of an albino ($\frac{1}{4}$) is so much higher than the population incidence (about $\frac{1}{15,000}$) that, apart from easily remediable effects of the method of ascertainment, sampling bias may be disregarded. This leaves the geneticist free to concentrate on the precision of the data, without the epidemiologist's fear of creating a bias by comparison with a less intensively studied, control sample. The genetic method demands, and usually obtains, a higher level of diagnostic precision than epidemiologic investigations. Thus the genetically heterogeneous rubrics of the standard disease classifications and the interview data of the United States National Health Survey are of value to epidemiologists, but are much too crude for segregation analysis, although they may be useful to obtain a sample for follow-up study. An appreciable frequency of false or missed diagnoses might be tolerable to an epidemiologist if they were random among groups, but such errors would be fatal to formal genetics.

However, if segregation analysis shows that the risk in relatives is uniformly low, of the same order as the incidence, and with no division into high- and low-risk cases, and if the diagnosis of affection is not accurate and objective, then the genetic method is definitely less reliable than the epidemiologic approach. In the absence of methods to distinguish sporadic and high-risk cases, a pedigree study of such conditions as mental retardation and the common types of cancer is at best sterile and likely to be grossly misleading.

THE PROPORTION OF SPORADIC CASES (x)

Even if many trait-bearers are due to simple genetic mechanisms and therefore occur in high-risk families, it is the rule rather than the exception for some cases to be *sporadic* due to mutations, phenocopies, technical errors, extramarital conceptions, rare instances of heterozygous expression of a recessive gene, chromosomal nondisjunction, polygenic complexes, etc. These sporadics, of different etiology from the high-risk cases, must be distinguished from *chance-isolated* cases whose siblings, although normal, have the same *a priori* risk, *p*, of being affected. Sometimes the distinction between sporadic and chance-isolated cases can be made phenotypically, for example by the use of a discriminant function between isolated and familial cases.[139] More commonly, a phenotypic distinction is difficult or impractical, but the proportion of sporadic cases can be determined.

The first and most general method of estimation uses the distribution of isolated and familial cases. If x is the proportion of cases that are sporadic among all cases in the population, then the probability that an isolated case with $s-1$ normal siblings is sporadic is $x/[x + (1-x)q^{s-1}]$, where $q = 1 - p$, and p is the segregation frequency in high-risk sibships. A family with an isolated case, either sporadic or chance isolated, is called *simplex*, and a family with two or more cases is called *multiplex*. The frequencies of simplex and multiplex families of size s lead to an estimate of x.[570, 573]

The second method to determine x is appropriate if high-risk cases are due to rare, recessive genes. Let y be the proportion of isolated cases that are sporadic, F_I be the inbreeding coefficient of isolated cases, F_F be the inbreeding coefficient of familial cases, and α be the inbreeding coefficient of the general population. Then if sporadic cases are not associated with inbreeding,

$$F_I = y\alpha + (1-y)F_F,$$

$$\text{or} \quad y = \frac{F_F - F_I}{F_F - \alpha}.$$

If the number of isolated probands is n and the number of familial probands is N, then $c = n/(n + N)$ is an estimate of the proportion of isolated cases in the population, and cy is an estimate of x.[142] Comparison of these two independent estimates of x tests whether sporadic cases are associated with inbreeding. There is no association for the three traits studied so far.[568]

Although the model of sporadic and high-risk cases has heuristic value and appears to be correct for many traits, it is clearly only a simple abstraction from a reality which may be much more complex. If the occurrence of sporadic cases is random among families, with incidence γ, then the condition for familial cases of sporadic origin to have negligible frequency is $x\gamma \ll (1-x)p$. More generally, independent sporadic cases may sometimes occur in the same sibship, or the segregation frequency, p, may be low or variable among families. Segregation methods could readily be developed to cover these cases, and, with modern computing equipment, the arithmetic would present no difficulty, but the analysis is bound to be of low resolving power. Extension of the segregation model can wait until a clear need is shown. Development of an ill-health registry and record linkage as part of our national vital-statistics system would generate sample sizes large enough to warrant application of more refined methods, if the data were reliable enough to justify analysis.[618]

THE PROPORTION OF PARENTS WHO CANNOT SEGREGATE (h)

Among parents of a particular phenotype there will be some who cannot produce affected children or do so only with the frequency of sporadic cases in the general population. Let the proportion of such parents be h. The probability of a segregating mating (with at least one affected child) is $(1 - h)(1 - q^s)$ in backcrosses and $(1 - h)^2(1 - q^s)$ in intercrosses.

There are two principal reasons for parents who cannot segregate. One is homozygosity for a dominant allele, the probability for which can be calculated from the genes frequencies in the population. For example, the probability for a person of blood type A to be homozygous is

$$h = f_A{}^2/(f_A{}^2 + 2f_A f_O) = f_A/(f_A + 2f_O)$$

in a randomly mating population, where f_A and f_O are the frequencies of the genes for A and O, respectively. Complexities introduced by multiple alleles have been considered by Morton.[570]

A second reason for parents who cannot segregate is *phenocopies* (nontransmissible, sporadic cases that simulate genetic ones) . For example, retinoblastoma is sometimes due to a rare, dominant gene with incomplete penetrance and sometimes to a phenocopy that may be a somatic mutation. Suppose an isolated proband has s children and the probability that the parent is a phenocopy is h, then the probability of at least one affected child is $(1 - h)(1 - q^s)$, where p is the probability of an affected child from a carrier parent and $q = 1 - p$ is the probability of a genetically normal child or an unaffected carrier. Analysis of this distribution permits estimation of h and p from families of any size, while other methods either confound h and p or use only exceptionally large families which can be classified as low- or high-risk by inspection.[887]   In other cases, h may represent diagnostic errors. For example, in the mating $MN \times M$, the probability of at least one M child is $(1 - h)(1 - q^s)$, where h is the probability that a parent of phenotype MN is actually of genotype NN but gives a false-positive reaction with an M antiserum. This distribution has been used to show that technical errors are not frequent enough to explain the excess of MN children observed in published family studies, which is therefore probably due to preferential survival of MN fetuses.[137, 574]   Segregation analysis should not be considered evidence for differential mortality or biased segregation unless it can be shown that the discrepancy from genetic expectation is in p and not in h. Properly applied, segregation analysis is a powerful instrument for studying selection in man.

THE ASCERTAINMENT PROBABILITY (π)

In human genetic-studies, some trait bearers are selected through hospital records, death certificates, inquiries to physicians, examination of a population sample, or other direct means of ascertainment. These cases found by direct methods are called *probands,* in contrast to secondary cases ascertained through investigation of the relatives of probands. If the children of probands or their affected relatives are sampled without regard to their own phenotypes, we speak of *complete selection* of these sibships. Selection is said to be *incomplete* if only sibships with an affected member are sampled. This type of selection generates a continuum of simple possibilities, ranging from selection so incomplete that two probands never occur in the same sibship *(single selection)* to selection so thorough that sibships with many cases are no more likely to be sampled than sibships with only one affected case *(truncate selection)*. These possibilities may be described in terms of the *ascertainment probability* (π), which equals 1 for truncate selection and approaches 0 for single selection, the intervening range constituting *multiple selection*. Since π represents the probability that an affected person in the population is a proband, estimation of π is necessary not only for genetic analysis but also for determining the incidence of cases in the population.

There are several sources of information about π. Even if probands are not identified, a small amount of information is contributed by the distribution of affected children among families. Haldane [315] has suggested that the hypotheses of single and truncate selection be used to set a lower and upper limit to the segregation frequency, p. Of course if probands are identified, this is very inefficient. If probands are not identified, the assumption of a uniform ascertainment probability may be questionable, since, with nonsystematic ascertainment procedures, isolated cases may be expected to be underrepresented because of greater medical interest in families with several affected. However, Haldane [311] found no evidence for such a bias in the literature on albinism.

More information can be extracted if probands are identified. A sibship with r affected may contain from 1 to r probands. Let the actual number of probands be a, then the distribution of a is

$$\frac{\binom{r}{a} \pi^a (1 - \pi)^{r-a}}{1 - (1 - \pi)^r}. \qquad (r, a > 0)$$

If most cases are sporadic, even a large sample will give little information about the segregation frequency. The geneticist may then concentrate on familial cases, selected either from the distribution of affected children including families with only one proband or because there were two or more probands. In the latter event, the distribution of a probands in selected families with r affected is

$$\frac{\binom{r}{a} \pi^a (1 - \pi)^{r-a}}{1 - (1 - \pi)^r - r\pi(1 - \pi)^{r-1}}. \qquad (r, a > 1)$$

Analysis of the appropriate distribution gives an estimate of π. It will be noticed that isolated cases yield no information by this method, since they are necessarily probands.

An even better estimate may be obtained from the number of ascertainments per proband, provided a proband can be ascertained in more than one independent way. The simplest case occurs if there are two independent modes of ascertainment with probabilities π_1 and π_2, so that the total ascertainment probability is

$$\pi = \pi_1 + \pi_2 - \pi_1\pi_2.$$

The probability of ascertainment by the first mode alone is

$$\pi_1(1 - \pi_2)/\pi = (\pi - \pi_2)/\pi.$$

The probability of ascertainment by the second mode alone is

$$\pi_2(1 - \pi_1)/\pi = \pi_2(1 - \pi)/\pi(1 - \pi_2),$$

and the probability of double ascertainment is

$$\pi_1\pi_2/\pi = (\pi - \pi_2)\pi_2/(1 - \pi_2)\pi.$$

From this distribution we can estimate π, together with the nuisance parameter, π_2. Extension to multiple modes of ascertainment is easy in principle, but introduces an additional nuisance parameter for each mode of ascertainment. A simple approximation, adequate if there are several modes of ascertainment, is provided by the truncate Poisson distribution. Let the mean number of independent ascertainments per proband be m. Then

$$\pi = 1 - e^{-m},$$

and the distribution of t ascertainments per proband is

$$\frac{m^t e^{-m}}{t!(1 - e^{-m})} = \frac{[-ln(1 - \pi)]^t(1 - \pi)}{t!\pi}.$$

The distribution of t gives the greatest amount of information about π, since both isolated and familial cases can be used. In two test cases, the distribution of probands among affected gave less than $\frac{1}{4}$ as much information as the distribution of ascertainments per proband.[570, 573] However, for limb-girdle muscular dystrophy the estimates were significantly heterogeneous. It was suggested that the ascertainment distribution may tend to underestimate π because ascertainment by one source tends to make another less likely, as one physician's referral tends to preclude another. On the other hand, the distribution of probands may overestimate π because two or more siblings may be examined and counted as probands, even though only one of them would have submitted to examination independently. Pooling the two independent sources of information gives an intermediate value that is closer to the lower estimate. Further experience with the same population has suggested that the ascertainment distribution was more reliable than the proband distribution and that the pooled estimate was substantially correct.

Unless the ascertainment model is clearly understood, probands and ascertainments will not be defined appropriately. Multiple referrals by the same physician are usually not independent. For example, if a proband was ascertained through a hospital record, the referring and consulting physicians should not be counted as separate modes of ascertainment unless they also report the proband independently of the hospital record. Since the sibship is the unit of analysis, a sibship with no proband, but one or more secondary cases ascertained through related probands, is obtained by truncate selection, as ascertainment is assured in such a sibship by presence of a single affected child. Such secondary sibships should be scored for $\pi = 1$ and appropriate values of p and x, but of course they give no information about the ascertainment probability. Unrelated affected individuals ascertained in the course of a family study are probands in their sibship, and such fortuitous detection is one mode of ascertainment.

The definition of probands may be difficult under certain methods of ascertainment, for example through membership in a society of families with affected children. If ascertainment is not sufficient to assure coöperation and inclusion in the sample, the crude ascertainment probability should be multiplied by the probability that a proband participates in the study. Exploitation of common polymorphisms, with selection complete or truncate, and of vital statistics or institutional registries to ascertain rare conditions facilitates identification of ascertainments and probands. The investigator should always remember that ascertainment is defined not by words but by conformity to a probability model. Fortunately, consideration of a number of published studies which used quite different definitions of probands indicates that the segregation analysis is reasonably robust, with a sufficiently small covariance between π and the other parameters so as not to be disturbed appreciably by small departures from the ascertainment model. Since even the incomplete and crude data provided by published genetic studies can apparently be tolerated, an enlightened effort to define probands and ascertainments appropriately seems all that is required to assure a valid segregation analysis.[573]

SEGREGATION MODELS

Because the uses and limitations of segregation analysis are implicit in the probability distributions, it may be helpful to set out here the models that have actually been used in human genetics.[570, 573, 574] In all cases the following assumptions are made:

1. The ascertainment probability, π, is constant; all probands in a sibship are ascertained independently; and all ascertainments of a proband are

independent. There are a probands among r affected in a sibship of size s.

2. In multiplex and simplex sibships of the same origin, there is a constant *a priori* probability p that a child be affected $(q = 1 - p)$.
3. For sibships of all sizes, sporadic cases make up a proportion x of all cases in the population.

Case 1. Complete selection with separation of homozygous and heterozygous parents. No sporadic cases.

$$P_{(r)} = \binom{s}{r} p^r q^{s-r}.$$

Case 2. Complete selection with separation of segregating and nonsegregating families. No sporadic cases.

$$P_{(r>0)} = (1 - h)(1 - q^s). \qquad \text{(backcrosses)}$$
$$= (1 - h)^2(1 - q^s). \qquad \text{(intercrosses)}$$

Case 3. Single selection $(\pi \to 0)$

$$P_{(r=1 \mid r>0)} = x + (1 - x)q^{s-1}$$
$$P_{(r \mid r>1)} = \binom{s-1}{r-1} p^{r-1} q^{s-r} / (1 - q^{s-1}).$$

Case 4. Multiple selection $(0 < \pi \leqslant 1)$

$$P_{(r=1 \mid r>0)} = \frac{sp\pi[x + (1 - x)q^{s-1}]}{xsp\pi + (1 - x)[1 - (1 - p\pi)^s]}$$

$$P_{(r \mid r>1)} = \frac{\binom{s}{r} p^r q^{s-r}[1 - (1 - \pi)^r]}{1 - (1 - p\pi)^s - \pi sp q^{s-1}}.$$

Case 5. Multiple selection with at least one affected girl.

This is used for rare, recessive traits that are a mixture of autosomal and sex-linked cases. Families with autosomal or sporadic cases can be recognized if they contain at least one affected girl.

$$P_{(r=1 \mid r>0)} = \frac{sp\pi[x + (1 - x)q^{s-1}]}{xsp\pi + 2(1 - x)[1 - (1 - p\pi)^s - (\tfrac{1}{2}p + q)^s + (\tfrac{1}{2}p + q - \tfrac{1}{2}p\pi)^s]}$$

$$P_{(r \mid r>1)} = \frac{\binom{s}{r} p^r q^{s-r}[1 - (1 - \pi)^r][1 - (\tfrac{1}{2})^r]}{1 - (1 - p\pi)^s - (\tfrac{1}{2}p + q)^s + (\tfrac{1}{2}p + q - \tfrac{1}{2}p\pi)^s - \tfrac{1}{2}\pi sp q^{s-1}}.$$

Case 6. Uncles of isolated probands of a rare, recessive, sex-linked or autosomal, dominant, male-limited gene.

Let x' be the probability that the brothers of a carrier female be at risk and r be the number of maternal uncles. Under sex-linkage $x' = \tfrac{1}{2}$, and under autosomal sex-limitation $x' = m/2$, where m is the selection coefficient against the trait in males. The number of children in the proband's sibship is s, and the probability that the isolated proband is sporadic is $x/[x + (1 - x)q^{s-1}]$, where $x = m/(1 + m)$ under autosomal sex-limitation and $x = mu/(2u + v)$ under sex-linkage and u and v are the mutation rates in egg and sperm, respectively.[573]

$$P_{(r=0)} = \{x + (1 - x)q^{n-1}[x' + (1 - x')q^s]\}/\{x + (1 - x)q^{n-1}\}.$$

Case 7. Multiple selection through two or more probands.

This may occur increasingly through selection of multiplex families for analysis. For example, familial cases of mental defect might be selected through two or more institutionalized siblings or through an ill-health registry, while affected siblings who were not institutionalized or registered would be secondary cases. Since at least two probands are required for selection of a sibship,

$$P_{(r \mid r>0)} = \frac{\binom{s}{r} p^r q^{s-r}[1 - (1 - \pi)^r - r\pi(1 - \pi)^{r-1}]}{1 - (1 - p\pi)^s - sp\pi(1 - p\pi)^{s-1}}.$$

Case 8. Complete selection with separation of segregating, A-nonsegregating, and a-nonsegregating families. No sporadic cases.

This distribution has been used to investigate technical error and selection hypotheses for the *MN* blood groups.[573]

$$P_{(r=0)} = h + (1 - h - y)q^s$$
$$P_{(r=s)} = y + (1 - y - h)p^s$$
$$P_{(0<r<s)} = (1 - h - y) \binom{s}{r} p^r q^{s-r}.$$

Clearly these models are only a few of the possible ones, and more will be developed as the need arises. They have in common the assumption that a sibship is not deliberately terminated because of the birth of an affected child. Some statisticians have expressed apprehension that neglect of such optional stoppage may introduce appreciable error into segregation analysis. There are several reasons why there has been delay in extending segregation methods to cover this situation. (1) Actual analyses of human material have not clearly revealed any discrepancy which could be assigned to optional stoppage. Because of small sample sizes and uncertainty about the ascertainment probability, it is doubtful whether such a discrepancy could be detected in the material currently available. (2) Stoppage rules are variable among families and are unenforceable by reason of delayed diagnosis, contraceptive failure, and compensation and overcompensation for defective children. Whatever the rules governing human fertility, they are much too complicated to be subsumed by simple models. (3) It is doubtful if correct information about stoppage rules could be elicited from many families in view of religious and other emotional factors surrounding such decisions.

(4) **Even** if some families revealed that they had terminated reproduction because of one or more defective children, any attempt to incorporate this information into the analysis would require knowledge of the distribution of incomplete family size as a function of parental age, since it would hardly be feasible to limit genetic analysis to completed families. There is a lack of such information in all countries, and its use would be complicated by the ascertainment problem that age at diagnosis and parental age are related. Thus a different distribution of family size would be expected for similar conditions such as infantile and juvenile amaurotic idiocy, while there could not be any effective optional stoppage rule for a condition with late onset, like Huntington's chorea. Attempts have been made to describe complete family size by a modified geometric or Poisson distribution,[277a] but in populations with mixtures of contraceptive and noncontraceptive groups and with biological variations in fertility, there may be no good approximation to the distribution of family size, especially incomplete size. Geneticists have preferred to avoid the unknown prior distribution of family size in favor of distributions conditional on fixed size, especially as a distribution of family size which must be estimated can provide only negligible and equivocal information about segregation. Until there is evidence that stoppage rules have detectable effects, these problems may well be ignored.

STATISTICAL ANALYSIS OF SEGREGATION MODELS

There have been four stages in the development of segregation analysis in man. At first, tests of significance based on complete selection were applied to rare dominant pedigrees and later to codominant factors. Next, the disturbing effects of truncate selection were recognized, leading to the development of the *a priori* methods of Bernstein and others and the *a posteriori* method of Haldane. Third, multiple selection was considered by Weinberg in the proband method, which is not fully efficient except in the limiting case of single selection, and more elaborately by later authors, none of whom used the large amount of information present in the number of ascertainments per proband. For a review of developments up to this point, see Smith [809] and Steinberg.[828] Finally, interest in mutations and other sources of sporadic cases have led to their inclusion in more general segregation models, with separation from incomplete penetrance and other superficially similar phenomena.[570] Because of the advantages of the maximum-likelihood theory of Fisher, this is the method of choice. Its theoretical credentials are discussed in any statistics text,[392, 441, 531, 547, 705] and the practical advantages include the simplicity and power of maximum-likelihood scores, which permit combination of data from many sources, tests of heterogeneity and goodness of fit, and iterative estimation.

The basis of this system is derived from the probability distribution as a score for each parameter to be studied, say U_p, U_x, and U_π for p, x, and π, respectively.

Associated with these scores are their variances, K_{pp}, K_{xx}, and so on, which are also the amounts of information about each parameter on the null hypothesis that all parameters are specified correctly. The scores may be arranged as a vector, U, and the variances and the covariances K_{px}, $K_{p\pi}$, et cetera, as a square, symmetrical matrix, K. Formulae for manipulation of these arrays to give estimates, tests, and standard errors are illustrated by Morton [570] and Morton and Chung.[573] * In a test case, 743 units of information were contributed by the distribution of segregating and nonsegregating families, 603 units by the distribution of simplex and multiplex families, and 1,724 units by the distribution of affected within multiplex families.[573] Methods which neglect any of these sources of information, such as the double truncate selection of Bennett and Brandt,[50] are grossly inefficient.[138]

EXAMPLES OF SEGREGATION ANALYSIS

Many rare errors of metabolism have been observed in only a few families. In such cases the occurrence of parental consanguinity or some other aspect of the family history may suggest a simple mode of inheritance, or a genetic etiology may be proposed solely from analogy with other metabolic defects. Thus tyrosinosis has been observed in only three patients and the etiology is unknown, but it may well be recessive like alkaptonuria, phenylketonuria, and other blocks in amino-acid metabolism. If the material is too scanty, there is no point in applying formal segregation analysis. One does not need an elephant gun to hunt mice.

With more adequate genetic data, the segregation analysis will often lead to discovery or verification of sporadic cases, incomplete penetrance, differential mortality, or other phenomena. For example, most investigators of limb-girdle muscular dystrophy were so impressed by the occurrence of parental consanguinity and familial cases as to conclude that the disease is due to an autosomal recessive gene. However, segregation analysis shows that about 41 per cent of limb-girdle cases are sporadic and of different etiology from the recessive group.[574] This is confirmed by the difference between the inbreeding coefficients of isolated and familial probands ($F_I = .00119$, $F_F = .00804$). There is evidence suggesting that some sporadic cases may represent rare instances of heterozygous expression of a gene which is usually recessive.[140]

---

* Evaluation of the scores and their variances involves tedious arithmetic, which can be eliminated by an electronic computer. Once programmed, almost no labor is required to tabulate scores for various values of the parameters and to perform the same type of analysis on other data, with incorporation of computing checks that insure accuracy. These methods have been programmed for the IBM 650 computer, checked exhaustively, and employed in many analyses. The SEGRAN program, written by Mrs. Nancy S. Jones, may be used by arrangement with the Department of Medical Genetics, University of Wisconsin Medical School.

Early investigators of Duchenne muscular dystrophy concluded that it is either due to a recessive sex-linked gene or to an autosomal gene dominant in males and not expressed in females. Since affected males seldom reproduce, the distinction between these two modes of inheritance did not at first seem feasible. However, all the tests that discriminate between the two mechanisms have now been applied and show that sex-linkage is the correct explanation. In particular, the proportion of sporadic cases is .355 ± .050, which is too small to be consistent with autosomal sex limitation. On the other hand, it is too high to support Haldane's contention that the gene for Duchenne dystrophy mutates more frequently in sperm than in eggs.[573] Thus segregation analysis of sex-linked genes leads to a critical distinction between mutation rates in eggs and sperm, a contrast that cannot be made in any other way in man, and gives an estimate of the mutation rate by the semidirect method.[574]

Stevenson and Cheeseman [842] reported on deaf mutism in Northern Ireland, and Krooth [464] and Slatis [805] commented on their analysis. The data have been interpreted in terms of heterozygote advantage, absence of dominant cases, synergistic effects of nonallelic recessive genes, and hypotruncate selection. However, Chung, Robison, and Morton [142] showed by segregation analysis that the material is consistent with only one simple model and that the proposed alternatives do not fit the observations. They concluded that the congenital deaf-mute population appears "to consist of three types:

1. Autosomal recessives, complete penetrance, 68 per cent of all cases, at least several loci, average mutation rate 1×10^{-5} per locus, carrier frequency 0.080 per gamete.
2. Autosomal dominants, high but not complete penetrance, 22 per cent of all cases, at least two loci, total mutation rate 5×10^{-5} per gamete.
   a) Genes inherited from affected parents, 7 per cent of all cases.
   b) New mutants inherited from normal parents, 15 per cent of all cases.
3. Sporadic cases, due to unrecognized infection or more complex genetic mechanisms, 9 per cent of all cases."

The existence of sporadic, nonrecessive cases was confirmed by the inbreeding coefficients of isolated and familial probands ($F_I = .00278$, $F_F = .00854$).

Recently, Dewey and Morton [178] reëxamined the classic Colchester survey of mental deficiency [649] as a pilot study for their own work. After exclusion of cases due to trauma, infection, neoplasm, mongolism, and hydrocephalus, they found that the low-grade cases from normal parents are of two types: a sporadic type amounting to 87 per cent of residual defect and a group of recessive entities estimated to be about 69 in number, with apparently high penetrance and no heterozygote advantage. Estimates of gene frequencies and mutation rates (2×10^{-5}

per locus) from these data are in agreement with other evidence.[568] The inbreeding coefficients of isolated and familial cases indicate no association between sporadics and inbreeding (F_I = .00097, F_F = .01349). The etiology of the sporadics is complex and includes unrecognized exogenous causes, polygenes, dominant mutations, and aneuploidy.

Only a beginning has been made in the application of segregation analysis to the dynamics of balanced polymorphisms. Morton and Chung[574] found that heterozygote excess in the *MN* system is distributed in such a way that they could not explain it in terms of technical errors, which would have to be postulated to occur only when the mother is *MN*, with the same frequency in *M* and *N* individuals, and only in children. These assumptions are inconsistent with available evidence, notably for the same frequency of the *MN* phenotype in parents and children. The most likely explanation appears to be heterozygote advantage, acting in early fetal stages. If this is correct, the *MN* locus makes a major contribution to the genetic load expressed as fetal death in human populations.[137]

Evidence of strong selective forces acting on the *ABO* system is more direct and manifold. Early claims of segregation disturbance due to *ABO* incompatibility [910] were severely criticized on statistical grounds, using methods that extract no information from backcross families of size 1 or 2, only $\frac{1}{4}$ of the information from families of size 3, $\frac{1}{2}$ from families of size 4, et cetera.[50] However, several studies in Caucasians and Japanese have revealed a large increase in abortions and stillbirths with maternal-fetal *ABO* incompatibility. The genetic evidence amply confirms this and, in principle, even offers the possibility of detecting (through disturbed segregation) the effects of early, unobserved abortions.[138] There are interesting parity effects on segregation, differing in some ways between Japanese and Caucasians, indicating the dual effects of heteroimmune and isoimmune mechanisms and perhaps the increasing relative importance of the isoimmune response.[137, 568] Unlike *ABO* hemolytic disease, which is almost limited to A_1 and *B* children of *O* mothers, abortions occur with the same frequency in incompatible children of *O, A,* and *B* mothers. There is also highly significant evidence for heterozygote (*AO, BO,* and *AB*) excess in compatible matings, not due to extramarital conceptions and most likely caused by heterozygote advantage, which is sufficiently strong to make the *ABO* system a balanced polymorphism despite the elimination of incompatible heterozygotes.

The exploration of this fascinating aspect of population genetics has only begun, but the present inadequate data are still sufficient to show that the *ABO* locus is a major selective force in man. It seems hardly too sanguine to predict that this generation will see a determined and successful attack on the problems of balanced polymorphisms in human populations, that they will be revealed as principal causes of differential morbidity and mortality, and that the advantages that man offers for detailed examination will make him the preferred organism for the study of population genetics through segregation analysis.

CONCORDANCE IN TWINS

Since the time of Francis Galton, twin studies have seemed to some students to hold forth the promise of separating the contributions made to family resemblance by heredity and environment. An entire journal is devoted to this so-called "twin-method," but the assumptions required for estimation of heritability from twin studies are so little subject to proof that they have taken on the character of articles of faith, to which few of the younger geneticists adhere. Nevertheless, concordance of twins is often a useful datum, especially when the cause of discordance in monozygotic twins can be identified. Even a single instance of discordance of monozygotic twins is critical evidence against the hypothesis that incomplete penetrance is due strictly to genetic modifiers of a major gene. Monozygotic twins provide an upper estimate of penetrance and valuable experimental material on the effects of environmental differences, whereas comparison of the concordance of dizygotic twins and ordinary siblings gives evidence of transient uterine and postnatal environmental differences within families. Dizygotic twins have not yet been used to full advantage to study the physiologic effects of genetic differences. In conjunction with other family data, twin concordances are of circumstantial and qualitative value in assessing the role of genetic factors.

To obtain comparability with segregation frequencies and empiric risks, concordance may be defined in a somewhat novel way as the probability that a twin be affected if the cotwin is. When we attempt to estimate concordance, a difficulty becomes apparent. The ascertainment probability, π, can be determined from data on probands and ascertainments; that twin investigators rarely do this, or even record these data, is in principle irrelevant, although it reduces the value of most twin studies in the literature. But when we come to estimate the concordance in high-risk twinships (p) and the frequency of sporadic cases (x), we find that this is impossible, since our only datum is the observed proportion of concordant twins, which is formally the same as the proportion of familial cases in sibships of size 2. Knowing x, we can estimate p, or knowing p we can estimate x, but we cannot determine both simultaneously. Without a segregation analysis of other kinds of family data, our only recourse is to assume that there are no sporadic cases.*

The consequences of such an assumption are best appreciated from an example. Limb-girdle muscular dystrophy occurs sporadically in about 40 per cent of cases and is due to highly penetrant recessive genes in the remainder.[573] Accordingly, we should expect the concordance for monozygotic twins to be at least

---

* If p and p' are the concordances for monozygotic and dizygotic twins, respectively, and ϕ is the ratio of monozygotics to dizygotics in the general population, then the ratio expected in a sample under incomplete ascertainment is

$$\phi[1 - (1 - p'\pi)^2]/[1 - (1 - p\pi)^2] = \phi p'(2 - p'\pi)/p(2 - p\pi).$$

$p = .60$, but for dizygotics to be only $p = (.60)(\frac{1}{4}) = .15$ or less, depending on the age at examination. From these data we would correctly conclude that genetic factors are important in this type of muscular dystrophy, although with less assurance than by analysis of segregating families. However, we would be unable to demonstrate from twin concordance the mixture of recessive and sporadic cases which is so easily demonstrated by family analysis. Not only are twin studies incapable of elucidating complex genetic situations, but they are of limited value in simple cases.

LINKAGE ANALYSIS

Segregation and linkage, the two aspects of formal genetics, pose contrasting analytical problems. Segregation relates to the distribution of progeny with respect to a single locus and opposes a variety of alternatives to the null hypothesis that specifies the segregation frequency, p. The true frequency may be larger or smaller than this, the ascertainment model may be incorrect, or segregation may be mixed with sporadic cases of different etiology, and to these alternatives no *a priori* probabilities can reasonably be assigned. To such problems, characteristic of scientific research, acceptance sampling theory provides no solution. Fortunately, even a relatively small number of families gives enough information about segregation for the large-sample maximum-likelihood theory to be appropriate.

Linkage relates to the association between segregation at two loci located on the same chromosome. The alternative to 50 per cent recombination is linkage with some smaller recombination value with *a priori* distribution which is specifiable in principle and can in fact be approximated. Usually no other parameter need be estimated, nor the ascertainment of the main locus specified. Even when penetrance and gene frequencies must be estimated, this can nearly always be done on other evidence and with such precision that covariance with the recombination fraction may be neglected.[625] Sample sizes are so small that large-sample theory is often unreliable. Families are collected over a period of time and may be analyzed as sequential samples. Here is a problem differing from that usually encountered in scientific research in being amenable to the acceptance sampling theory of Neyman and Wald, which assumes a formal alternative to the null hypothesis. These small-sample methods, adapted to sequential analysis, are the most precise, powerful, and utilitarian of all linkage tests in man.

DETECTION OF LINKAGE

Consider two gene loci, G and T, not necessarily on the same chromosome. Information about linkage is given only if at least one parent is doubly heterozygous, $GgTt$. An individual of this genotype may be of either of two possible phases, GT/gt or Gt/gT, corresponding to his formation by the union of GT and gt gametes or of Gt and gT gametes.   If the G and T loci happen to be on the same chromosome, these two phases correspond to the usual meanings of coupling and repulsion.   In any case, the frequencies of the four types of gametes produced by this individual, if he is GT/gt, will be

$$(1 - \theta)/2 \quad GT, \quad \theta/2 \quad Gt, \quad \theta/2 \quad gT, \quad (1 - \theta)/2 \quad gt,$$

whereas, if he is Gt/gT, they will be

$$\theta/2 \quad GT, \quad (1 - \theta)/2 \quad Gt, \quad (1 - \theta)/2 \quad gT, \quad \theta/2 \quad gt,$$

where θ is the probability of recombination between the two loci (nearly always, $\theta \leqslant \frac{1}{2}$).

If the population is at equilibrium, coupling and repulsion are equally likely and the pooled segregation frequency is 1:1:1:1, so there is no association between linked genes in the general population.   Within a sibship, in the absence of information about the linkage phase, the probability of a, b, c, d gametes of types GT, Gt, gT, and gt, respectively, from a doubly heterozygous parent is

$$(\tfrac{1}{2})^{s+1}\{\theta^{a+d}(1 - \theta)^{b+c} + \theta^{b+c}(1 - \theta)^{a+d}\}$$

where $a + b + c + d = s$.   The simplest type of family is a double backcross $(GgTt \times ggtt)$, in which the gametic constitution is recognizable by inspection. Complexities introduced by single backcrosses $(GgTt \times Ggtt)$ and a variety of double intercrosses (such as $GgTt \times GgTt)$, as well as by dominance, ascertainment, incomplete parental testing, partial sex-linkage, and multiple alleles and pseudoalleles have been treated by Morton.[571, 572]   In all cases, the probability is a function of θ which consists of the sum of as many terms as there are possible combinations of phase in the parents.   Pedigrees of three or more generations lead to complex probabilities which, in general, cannot be decomposed efficiently into the contributions of individual sibships.

All these cases are subsumed by the following theory.   Let $f(i; \theta_1)$ denote the probability of the ith pedigree evaluated at $\theta = \theta_1$, and $f(i; \frac{1}{2})$ be the probability of this pedigree under random recombination $(\theta = \frac{1}{2})$.   Then

$$z_i = \log \frac{f(i; \theta_1)}{f(i; \frac{1}{2})}$$

is called the *lod score* (by elision of *logarithm of the odds*) for testing the null hypothesis H_0 that $\theta = \frac{1}{2}$ against the alternative hypothesis H_1 that $\theta = \theta_1$.   Under the assumptions of the test with respect to ascertainment and the segregation frequency at each locus, rejection of H_0 at a specified significance level implies acceptance of the conclusion that G and T are linked with a posterior probability that can, in principle, be specified.   Rejection of H_1 implies acceptance of the hypothesis that the true value of θ is greater than θ_1.   Thus the investigator may test with great efficiency for close linkage (say, $\theta < .1$) in a small sample or for loose linkage ($\theta > .1$) in a larger sample.

Wald [904] and Haldane and Smith [316] elaborated a fixed-sample-size theory. Let Z denote the cumulative score Σz_i. Then if $Z > \log A$, this provides evidence for linkage significant at the $1/A$ level (at least). The proof of this is simply that the expected value of antilog Z is 1 on the null hypothesis. Suppose the inequality antilog $Z > A$ occurs with probability α. Then under the null hypothesis $A\alpha < E(\text{antilog } Z) = 1$, which implies $\alpha < 1/A$. Similarly, if $Z < \log B$ this is evidence that θ is greater than θ_1, significant at the B level (at least).

Wald [904] made an outstanding contribution to this subject by the theory of sequential analysis. Because families that give information about a particular pair of genes G and T are usually collected over a period of time and often by several investigators, it is convenient to have a test that will permit detection or exclusion of close linkage (or any specified value of linkage) with the smallest possible number of families. Under such circumstances, it is difficult to see how the conventional fixed-sample-size theory could be efficient or even valid. Sequential analysis, on the other hand, sets the boundary conditions $\log B < Z < \log A$. If at any point in the sampling Z exceeds $\log A$, the evidence for linkage is significant. If at any point Z becomes less than $\log B$, the evidence that the recombination fraction is greater than θ_1 is significant. In the intervening range between $\log B$ and $\log A$, the evidence is indecisive at the preassigned significance levels, and more data must be collected to reach a decision.

Since there are 22 pairs of autosomes in man, the chance that two random loci are linked is only about .05, so a conservative value of A (say 1,000) should be used to give a significance level of .001. At this level, more than 95 per cent of significant linkages are expected to be genuine.[572] The value of B is less critical, but .01 or more has been suggested, depending on the assurance with which it is desired to detect or exclude close linkage. This procedure leads to detection of close linkage with sample sizes that are less than $\frac{1}{3}$ as large as for fixed-sample-size tests with the same probabilities of error.[572]

ALTERNATIVE LINKAGE TESTS

Until lod scores were developed, linkage tests consisted in comparison of the deviation of a total score with its standard error, both calculated on the hypothesis of no linkage. It was assumed that the amount of data was sufficiently large that the distribution of the total score in the absence of linkage would be approximately normal. This large-sample theory was refined by Bernstein, Hogben, Haldane, Fisher, Finney, Bailey, and Smith, until it seemed that every bit of information could be wrung from human pedigrees. Penrose extended these applications to siblings from unspecified parents, which increased the generality but decreased the efficiency and reliability of the linkage test.

Despite these developments, no linkages were discovered in man for twenty years after the possibility of linkage detection was first explored. In fact, a number of claims put forward were subsequently proved false. As the available methods were being used, they lacked precision and specificity. Smith [807] in a critical review summarized the disadvantages of these methods as follows:

1. Calculation of the variance of these scores may be intractable when the parental genotypes are unknown.

2. They are efficient only in the limit for loose linkage, which it is not practicable to detect. An ideal test would be efficient for close or moderate, rather than loose, linkage.

3. Information about linkage can be greatly increased by using data from three or more generations. It is not feasible to combine this information with families of unknown phase on the null hypothesis.

4. The assumption of normality may be far from true for moderate sample sizes. Exact tests fail to confirm evidence for linkage by large-sample tests.[571]

5. The sib-pair method of Penrose is inefficient, sensitive to pleiotropy and heterogeneity in gene frequencies, and inexact for families of size greater than 2.

6. Because of the low prior probability of linkage, usual significance levels provide inadequate protection against false claims of linkage. For example, most assertions of linkage based on the conventional .05 significance level are expected to be incorrect, even in samples large enough for the calculated significance level to be valid. However, a significance level of .001 or less provides adequate grounds for claiming linkage, if the assumptions of the test are appropriate.

Morton [572] has stressed the contrasting advantages of the sequential linkage test:

1. The type I and II errors $(1/A$ and $B)$ are reliably specified.

2. No variances need be calculated.

3. The average sample size for given type I and II errors is smaller than for fixed-sample-size tests, with fewer than one-third as many observations required in typical applications.

4. Data from different mating types, with phase and parental genotypes known or unknown, may be combined simply by adding the lod scores.

5. Given the scores for various values of θ, the maximum likelihood estimate, its standard error, and tests of homogeneity can easily be obtained.

6. Calculation of the lod score is simple in principle and can be programmed for an electronic computer, relieving the investigator of uninspiring arithmetic.[801]

Recently, Smith [808] has proposed that posterior probability (Bayes' theorem) be substituted for lod scores as the linkage test. This suggestion was prompted by the observation of Haldane and Smith [316] that "There is some evidence, chiefly from a comparison with the known linkage values of *Drosophila*, that it may not be a bad approximation" to assume that the recombination fraction for linked genes has a uniform distribution from 0 to $\frac{1}{2}$, and the demonstration by Morton [572] that a chromosome with a genetic map length of 100 units, along which gene loci are distributed uniformly and recombine in accordance with Kosambi's mapping function, has a nearly uniform distribution of recombination fractions. *Drosophila*, corn, and the mouse were shown to conform approximately to this distribution. For man, there was then only the unpublished evidence of Schultz that the mean map length of a human chromosome, based on chiasma frequency, is about 100 units,[604] but to this may now be added the observation of Ford and Hamerton [237] of a mean chiasma frequency of 56 per cell, corresponding to 120 units per chromosome. At least in the mouse, estimates of map length from genetic data and chiasma frequency are in good agreement.[118]

Given these results, it would seem that a uniform distribution is likely to be a good approximation in man. On this assumption, average sample sizes and powers for sequential tests were calculated, and the significance level $1/A$ was chosen with these parameters in mind.[572] The posterior probability of no linkage is $\rho = 1/(1 + \Lambda/\phi)$, where ϕ is the prior odds against linkage (about 19) and Λ is the probability ratio score on the hypothesis of linkage, or

$$\Lambda = 2 \int_0^{\frac{1}{2}} (\text{antilog } Z) \, d\theta$$

If Λ is sufficiently large, the evidence for linkage is significant.

However, I have been unable to accept the further step, proposed by Dr. Smith, of incorporating the assumed prior distribution into the linkage test. The reasons for reluctance are as follows:

1. The lod score can be calculated exactly, but the posterior probability requires numerical integration of probability ratios for at least several values of θ. Smith gives convenient approximation formulae for Λ, but the computations are longer and less accurate than that for the sequential test.

2. Since the posterior probability is a monotonic function of the probability ratio, Λ, any property of one must apply to the other. In particular, Smith's statement that the validity of a probability ratio depends on the stop rule but the validity of Bayes' theorem does not, is obviously incorrect. In both cases, the condition for the validity to be independent of the stop rule in repeated sampling is that the difference between θ_1 and $\frac{1}{2}$ be large relative to the standard error of θ from a single sample.[17, 18] This is true for a sequential test of close linkage but is not true for a test based on Λ, which allows θ_1 to approach $\frac{1}{2}$. Thus the stop rule cannot be ignored in using Bayes' theorem.

3. Although a uniform distribution of θ is plausible for two random loci, it is inaccurate in two respects. First, the distribution can at best be only an approximation, as it is based on a uniform density of loci, constant chromosomal length, and Kosambi's mapping function. Second, in human genetics it is not generally possible to choose two loci, but only two traits. Thus, genetic heterogeneity of one or both traits hangs over a linkage study like the sword of Damocles, and when there is heterogeneity the uniform distribution of θ will no longer be even ap-

proximate. Detection of genetic heterogeneity is the principal purpose of a linkage test in man, and there can be little merit in assuming a prior distribution that is grossly incorrect in the cases of greatest interest.

4. A sound theory of statistical inference should distinguish prior evidence and evidence in the sample under investigation. As Fisher [232] has argued with great cogency, the implication of statistical significance is "logically that of the simple disjunction: *Either* an exceptionally rare chance has occurred, *or* the theory . . . is not true." An example of the confusion that can result when prior probability and the information in the sample are confounded by Bayes' theorem is afforded by Smith's treatment of data of Steinberg and Morton [831] on linkage between cystic fibrosis and the *MN* blood groups. He finds that the "probability of the loci being unlinked" is .89. A novice in human genetics might suppose that this provides good evidence against linkage. On the contrary, examination of the lod scores shows that only close linkage can be ruled out on these data, the lods for larger values of θ being equally consistent with loose linkage or no linkage. In other words, the sample is small and indecisive, except for close linkage, and the "evidence" against loose linkage is furnished entirely by the prior expectation that any two loci are unlikely to be linked. In fact, had no data at all been collected, the statement could have been made that "the probability of the loci being unlinked is about .95," which is not appreciably different from the statement made on the evidence. The conclusion (based directly on the lod scores) that "close linkage is reasonably excluded by this sample" is more informative than an equivocal statement about posterior probability.

5. Smith's objections to sequential analysis seem to be largely on semantic grounds. Words like "accept," "reject," and "terminal decision" are used for brevity by mathematical statisticians to indicate strength of evidence. If "terminal" is replaced by "tentative," the nature of a "decision" in sequential analysis is not different in any respect from a conventional statement of statistical significance in fixed-sample-size theory.

6. One advantage of sequential tests is that they permit the investigator to concentrate on detecting or excluding close linkage. This is the only kind of linkage that can be detected in man and certainly the only kind that is of value either in resolving genetic heterogeneity or predicting the outcome of human matings. However desirable the detection of loose linkage may seem, it is so far beyond the bounds of reasonable effort with present methods of human genetics that the choice among competing linkage tests must be based on their power to detect close linkage. Accordingly, it is appropriate to compare sequential and posterior probability tests in the limiting case for close linkage, when $\theta = 0$. All progeny will then be nonrecombinants and the probability ratio for n double-backcross progeny of known phase is

$$\Lambda_n = 2^{n+1} \int_0^{\frac{1}{2}} (1 - \theta)^n \, d\theta = \frac{1}{n+1} (2^{n+1} - 1)$$

For the sequential test, the posterior probability that a case of apparent linkage be a type I error is $1/(1 + A\overline{P}/\phi)$, where A is the upper boundary of the test and $\overline{P}$ is the average power. Comparing this with the corresponding probability, $1/(1 + \Lambda/\phi)$, for the posterior probability test, we see that $\Lambda = A\overline{P}$ is a critical value at which the two tests give the same posterior probability of linkage. For the sequential test, the average sample number n when $\theta = 0$ is $\log A/\log 2(1 - \theta_1)$. Table 6 compares the two tests.

Table 6

COMPARISON OF THE SEQUENTIAL AND THE BAYES' TESTS IN THE LIMITING CASE FOR CLOSE LINKAGE ($\theta = 0$), ASSUMING A UNIFORM DISTRIBUTION OF THE RECOMBINATION FRACTION θ BETWEEN 0 AND $\frac{1}{2}$

Characteristics of sequential test [572]						Corresponding Bayes' test	
θ_1	A	B	$\overline{P}$	$\log 2(1 - \theta_1)$	$n(\theta = 0)$	$\log A\overline{P}$	$\log \Lambda_n$
.05	2000	.01	.28	.2788	11.8	2.75	2.75
.10	1000	.01	.39	.2553	11.8	2.59	2.75
.20	1000	.01	.56	.2041	14.7	2.75	3.53
.30	1000	.01	.71	.1461	20.5	2.85	5.14

Since

$$\log A\overline{P} = \log \Lambda_n$$

in the first row, a sequential test with $\theta_1 = .05$ is as good as the Bayes' test in the limit for close linkage and better in the neighborhood of .05.[905] A sequential test with $\theta_1 = .10$ is not quite so good as the Bayes' test when $\theta = 0$, the difference between $\log A\overline{P}$ and $\log \Lambda_n$ corresponding to less than one observation. Because the sequential test is necessarily better in the neighborhood of θ_1, it is clear that the Bayes' solution can lead to no appreciable gain in power compared to a sequential test of close linkage. (These calculations do not tell whether the Bayes' test has as good power for close linkage as the sequential test, only that it is no better.) Table 6 shows that the Bayes' solution recognizes close linkage more efficiently than a sequential test of loose linkage ($\theta_1 = .2$ or .3). This is of little moment, since no reasonable investigator would undertake a test of loose linkage unless he could command an amount of data much more than sufficient to detect close linkage. If human geneticists follow the practice of testing every pair of loci first for close linkage ($\theta_1 = .05$ or .10) and only with exceptionally large samples extend the test to loose linkage, the efficiency of simple sequential tests to detect close linkage cannot be exceeded by Bayes' theorem. Thus, the investigator who ignores the weight of statistical opinion and embraces Bayes' theorem must expect to make more calculations, but to detect no more close linkages (perhaps fewer) than if he had used a sequential test.

In summary, advocates of posterior probability have not been able to adduce evidence that the extra labor of calculation involved will lead to any improvement over the sequential test, and there are logical arguments against the use of Bayes' theorem even if, as for linkage, the assumed prior distribution has some plausibility. However, the composite probability ratio, Λ, and the posterior probability, first used by Haldane and Smith,[316] are still useful statistics for summarizing evidence on linkage and confirming the significance of a sequential or other test. Their use will generally lead the investigator to be properly cautious in claiming a barely significant linkage,[945] and, like sequential tests, they have the advantage of being as reliable in small as in large samples. There is reason to think that the posterior-probability test is of high power, and its use in sequential sampling is appealing on theoretical grounds,* although extremely laborious in practice. Thus my preference for simple sequential tests is pragmatic, and the difference in viewpoint between Smith and me, since we both use probability ratios either as lods or posterior probabilities, is much less than between us and the proponents of large-sample tests a generation ago.

ESTIMATION OF LINKAGE AND TESTS OF HETEROGENEITY

Because of small sample size and the low prior probability of two loci being on the same chromosome among 22 pairs of autosomes, most linkage tests in man are either negative or indecisive. Morton [571] concluded that, with a relatively large amount of data, the chance of detecting linkage to a particular blood-group factor is about .02–.03 for a rare dominant and .01–.02 for a rare recessive. Thus, it is convenient to have separate theories for detection and estimation of linkage. Fixed-sample—size-estimation theory is appropriate for close linkage, ignoring the sequential nature of the observations.[18]

Once linkage has been recognized, effort will usually be made to extract all information from the record and to collect additional pedigrees. More than half the data from three or more generations may be lost if the pedigrees are decomposed into individual sibships, because of the information each individual gives about the phase of close linkage in relatives. Calculation of the full probabilities for large pedigrees can be extremely laborious, but electronic computers promise to eliminate this nuisance.[801]

---

* By the same argument as for sequential tests based on probability ratios, it can be shown that a sequential test based on the posterior probability of linkage, $\Lambda/(\phi + \Lambda)$, with stop-rule

$$B < \Lambda/(\phi + \Lambda) < A$$

is valid in the sense that the probability of rejecting linkage when it is present is less than B and the probability of accepting linkage when it is absent is less than $1/A$. The principal objections to such a test are the uncertainty of the prior distribution, the heavy computations required, and lack of evidence that any increase in power to detect close linkage will result. Clearly, weighting the probability ratio by a uniform distribution between 0 and $\frac{1}{2}$ cannot give a most powerful test with respect to close linkage ($\theta \to 0$), although Neyman-Pearson theory shows Λ to be a most powerful test over the whole uniform distribution.

Given the probabilities, the maximum-likelihood estimate of linkage still poses a problem because of the compounding of many binomial terms in pedigrees with unknown linkage phase. The usual maximum-likelihood device of differentiating the logarithm of the likelihood founders in forbidding algebra, but several shortcuts have been proposed. The first finds the maximum of Z either graphically or by fitting a polynomial. In large samples, the central-limit theorem makes Z a parabola with the vertex at the maximum-likelihood estimate. Therefore any three points, say (θ_1, Z_1), (θ_2, Z_2), and (θ_3, Z_3), determine the large-sample estimates, which from the normal distribution are

$$\hat{\theta} = \theta_3 + R/2S, \qquad \sigma^2 = MQ/2S, \qquad \hat{Z} = Z_3 + R^2/4SQ,$$

$$Q = \begin{vmatrix} \dot{\theta}_1 & \dot{\theta}_2 \\ \dot{\theta}_1{}^2 & \dot{\theta}_2{}^2 \end{vmatrix}, \qquad R = \begin{vmatrix} \dot{Z}_1 & \dot{Z}_2 \\ \dot{\theta}_1{}^2 & \dot{\theta}_2{}^2 \end{vmatrix}, \qquad S = \begin{vmatrix} \dot{Z}_1 & \dot{Z}_2 \\ \dot{\theta}_1 & \dot{\theta}_2 \end{vmatrix}$$

and $M = \log_{10} e = .4543$, $\dot{\theta}_i = \theta_i - \theta_3$, $\dot{Z}_i = Z_i - Z_3$.

By using a transformation to make Z more nearly parabolic or by choosing the three points sufficiently close to the vertex, a good estimate of the three parameters can be obtained.[567]

The second method for finding $\hat{\theta}$ is appropriate if $\hat{\theta}$ is small. Then the frequency of certain recombinations is very nearly an unbiased estimate, with approximately a binomial distribution. (If θ^n is the highest power of θ that is a factor of the probability, then n is the number of certain recombinants.)

The third method is called "counting" recombinants.[810] It depends on treating θ and $1 - \theta$ as independent in first differentiating the probability. If $\partial P/\partial \theta$ and $\partial P/\partial(1 - \theta)$ represent these partial derivatives, and if P is of the form

$$P = \Sigma C_{ij}\theta^i(1 - \theta)^j,$$

where the C_{ij}'s are constants, then

$$dP/d\theta = \partial P/\partial\theta - \partial P/\partial(1 - \theta)$$
$$= \frac{\Sigma i C_{ij}\theta^i(1 - \theta)^j}{\theta} - \frac{\Sigma j C_{ij}\theta^i(1 - \theta)^j}{1 - \theta}.$$

Since $dP/d\theta = 0$ at the maximum-likelihood estimate, we obtain the iterative solution

$$\theta = \frac{\Sigma i C_{ij}\theta^i(1 - \theta)^j}{\Sigma i C_{ij}\theta^i(1 - \theta)^j + \Sigma j C_{ij}\theta^i(1 - \theta)^j},$$

where the numerator may be regarded as proportional to a "count" of the number of recombinants.* Starting with a trial value of θ, an improved estimate is ob-

---

* If the phase of linkage were known, the probability of the sample would be

$$P = \theta^x(1 - \theta)^y,$$

tained and substituted in the equation. Continuing in this way, the maximum-likelihood solution $\hat{\theta}$ is finally obtained to as many decimal places as desired without the labor of evaluating the information as in the conventional method. Smith [810] shows how the counting method can be extended to give the standard error and tests of heterogeneity and how it simplifies the estimation of gene frequencies with multiple alleles.

Smith [808] has given formulae for numerical integration by Simpson's rule to obtain Λ from several values of Z. In large-sample theory,

$$\log \Lambda = \hat{Z} + \log \sigma + .7001,$$

provided linkage is highly significant and there is at least one certain recombinant.[571]

A confidence interval for θ may be obtained from the standard error σ by large-sample theory, from the binomial or Poisson distribution if $\hat{\theta}$ is small (as it usually will be), or by a method of Haldane and Smith [316] if θ is constant. They showed that the inequality

$$Z' > \log \Lambda - \log A$$

provides, by projection on the θ axis, a confidence interval of strength at least $1 - 1/A$. Upper confidence limits for the case of no certain recombinant, appropriate to alleles or pseudoalleles, were given by Morton.[571]

Tests of heterogeneity of θ are extremely important, and may well be considered the *raison d'être* of linkage detection in man. A convenient large-sample criterion is provided by the maxima of the lod scores as

$$\mathcal{L} = 4.605 \left(\sum_{i=1}^{n} \hat{z}_i - \hat{Z} \right),$$

where $\hat{z}_i$ denotes the maximum of z_i in the interval.

$$0 \leqslant \theta \leqslant \tfrac{1}{2},$$

_____

where x is the number of recombinants and y the number of nonrecombinants. The maximum-likelihood score would be

$$dL/d\theta = x/\theta - y/(1 - \theta).$$

When the linkage phase is unknown and the probability is therefore compound,

$$P = \Sigma C_{ij}\theta^i(1 - \theta)^j,$$

the quantity $\Sigma i C_{ij}\theta^i(1 - \theta)^j/P$ enters into the maximum likelihood score in the same way as x, and therefore may be called the (likely) number of recombinants, and similarly,

$$\Sigma j C_{ij}\theta^i(1 - \theta)^j/P$$

is the (likely) number of nonrecombinants. Because the true number of recombinants is unknown if the probability is compound, this analogy is not exact but it leads to the maximum-likelihood solution.

$\hat{Z}$ is the maximum of Z in the same interval, n is the number of pedigrees or samples, and logarithms to the base 10 are used for the lods. This likelihood-ratio criterion is distributed in large-sample theory as χ^2 with $n - 1$ degrees of freedom on the hypothesis of no heterogeneity. There is some reason to believe that this may be preferable to other large-sample heterogeneity tests in samples of the small size encountered in human linkage data.[567] Smith's counting method provides, with rather more labor, a maximum-likelihood heterogeneity test that converges asymptotically to the likelihood ratio, $\mathcal{L}$. No practical small-sample heterogeneity test has been proposed.

In recent years three cases of close autosomal linkage have been established in man—between the nail-patella syndrome and the ABO blood groups, elliptocytosis and Rh, and the Lutheran antigen and ABO secretor.[481] Recombination takes place in both sexes with approximately equal frequency. The first of these linkages is homogeneous, indicating that the nail-patella syndrome is determined by the same locus in different pedigrees. Recombination between elliptocytosis and the Rh locus is heterogeneous among pedigrees, and analysis demonstrates that hereditary elliptocytosis may be determined by either of two dominant factors, at present distinguishable only by the fact that one of them is closely linked to the Rh locus.[567] This is the first instance in man of resolution of genetic entities by linkage to another marker, a technique that has great promise for the incisive analysis of rare factors. Resolution of loci and alleles has the same fundamental importance to the geneticist as purity to the chemist, recognition of clinical entities to the physician, or fractionation of antibodies to the immunologist. Linkage analysis provides the ultimate criterion of genetic unity.

EXPERIMENTAL ANALYSIS OF SEGREGATION AND LINKAGE IN MAN

In recent years, preliminary experiments have raised the hope that linkage and segregation can be studied in the somatic cells of man by the methods developed for microbial genetics.[484] Somatic segregation and crossing-over and genetic exchange through transduction and cellular fusion have been studied in lower organisms. Mosaicism of possible segregational origin has been observed in man;[23, 152] and recent discoveries of several viable types of human aneuploidy suggest that a wide range of chromosomal variation may be observed in somatic cells and provide a means for chromosome mapping without meiosis.[236] The remarkable cytologic observations of inversion heterozygotes in *Drosophila* salivary chromosomes may have a parallel at pachytene in human chromosomes, which Yerganian[952] considers to be more favorable for study than rodent material. The complementarity test of allelism has so far only been applied to abnormal hemoglobins and genetic coagulation defects in man, but it can undoubtedly be extended to other systems. If an artificial mixture of two defective plasmas has a normal coagulation time, the genes are shown to be members of different cistrons, and therefore to be nonallelic.[51]   Determination of the site of amino-acid substitution in a protein is *ipso facto* a test for allelism, genes acting at the same site being allelic.[397]   There is even hope that genetic analysis of the human haplophase may be feasible, if unconvincing reports of phenotypic expression of segregation in sperm are confirmed.[309]   Prospects for an experimental genetics of human cells which now seem speculative or even fantastic may be realized within the next few years, to match the strides that are being made by nonexperimental methods.

CONCLUSION

There is a tendency among experimental scientists to assume that observational disciplines are inherently less accurate than those that can manipulate the objects of study in controlled experiments.   However, a little reflection will show that this is by no means the rule, and it would be both false and invidious to assert that astronomy is less precise than physics, or geology than chemistry, despite the pronounced inequalities of experimental opportunity in these areas. Human genetics, still in its infancy compared with astronomy or geology, began with a search for examples of Mendelian inheritance, progressed to precise analysis under restrictive assumptions, and is now entering the third stage of generalizing by analysis and experimentation the conditions under which similar phenomena, genetic or environmental, can be incisively resolved.   That human genetics has the brashness and excitement of a young science is because scarcely fifty years have elapsed since its beginning, and there has not been time enough to develop a great body of knowledge or to attract more than a few men of genius.   The tempo of discovery is accelerating, and if we do not have our Galileo, it may be that this generation will produce him.

DISCUSSION

DR. BURDETTE: Dr. James Neel will open the discussion of Dr. Morton's paper.

DR. NEEL: Dr. Morton's paper is a complex and comprehensive presentation which obviously deserves intensive study.   He was kind enough to make it available to me as we traveled together en route to this meeting, but that was obviously scarcely long enough to digest its contents and my comments are of necessity rather superficial.   The paper is an excellent summary of the recent developments in segregation and linkage analysis.   These are topics that Dr. Morton is certainly eminently qualified to discuss, in view of his own many contributions to both the theory and the techniques.   Of the various aspects of his presentation which might be singled out for comment, I have chosen to say a few words about the matter of ascertainment probability because of some of our own recent experiences.

As mentioned in his presentation, satisfactory estimates of ascertainment probability are necessary for accurate segregation analysis.   Dr. Morton has demonstrated that the number of ascertainments per affected individual can be utilized to estimate the total ascertainment probability, provided we make the assumption that the different ascertainments of a single individual are independent of one another. However, this is an assumption that those of us who labor in the field must examine with some care. Dr. Shaw, Dr. Falls, and I have recently completed a study of congenital aniridia which illustrates precisely the questions that arise when one assumes the independence of multiple ascertainment of a single individual.[794]

Congenital aniridia is a dominantly inherited absence of the iris of the eye. In the young, it is accompanied by a visual handicap. Later, when affected individuals reach their fourth or fifth decades, one frequently observes certain complications of the disease. These include glaucoma and cataract, and an affected individual may become almost or completely blind as a consequence of either or both of these. We wished to set up a roster of all the cases of this disease occurring in the lower peninsula of the state of Michigan. The procedures for establishing such a roster are well known. We turned, for instance, to the practicing ophthalmologist. We turned to the schools for the visually handicapped, particularly the school at Lansing. We turned to the state agencies that are responsible for the state-aid programs for the visually handicapped and for the blind. In brief, as one does when one attempts to establish a complete roster, we attempted to follow every lead.

Now consider the case of the individual who, as a youth, is visually handicapped because of his aniridia and then, at the age of twenty-five, develops one of the complications of the disease which I have just mentioned. If he has not previously consulted an ophthalmologist, he will do so at this time. The ophthalmologist is quite aware of the state facilities for the blind and visually defective and will first put the individual in contact with the program for the visually handicapped if the vision deteriorates, and then in contact with the program for the blind if there is complete loss of vision. The three ascertainments that we might make of these individuals are obviously not independent of one another. There are, of course, ways of setting up rosters whereby one can minimize this matter of overlapping ascertainment. For example, one's ascertainments can be restricted only to ophthalmologists and avoid the aid programs for the blind. However, once an individual is given a diagnosis with a very bad prognosis in medicine it is not at all uncommon for that individual to begin to shop around, looking for a ray of hope. There is probably a positive association between being seen by one competent specialist and being seen by several in the case of a disease with a poor medical prognosis.

Dr. Morton has stated that segregation analysis, in his opinion, is sufficiently robust to withstand a considerable error in the estimate of ascertainment probabilities. However, I note that some of his formulae entail exponential functions of ascertainment probabilities, so where multiple ascertainments are the rule rather than the exception, errors in estimating ascertainment probabilities may be exaggerated to the point that they pose a real problem for the analysis.

These remarks are not meant in any way to detract from the recent mathematical developments in this field which we owe to a number of individuals, and particularly to Dr. Morton. They are meant to draw attention to some of the problems involved in working with multiple ascertainments of a single individual in an effort to arrive at over-all ascertainment probabilities. I am sure the points I make do not constitute in any way insurmountable obstacles in that one can correct for nonindependent or overlapping ascertainment, and I am perfectly confident that Dr. Morton is capable of developing the necessary refinements of the formulae.

DR. MORTON: The ascertainment probability is a nuisance parameter, of no intrinsic genetic interest, but there are two reasons why a human geneticist must be concerned with it. The first concerns the determination of prevalence by dividing the number of living probands (n) by the ascertainment probability (π). This is fairly insensitive to errors in estimating the ascertainment probability, since errors in n and π tend to be compensatory. However, we could use help from the epidemiologist to get an independent estimate of the prevalence by standard sampling methods, if their use is not prohibitively costly for rare traits. The second reason for interest in the ascertainment probability is to test hypotheses about genetic parameters, such as segregation frequency and proportion of sporadic cases. Fortunately, these tests also appear to be fairly insensitive to errors in estimating ascertainment probability.

The ascertainment problem for dominant genes is much more complex in principle than for recessive genes, since the more affected there are in previous generations, the more likely the kindred is to be ascertained in this generation. However, the practical difficulties are reduced in large kindred, because ascertainment of the kindred is almost certain even if the ascertainment probability of an affected individual is low.

Dr. Neel raised the problem of variable ascertainment probability, dependent on age, severity, or other circumstances. This is relevant only to the extent that the ascertainment probability is variable among families. This can be allowed for very easily if the number of independent ascertainments per proband is recorded, from which the ascertainment probability can be determined as a function of age, severity, et cetera. Again I would stress that modern methods of human genetics appear to be quite insensitive to small errors in estimating ascertainment probability.

DR. LILIENFELD: Dr. Morton, you mentioned use of the epidemiologic method as being distinct from the genetic method. I do not think we are dealing with two different methods, since both are very much alike in that both the epidemiologist and human geneticist are concerned with studies of natural phenomena as they occur in human-population groups. Most of the time we are unable to conduct experiments among humans to test various hypotheses of interest. The epidemiologist is interested in studying disease or other conditions in human populations from a broad variety of viewpoints, including environmental and genetic aspects. The approaches used for these two aspects are quite similar.

You also made the statement that the epidemiologist is satisfied with a low degree of diagnostic accuracy of a human disease in contrast to the human geneticist who desires a higher degree of accuracy. I do not think this desire is related to a person's disciplinary background. The epidemiologist and geneticist, as well as any other student of human disease, is interested in obtaining diagnoses of the highest possible accuracy. I think the question of whether someone is satisfied with a low degree of diagnostic accuracy does not depend upon his approach, but upon the kinds of data with which he is dealing, the level of development of the hypothesis he is testing, and so on. Your statement is based on the fact that a great deal of epidemiologic data is obtained by interviews, which do not provide the best quality of diagnostic information. However, the epidemiologist may be satisfied with this at an exploratory level in a study. As soon as he arrives at a higher level of development of an hypothesis, he begins to look for more refined methods of measuring the disease he is studying.

I have one question I would like to ask with respect to one of the methods you discuss in your paper. Apparently, in the tests you have developed, you have added an additional parameter for sporadic cases. It is not clear to me how you actually decide which are the sporadic cases. I have the feeling there may be some sort of circular reasoning involved, but I am really not sure.

Dr. Burdette: In a statistical discussion, recognition of the fact that the statistical approach is limited by the degree of accuracy of the original observations is refreshing.

Dr. Morton: Sporadic cases present no difficulty if they can be recognized phenotypically. If they are not phenotypically distinct, a useful probability statement can still be made. Isolated cases can be regarded as two different kinds: chance-isolated cases with the same etiology as familial cases and sporadic cases with a different etiology and with a negligible risk of recurrence in the sibship. If we determine the proportion of cases that are sporadic and the risk (segregation frequency) in high-risk families with familial or chance-isolated cases, then we can specify exactly the probability that an isolated case with a given number of normal siblings is sporadic.

This turns out to be quite a powerful method because we have at least two independent ways of determining the proportion of sporadic cases. The inbreeding coefficients of isolated and familial probands give one estimate when familial cases are due to rare recessive genes, and the excess of isolated cases gives another; so I think the definition of sporadic cases is quite clear and useful when the risk in high-risk families is several orders of magnitude greater than the frequency of sporadic cases.

With respect to the objectives of epidemiologists and geneticists, there is this important difference: the geneticist is usually interested in the mode of inheritance but not the fact of inheritance. He is investigating single genes, not "hereditary influences." Few geneticists are concerned with conditions like breast cancer, which present essentially epidemiologic problems. I am sure that methods can be improved in this area, but I doubt that geneticists can make a real contribution unless some of the new genetic techniques turn out to be unexpectedly useful.

DR. LILIENFELD: In Copenhagen, there is an Institute of Human Genetics from which studies of familial aggregation of various types of cancers have been reported. A few examples are given in the references.[334, 576, 895] There is a great deal of published material by human geneticists on studies of familial aggregation of various other diseases. This indicates the existence of much interest in such matters on the part of human geneticists, and I think human geneticists should be interested in studying familial aggregation of human disease.

DR. MORTON: My contention about geneticists is based on a biased sample, but I will stay with it. There are differences between genetics and epidemiology in methods and objectives from which both sides can learn. I think the epidemiologist working with rare diseases would find it useful to use methods based on incomplete ascertainment instead of aiming at a complete roster. Here the geneticist is ahead of him.

That there is still a rather wide gulf between genetics and epidemiology is illustrated by the National Health Survey.[892] I have yet to find a geneticist who could see any possibility of using information obtained under such inadequate diagnostic standards, yet many epidemiologists are hopeful about this. Epidemiologists have been able to make considerable use of the standard classifications of disease and causes of death. I do not think a geneticist can use these for any purpose except ascertainment of cases, which must then be studied intensively to confirm the diagnosis and obtain family material. The geneticist is usually trying to separate a genetic entity from an heterogeneous clinical category, and therefore he tries to operate at the forefront of methods of diagnosis and testing. The epidemiologist is much more conservative, retaining heterogeneous categories for the sake of uniformity of statistical tabulation and to generate an incidence large enough to be studied by his random-sampling techniques. Random samples of the general population are like Procrustes' bed; they provide a solution for all problems only at a prohibitive price.

DR. NEEL: The antithesis that has been posed between the epidemiologic and the genetic approaches may have been a valid thesis thirty years ago when the epidemiologist was concerned with bacteria and the geneticist with genes, but I see no place for an antithesis in the context of the times.

DR. MORTON *: The genetic and epidemiologic "approaches" can be defined in many ways, for some of which there is no valid antithesis. Please note that for

---

* Addendum (written communication).

the purposes of my paper I made a precise definition of the genetic and epidemiologic methods. One may quarrel with my use of these terms, but there is no question that, as defined, they are antithetical.

Since Dr. Neel has presented aniridia as an example of ascertainment procedures which may not fit my models, Dr. Italo Barrai in my laboratory has examined the data in more detail. In the published paper of Shaw et al. it is reported that of 95 probands, 54 were ascertained once, 29 twice, 9 three times, and 4 four times. In addition, 23 were ascertained only through family history. Making the (perhaps) incorrect assumptions that all familial cases were ascertained and that ascertainment of isolated and familial cases was equally likely, the authors estimated that about 18 isolated cases were missed, giving 136 as the minimum number of cases living in Michigan.

If ascertainments were independent, the distribution of number of ascertainments (t) per proband would be a truncated Poisson, and the equation to be solved to give the maximum likelihood estimate of π would be

$$U_\pi = \frac{-\Sigma t}{(1 - \pi)\ln(1 - \pi)} - \frac{\Sigma a}{\pi(1 - \pi)} = 0,$$

where the number of probands is $\Sigma a = 95$ and the number of ascertainments is $\Sigma t = 152$. We obtain $\pi = .642$, giving $95./642 = 148$ as the estimated number of aniridiacs living in Michigan. Clearly this is substantially the same as Dr. Neel's minimum estimate, thus providing a basis for his assumpton that nearly all familial cases were ascertained. Furthermore, a χ^2 test shows that the numbers of 1, 2, 3, and 4 or more ascertainments give an extremely close fit to a truncated Poisson distribution. Thus there is no support for Dr. Neel's apprehension that "errors in estimating ascertainment probabilities may be exaggerated to the point that they pose a real problem for the analysis." Undoubtedly ascertainments could be defined so as to violate the assumption of independence, the validity of which can always be tested as in this example. In the hands of a careful worker like Dr. Neel, ascertainments and probands will be defined so as to be amenable to these modern methods of analysis.

James Crow, Ph.D.

POPULATION GENETICS: SELECTION

INTRODUCTION

In this chapter, selection in man will be considered from three standpoints:

1. The *mechanics* of selection. In analogy with physical mechanics, evolutionary mechanics may be divided into statics and dynamics.[308] Dynamics is concerned with the rate at which selection changes the phenotype, the genotype, or, at the most elementary level, the frequencies of individual genes. Statics deals with the variety of factors that maintain the population in equilibrium, and the conditions for stability of such equilibria.

2. The *amount* of selection. Under this heading, questions such as the following arise. How much differential mortality and fertility is there, and how is this measured? How much of the differential is associated with differences in phenotypes or genotypes? How much could the population be changed if all differential viability and fertility were directly applied to a systematic change in gene frequencies? How has the intensity of selection changed with changes in the environment?

3. The *cost* of selection. Such processes as mutation, Mendelian segregation and recombination, and changes in the environment have the effect of reducing the population fitness from what it would otherwise be. Of course these processes are important and necessary for long-range evolution, but, from a short-term view, the population fitness is impaired by such processes. The reduction in fitness or the increase in concrete measures such as specific types of morbidity, mortality, or sterility has been called the genetic load.[159, 181, 575] The problems are to devise operational definitions of the different components of the over-all genetic load, criteria by which these may be distinguished, and ways of measuring them.

53

THE DEFINITION AND MEASUREMENT OF FITNESS

In a discussion of selection, an immediate question is: How is fitness to be defined and how measured? Fitness is the capacity to survive and to leave descendants. It may be measured therefore by the number of descendants per parent; but how far in the future are the descendants to be counted? Thoday [872] has suggested that representation by descendants in the distant future is the most important criterion of fitness; in this sense, the ancestors of the mammals were more fit than the dinosaurs. Fitness, in the broadest sense, should include both adaptedness and adaptability as indicated by the ability of successive generations to respond to changes in the environment; but such considerations in man do not lead to any quantitative treatment, nor do they have any predictive value. So we shall be content with more prosaic models in which fitness is measured by current viability and fertility. Usually only a single generation of selection is measured, although a later example will show that consideration of more than one generation is sometimes necessary for a clear formulation of the problem.

Throughout this chapter fitness will be used in the strict Darwinian sense of expectation of descendants and, as Haldane [312] has cautioned, "not fitness for football, industry, music, self-government, or any other activity."

The simplest measure of the fitness, W, of a genotype, or of a population, is the average number of children per parent, the parents and children being counted at the same age. This measure of fitness has been used by Haldane,[313] Wright,[943] and many others. Strictly, it is true only for populations with discrete, non-overlapping generations such as annual plants, but, when the differences in the average age of reproduction in the two groups being compared is negligible, this measure is a satisfactory approximation. In demographic analyses, it is frequently convenient to avoid some of the problems introduced by minor departures from complete monogamy by measuring the number of daughters per mother. On the other hand, for genetic conditions the effect on fitness is likely to be different in the two sexes, so that separate consideration of males and females is necessary.

It is important to realize that relatively minor differences in the average age of reproduction may be of great importance in determining the composition of future populations. In human disease there is likely to be a changed pattern of reproductive ages. For example, with a disease that causes death or sterility during the normal reproductive ages, an estimate of fitness based only on total number of children would underestimate the true fitness because the average age of reproduction of the diseased group would be younger than the population average and therefore the contribution to future populations greater than the total number of children would indicate.

When there are overlapping generations and variable ages of reproduction, a measure that takes account of the continuous and exponential growth of the population is desirable. Such a measure of fitness is provided by a quantity that was introduced to demographic problems by Lotka [573] and has been known by various names such as "true rate of natural increase," "intrinsic rate of increase," and "innate capacity for increase." I shall use Fisher's [230] vivid term, the "Malthusian parameter." This quantity, designated by m, is defined by the equation

$$\int_0^\infty l_x b_x e^{-mx}\, dx = 1. \tag{1}$$

where x is the age, l_x is the probability of surviving to age x, b_x is the probability of giving birth during the infinitesimal interval x to $x + dx$, and e is the base of natural logarithms.

The Malthusian parameter has the following important property: if a system of death and birth rates has been in operation for a sufficient length of time, the population will reach a stable age distribution. When this distribution has been attained, the proportional rate of increase of the population is measured by m. That is, if N is the population number and t is time,

$$\frac{dN}{dt} = mN. \tag{2}$$

The time, t, may be measured in any desired units. If x in equation (1) is given in years, as it usually is, then t is measured in years. However, to change the scale of measurement simple multiplication is correct. For example, if the average age at reproduction is twenty-eight years, the Malthusian parameter for time units measured in generations is simply 28 times that for units measured in years.

This formulation offers one difficulty. In the human population at present the age-specific birth and death rates are changing so rapidly that the present age distribution in most countries is far from the stable age distribution that this schedule of births and deaths would lead to if they remained unchanged for many generations. Thus, it is possible for a population with a negative Malthusian parameter actually to increase if the age distribution is such that a disproportionate share of the population is in the ages of maximum reproduction. One can, of course, determine the fitness in such a way as to predict directly the change in population numbers in the next generation. However, the Malthusian parameter has the merit of being calculable solely from the age-specific birth and death rates, and, for most purposes, it is the quantity of greatest interest, since it gives the long-time effect of these birth and death rates.

A more sophisticated approach to this problem, given by R. A. Fisher,[230] defines the *reproductive value* for each age. The reproductive value is analogous to the present value of an annuity; for example, an adolescent has a higher value than an infant not only because he is more likely to survive until the reproductive age, but also because he will reach that age sooner and therefore begin to contribute to future generations earlier. If the reproductive value is determined for each age, and if the average reproductive value of the population is determined by weighting each individual by the reproductive value appropriate to his age, then the total reproductive value of the population will increase at a rate given by the Malthusian parameter irrespective of the age distribution of the population.

However, for most problems in human genetics, the most suitable measure of comparative fitness of two genotypes is the Malthusian parameter. Although this may not give a correct prediction of the gene frequencies next generation in a population with a nonequilibrium age distribution, it will tell what this schedule of genotypic birth and death rates would lead to if continued for a long time.

A final problem remains. Equation (1) is not easily solved. Solutions by iteration are time consuming, although the problem should be simple for modern computing machines. Several authors have proposed rapid methods that give reasonably good approximations.[336] Most of the related work has been done by demographers not interested in genetic problems. Possibly some of the computation procedures worked out by animal ecologists may be useful in problems in human genetics.[3, 14]

Reed [710] has presented a different definition of the relative fitness of two genotypes. He sums the annual birth rate, divided by the age, for all survivors. The ratio of these quantities is used as a measure of relative fitness of two genotypes. This has the merit of taking age of reproduction into account. However, it requires the same kinds of data as the Malthusian parameter without having its theoretic advantages.

When fitness is measured in Malthusian parameters, it may be positive or negative. If a measure that is always positive and counts descendants at some future date is desired, it is easily obtained as follows: If m is the Malthusian parameter with time measured in years and g is the average age of reproduction, fitness can be measured by

$$W = e^{mg}.\tag{3}$$

If two genotypes, a and b, are to be compared their relative fitness is given by

$$W_a/W_b = e^{(m_a - m_b)g}.\tag{4}$$

When there is no difference in the distribution of reproductive ages, this is the same as the W mentioned earlier for discrete generations. It is frequently convenient to take the fitness of the control population as 1, that is, $mg = 0$.

In the determination of the fitness of a certain genotype, the question of a proper control for comparison always arises. Ideally one would prefer as a control group a sample chosen so as to be representative of the general population. This topic has been extensively discussed, for example by Macklin.[519] If the fitness of a particular genotype is to be measured, it should be compared with individuals who are similar in all ways except for the locus or loci being studied. The selection of a suitable control group composed of normal siblings has been suggested, but there are many sources of bias in such a procedure. Ways of eliminating many of these have been worked out by Krooth,[463] but the procedure is still doubtful and a general-population sample is preferable.

There are two indirect methods for measuring selection, both involving the measurement of selective values over a period of time. One method is to count the frequency of a trait at different ages. Preferably, one would start with a co-hort of individuals and measure the frequency of the trait in successive years; for example, ten-year-olds could be counted from 1910 records, twenty-year-olds from 1920 records, and so on. The changed incidences with age would, when compared with controls, provide a measure of selective mortality. If this is not feasible, comparisons of persons of different ages can be made at the same time. This is a valid measure of differential mortality, provided the initial incidence has not changed during the time covered by the age differences.

Another procedure, useful for traits whose exact heredity is known, has been applied in indirect estimates of human mutation rates.[643] If all instances of the disease can be attributed to a dominant allele with 100 per cent penetrance, the cases can be divided into hereditary and sporadic. The ratio of nonhereditary to total cases is a measure of the selective disadvantage of the condition. For exam-ple, Verschuer [893] reports from Nachtsheim's data that of 56 cases of Pelger anom-aly, 3 were from parents who were both free of the disease. Thus, the reduction in fitness due to this condition may be estimated as $\frac{3}{56}$. The procedure is most useful for deleterious conditions that reduce fitness by a large amount.

THE MECHANICS OF SELECTION

As stated in the introduction, evolutionary mechanics is usefully divided into dynamics and statics. The rates at which population attributes, such as gene fre-quencies, genotypes, or phenotypes, are being changed by selection (dynamics) can be considered as well as those mechanisms that lead to equilibria because of some sort of balance between the selective forces (statics).

If one knows the genotypic composition of a population and the fitnesses of the various genotypes comprising it, he can predict the changes that will occur in the composition of the next generation. To the extent that the environment stays constant and the genes do not interact in unforeseen ways, the same equations may be used to derive the composition of the population in generations farther re-moved. Most considerations of selection contain the tacit assumption that the fitness of an offspring is independent of his parents' fitness, except insofar as they share common genes. Of course, this is not always true, as, for example, when the genotype of the parent conditions the environment of the child so as to change thereby the selective value of the child's genotype; but such indirect effects are in many cases unimportant, and, in our present stage of knowledge of human selec-tion intensities, the error made by ignoring such factors is probably less than that due to other uncertainties.

The rate of change of gene (or genotype) frequencies is usually written on a per-generation basis. One assumes that selection is Markovian in the sense that the selective forces acting on a generation are determined solely by the genotype frequencies in that generation and not by the historical path by which these frequencies were reached. Any other assumption is too complex to handle except in particular cases, so I shall consider only selection models based on this assumption.

It may, nevertheless, be of interest to give a simple example of a situation in which consideration of more than one generation is helpful in understanding selection. The example is selection determining the sex ratio, studied by Fisher [230] and by Shaw and Mohler.[795]

Selection for the Sex Ratio.—It is not immediately apparent how selection acts to change the sex ratio, since a parent that produces offspring with an unusual sex ratio still leaves the same total number of progeny and thus makes the same gene contribution to the next generation as one producing a normal ratio. Nevertheless, consideration of two generations affords a clear basis for selection. The method followed here is that of Shaw and Mohler.

Consider a population that for some reason has an excess of males at the time of conception. In such a population a female zygote, on the average, leaves a larger number of children than a male, since each child must have one male and one female parent. A parent that produces a higher proportion of females will be producing more children of the more fertile sex, and hence will have more grandchildren. Thus a gene that alters the sex ratio in the direction of a 1:1 ratio at conception will be favored because of making a larger contribution to the gene pool two generations later. As a result, assuming of course some genetic variance in the sex ratio, the ratio will tend toward 1:1.

This is the ratio at conception. Clearly, post-zygotic mortality has no effect on the argument. Therefore this kind of selection should result in equal proportions of the two sexes at conception, but with departures from equality at later ages if there is differential mortality. This may explain why many polygamous species of animals, which would appear to need only a small proportion of males, still produce the two sexes in approximate equality.

Fisher [230] has noted, in the human species, that because of the greater male death rate both pre- and post-natally, it is likely that a female producing an excessive proportion of sons will therefore produce more zygotes. The reason is that a lost infant or embryo may be partially compensated by the next child being conceived somewhat earlier than it would otherwise have been. Thus, a mother producing an excess of males will on the average have a somewhat larger number of children conceived, although fewer born, than a mother who produces children with both sexes in equal frequencies. The population should therefore come to equilibrium with a slight excess of male zygotes, although the greater mortality of males would equalize the ratio at some later age. The fact that the human sex ratio appears to conform to this model offers some evidence for its validity.

The Rate of Gene-frequency Change.—The elementary quantities on which selection acts are the frequencies of the various alleles. A knowledge of the allele frequencies and the mating system is sufficient to specify the expected frequencies of the various genotypes. Usually, human populations have so little inbreeding that, for most purposes, random mating and the Hardy-Weinberg relations can be assumed. A review of the methods of population genetics as applied to demography and physical anthropology has been written by Spuhler.[825] For a general review of the elementary theory of population genetics see Li.[502]

As mentioned earlier, a model of gene-frequency change may be formulated either for discrete generations or for continuous change with overlapping generations. In many cases it makes no practical difference which model is used. Haldane[306] showed that the effect of overlapping generations is usually small, unless there is a difference in the average age of reproduction in the groups being compared. It is convenient to have both sets of formulae available in theoretical calculations because in particular cases one or the other leads to more manageable algebra.

With discrete generations, fitness is measured by W, defined as the expected number of progeny per parent. The two generations must, of course, be counted at the same age. To assign mortality and fertility effects to the same generation, counts may be made at the zygote stage. In the human population this is not possible, and the time of birth may be conveniently used instead. On the other hand, in certain circumstances it is more convenient to count individuals at some other age. Sometimes this is dictated by the way the data are made available. There is also an advantage of measuring the population at the beginning of the reproductive period if the effects of random fluctuations of gene frequencies are to be considered. As Fisher has shown, the variance is minimum if the populations are counted at this stage and the effects of random drift can best be assessed for data gathered in this way.[162, 231]

Throughout this discussion, it will be assumed that the parent and offspring generations are counted at birth. The change in frequency, q_i, of the allele A_i is given by Wright[943] as

$$\Delta q_i = q_i \frac{W_i - \overline{W}}{\overline{W}} \tag{5}$$

where W_i is the average fitness of all the genotypes containing allele A_i weighted in accordance with the number of A_i alleles (that is, twice for the A_i homozygote and once for each heterozygote). $\overline{W}$ is the average fitness of the population. The quantity, $W_i - \overline{W}$, is called[230, 233] the *average excess* of the allele A_i.

If fitness is measured in terms of Lotka's and Fisher's Malthusian parameter, m, the formula becomes

$$\frac{dp_i}{dt} = p_i(m_i - \overline{m})$$ (6)

or, in words: the momentary rate of increase in the frequency of an allele is equal to the allele frequency multiplied by the average excess in fitness of that allele. For a derivation of this see Crow and Kimura,[161] Kimura,[449] or Fisher.[230] The Malthusian parameter, m, may be measured either in years (or some other arbitrary time unit) or in generations.

Alternatively,

$$\frac{d \log p_i}{dt} = m_i - \overline{m}.$$ (7)

If m is measured in terms of generations (that is, the average age of reproduction, which is approximately 30 years), it is easy to show the near-equivalence of the two formulations.[943] In an infinitesimal time interval, Δt, the gene frequency changes from p_i to $p_i(1 + m\Delta t)$; then, in a period corresponding to an average generation, the frequency changes to

$$p_i(1 + m\Delta t)^{1/\Delta t},$$

which, as Δt becomes small, becomes e^m. Therefore, W for discrete generations corresponds to e^m for continuous change when time is measured in generations.

From these basic formulae the changes in gene frequencies in future generations can be derived from knowledge of the fitnesses of the various genotypes. Also, from the changes in gene frequencies the changes in genotype frequencies or changes in measurable attributes of the population, such as a metric character, or fitness, can be derived. Such formulae have been given for a great number of types of inheritance and for various assumptions about the mating system and the nature of the action of selection. Nature is, of course, more complex than even the most troublesome formulae, so that these have little long-term predictive value. Nevertheless, the theoretical studies of Haldane, Wright, Fisher, and others are of great value in showing what can and cannot happen under certain assumptions and provide a set of predictions with which the real situation can be compared.

For an elementary discussion of this subject and a review of the classical results, see Li[502] and also the appendix to Haldane's[309] early book. More recent and general treatments are given by Wright.[943, 944] For a discussion of the rate of change of metric traits, and especially the change in fitness itself, see Fisher,[230] Kimura,[449] and Kempthorne.[441]

Gene-frequency Equilibria.—In many respects, the statics of human selection is more interesting and important than the dynamics. It is likely that only a very small amount of the total selection has any permanent effect in changing gene frequencies. Much more selection probably is directed toward maintaining the status quo: eliminating recurrent mutations, maintaining polymorphisms, adjusting to momentary changes in the environment, and so on. Thus, much of the interest in selection is in determining what systems lead to stable equilibria.

A theoretically simple equilibrium occurs when there is recurrent mutation to a harmful allele. Every generation a certain number of mutant alleles arise, and these are eliminated by selection in later generations. It is clear that the two processes will balance when the number of new mutants arising is exactly equal to the number of old mutants eliminated from the population by selection during the same period. Thus the intensity of selection required by the fact of mutation must be equal to the mutation rate, except insofar as two or more mutants may be eliminated by the premature death or infertility of the same individual.

Also, a number of equilibria are possible under selection alone. The simplest occurs when the heterozygote for a pair of alleles is superior in fitness to either homozygote. Another possibility is an allele that is favored when rare, but selected against when common, as might be true of an allele producing some specialized ability that is important, but for which a small number of such individuals suffice.

It is sometimes quite difficult to determine whether a system of hypothetical genotypic fitnesses will lead to a stable equilibrium. A necessary condition for equilibrium is that

$$\Delta p_i \text{ or } dp_i/dt = 0$$

for all alleles. I am ignoring the possibility of recurrent cyclical changes, as may sometimes occur when the fitnesses of different genotypes have certain complex interdependencies [943] or in host-parasite relations.[513]

An equilibrium may be unstable as well as stable. The unstable type is of little interest; the slightest perturbation in the gene frequency away from the equilibrium point leads to a further departure, so that the gene frequency tends to drift away from the equilibrium value. If an equilibrium value is found, it is transitory. On the other hand, if the equilibrium is stable, any departure from this value is followed by an immediate return, so the equilibrium value persists in the population. For this reason, such an equilibrium situation is nevertheless likely to be encountered because of its persisting nature, even if its probability of arising is low *a priori*.

With multiple alleles or multiple loci, the conditions determining whether the equilibrium is stable are sometimes quite complex. Of the several criteria for stability, two are particularly useful. The first is applicable to any situation in which the genotypic fitnesses can be regarded as constants; that is, it would not necessarily be applicable if the fitness of a genotype depended on its frequency. With constant genotypic fitnesses, selection acts in such a way as to maximize the average population fitness.[448] Therefore the necessary and sufficient conditions for stability are that the equilibrium point be a relative maximum when the fitness is expressed as a function of the allele frequencies. The detailed application of these criteria has been given by Kimura [448] for either multiple alleles at a single locus or multiple independent loci. For two alleles, it is necessary that the heterozygote be more fit than either homozygote. Nevertheless, it is not necessary in a multiple allele system for all heterozygotes to be more fit than any homozygote; in some stable systems this is not true. There are also systems that are stable but which become unstable when one of the elements is removed, as the following example from Kimura shows:

Genotype	A_1A_1	A_1A_2	A_1A_3	A_2A_2	A_2A_3	A_3A_3
Fitness	$-.1$	0	$.15$	0	$.1$	$-.2$

The fitnesses are given in Malthusian parameters. This system leads to a stable equilibrium, with allele frequencies

$$p_1 = \tfrac{2}{15}, \ p_2 = \tfrac{9}{15}, \text{ and } p_3 = \tfrac{4}{15}.$$

Yet, if A_3 is somehow eliminated from the population, A_2 will quickly replace A_1, so the stability depends on the presence of all three alleles.

If the genotypic fitnesses are not independent of their frequencies, selection does not necessarily lead to maximum population fitness and another criterion for stability is needed. The usual procedure is to determine the equilibrium values by setting the time derivative of the gene frequency equal to zero. Then, the stability is determined by whether an infinitesimal perturbation is followed by a return toward the equilibrium point. This procedure has been used by Penrose et al.,[653] and an explicit set of rules has been given for more complex cases by Kimura.[446] Geometrically, the array of gene frequencies may be thought of as a point in n-dimensional space. If this point is displaced, one asks, crudely speaking, whether the vector for gene-frequency change points in the general direction of the equilibrium value. The analytical procedure is given in Kimura's article. See also Lewontin.[500]

A situation of importance exists when the phenotype of optimum fitness is intermediate for some metric character, such as height, weight, and probably intelligence. In most instances, there is no stability at intermediate gene frequencies under selection alone,[230, 946] but, under some special circumstances, there may be.[460] The mathematical procedures have been given by Kojima.[460]

High-speed computing machines have frequently been useful in testing the consequences of hypothetical schemes of selection and gene action that are not amenable to mathematical analysis. It is frequently possible to determine whether or not stable equilibria exist by letting the machine try a large number of special cases. Sometimes a general solution is suggested by such numerical results.

In any particular instance it is usually very difficult to determine the detailed way in which selection is operating. For some near-lethal or otherwise highly deleterious factors, a reasonable estimate of the fitness may be obtained. But for common factors, whose effect on fitness must therefore be relatively mild or somehow balanced, the measurement is frequently very difficult. (See Haldane.[304])

For example, it is likely that strong selection is acting on the blood-group genes. The *ABO* locus is known to be associated with a number of effects of maternal-fetal incompatibility. There are also strong associations between the blood groups and adult diseases, as for example between group *O* and duodenal ulcer; but which of these is more important demands a careful analysis. One study [138] suggests that the most important factors are those operating during fetal and early postnatal stages, and that the associations between blood groups and adult diseases are second-order effects from the standpoint of influence on allele frequencies. In addition to fetal deaths due to incompatible matings, there is a significant excess of heterozygotes, in compatible matings, of a magnitude probably sufficient to maintain the polymorphism.

A similar excess of heterozygotes is found in certain matings in the *MN* system. An analysis by Morton and Chung [574] illustrates the value of looking into as many sources of error as possible. In this case, the evidence is against the *MN* excess being due to false-positive reactions, illegitimacy, ambivalent *M–N* alleles, nondisjunction, preferential fertilization, or meiotic drive. After eliminating these as likely possibilities, the preferable remaining hypothesis is heterosis. Yet this must be of a complex sort, for it is apparent only in *MN* × *MN* matings.

It is evident that any instance of human polymorphism demands a detailed study, making use not only of genetic information but a full knowledge of the entity being studied and the possible technical errors. The importance of the use of genetic methods rather than those purely epidemiologic has been emphasized elsewhere in this volume. Detailed analysis of segregation ratios offers perhaps the best single source of information in human genetics.[569] Although gene-frequency analysis and epidemiologic methods are useful, they give the most unequivocal information when combined with a study of family data.

THE AMOUNT OF SELECTION

The intensity of selection can be considered at three levels. First is *total* selection. This is determined by the total prereproductive mortality and by the fertility differences in the population. A substantial fraction of such differences are neither related to the genotype nor to the phenotype, being determined by geography, epidemics, accidents, and other genetically irrelevant factors; but some of the differential mortality and fertility is associated with observable differences in phenotype, so the second level is *phenotypic* selection. Not all phenotypic differences are associated with genotypic differences, but to the extent that survival and fertility are determined by genotype, there is the third level, *genotypic* selection.

Although the third level of selection is the one of main interest to geneticists, it is also useful and important to consider the other two levels.

Total Selection.—A procedure for investigating the total amount of selection that is taking place in the human population was given by Crow.[159] It was suggested that the relevant quantity to measure is the *index of total selection, I,* defined by the formula

$$I = \frac{V}{\overline{W}^2} \tag{8}$$

where V is the variance in the number of children per parent and $\overline{W}$ is the mean number. Parents who die before reproduction may be counted as leaving zero descendants. This is a rather crude procedure, since it takes no account of differences in the age of reproduction; but it can serve as a first approximation, and better methods can be developed as the data warrant. The rationale for this choice of a measure of total selection intensity and the derivation of the formulae were given earlier and will not be repeated here.[159] In the same paper it was shown how selection due to prereproductive mortality and to fertility differences can be separated. This index has the meaning that if fitness were completely heritable, that is, if each offspring had exactly the average of his parents' fitnesses, the fitness of the population next generation would be that of the present generation multiplied by I. A trait or a gene that is genetically correlated with fitness will increase in proportion to this correlation.

The contribution to the index from differential mortality and differential fertility can be separated into I_m, due to mortality and I_f, due to fertility differences. The total index is then

$$I = I_m + I_f/p_s \tag{9}$$

where $\qquad\qquad I_m = p_d/p_s \text{ and } I_f = V_f/\bar{x}_s^2.$

In this formula, p_d and p_s are the proportions of infants that die before the reproductive period and that survive, respectively; $\bar{x}_s$ is the mean number and V_f is the variance in the number of births per survivor. The application of equation (9) to actual census data is complicated by deaths during the reproductive period. Fortunately, the death rate is low at this time and the error is not large.

These formulae have been applied to mortality and fertility data from United States Census reports [159] and from studies of the Ramah Navaho Indians.[825] It is not surprising to find that the decrease in death rates in recent years has reduced the index of total selection due to mortality, on the other hand, it appears that the component of the index due to fertility increases with decreasing family size. This trend shows clearly in the data in table 7. These values are obtained from various sources of various accuracy for the determination of I_f. In more recent

Table 7

INDEX OF SELECTION DUE TO DIFFERENCES IN FERTILITY, I_f, IN VARIOUS POPULATIONS *

Population	Mean number of children	Index I_f	Source of Data
Rural Quebec	9.9	.20	Keyfitz [445]
Hutterite	9.0	.17	Eaton and Mayer [197]
Gold Coast, 1945	6.5	.23	Fortes [250]
New South Wales, 1898–1902	6.2	.42	Powys [674]
United States, born 1839	5.5	.23	Baber and Ross [27]
United States, born 1866	3.0	.64	Baber and Ross [27]
United States, 1910	3.9	.78	U.S. Census
United States, 1950	2.3	1.14	U.S. Census
Ramah Navaho Indians	2.1	1.57	Spuhler [825]

* The figures are based on total children born to women who have survived to the end of the reproductive period.

U.-S. data, not shown in the table, the trend is reversed. The mean has increased while the variance has decreased, so that the index has again become smaller.

Clearly there is considerable differential fertility at the present time, although not much opportunity for selection by postnatal mortality. If all the fertility differences were associated with genotypic differences, a very rapid change in the genic make-up of the population could take place, but probably very little of the difference is actually genotypically determined at present.

Intensity of Phenotypic Selection.—The measurement of selection intensity for a phenotype that is unequivocally harmful is in principle not difficult. For example, the selection intensity against low-grade mental deficiency or muscular dystrophy could be determined from the average fitness of this phenotype in comparison with the population as a whole.

For many characters of interest the optimum is not extreme but intermediate. For example, persons who are very tall or very short have lower viability than persons of intermediate height. Haldane [303] has suggested ways in which the intensity of such phenotypic natural selection can be measured.

Haldane considers in particular an example of human birth weight. Here, too, the probability of survival is highest for intermediate values. Suppose that of those with the optimum weight a fraction w_o survive, whereas the over-all survival weight is $\overline{w}$. Haldane defines the intensity of natural selection for birth weight as $\log_e (w_o/\overline{w})$, or roughly $w_o - \overline{w}$. For example, when w_o is .983 and $\overline{w}$ is .959, as in the data he gives, the intensity of selection is $\log_e (.983/.959)$ or .024. Thus, the fitness of the population is about 2.4 per cent less than it would be if all births were of the optimum weight; that is, the *phenotypic* load due to variability in birth weight is .024. Haldane's article may be consulted for a detailed treatment of this subject.

Genotypic Selection.—The final aim of the study of selection in man is to determine the extent to which it is genotypic. This can be approached in a number of ways, most satisfactorily when the basis of the inheritance is known and the genotypes can be identified. Or, if the individual genotypes cannot be identified as with dominance, it is sometimes possible to make use of gene-frequency information and knowledge of the mating system to determine the frequencies of the various genotypes. Then the problem is one of measuring the fitness of the various genotypes, as has been discussed earlier.

Except for a few conditions, such as amaurotic idiocy and achondroplasia, with fitness greatly reduced, very little empiric information exists at present. Recently Spuhler [825] said: "We have practically no reliable information on selective values for the more common genotypes in man. Unless some mechanism like balanced polymorphism is involved, we must suppose that the selective differentials for such characters will be relatively low, probably less than 1 per cent, and thus extremely difficult to demonstrate statistically." There is clearly need for more measures of the fitnesses of known genotypes if the working of selection in the human population is to be understood. In a later section, the selective mechanisms involving the blood-group polymorphisms will be discussed.

For traits with unknown inheritance, not very much in the way of detailed information about genotypic selection can be obtained. In animal and plant breeding, it is conventional to determine the *heritability* for traits of economic importance.[214, 441, 942] This is estimated from the degree of similarity of individuals of different degrees of relationship, but such a measure, in order to have any genetic meaning, must have the genotypic and environmental factors independent. In experimental animals and plants this can be arranged by suitable randomization; in the human population, it is rarely even approached.

Correlations between identical twins suffer from the fact that they are reared in an environment at least as similar as that of ordinary siblings, and there are too few instances of twins reared apart to offer any quantitative information. Furthermore, the heritability determinations from identical twins are based on the total genotypic variance, much of which is due to dominant and epistatic factors that are not responsive to selection. Correlations between siblings or other relatives ordinarily have genetic and environmental factors so thoroughly confounded that it is not possible to assign any selective differences to genotypic differences. Thus, for the moment, measurement of genotypic selection intensities for characters whose genetic basis is not precisely known is a difficult, if not dubious, undertaking.

THE COST OF SELECTION: GENETIC LOADS

The final question is: What is the cost of evolution by natural selection? One way of approaching this is through the concept of a genetic load. The genetic load is a measure of the extent to which the population is impaired by the factor under consideration. The population impairment may be measured in terms of decreased average fitness, or more concretely in terms of mortality, sterility, or morbidity due to specific causes.

The genetic load is measured by the extent to which the average population fitness, or whatever trait is of interest, differs from that of the best genotype in that population. For example, if the best genotype at some particular locus is a heterozygote, there is a load because of the presence of homozygous segregants.

This procedure carries no implication that the optimum population structure is one in which all individuals have the same genotype. Clearly, in a complex society with division of labor the optimum structure will include a variety of mutually advantageous genotypes carrying out different functions. The idea behind the definition of a genetic load is to measure the change that would result if the phenomenon whose effect is being studied (for example, mutation) could be suspended with everything else left unchanged.

The load can be measured in various units. In its most general use, the load is measured by reduction in fitness. If death rates are studied, the most convenient unit is the lethal equivalent.[575] The unit is usually the incidence in the case of diseases. It should be emphasized that a unit amount of reduction in fitness or a lethal equivalent need not have a constant effect from the standpoint of human welfare. Needless to say, an early embryonic death is quite different from a childhood or early adult death in this respect.

Genetic loads have several bases of classification. It might be advisable to attempt to classify them into mutually exclusive and collectively exhaustive categories, but it is probably more useful in the present state of knowledge to devise components of the total load that can, in principle at least, be separated and measured. Therefore the following classification is an attempt to separate the total load into operationally definable categories.

1. *Mutation Load.* This is the extent to which the population is impaired by recurrent mutation. It can be measured in principle by comparison of the population with what it would be if the mutation rate were zero and the gene frequencies were permitted to come to a new equilibrium under this condition. Because mutation is a *sine qua non* for evolution, the mutation load may be regarded, as Haldane[310] has suggested, as the price the species pays for the privilege of evolution.

2. *Segregation and Recombination Loads.* The best genotypes may be heterozygotes, in which case inferior types will recur by segregation. Likewise, certain linkage combinations may be more advantageous than others, in which case recombination may lead to impairment. The loads due to these phenomena are the segregation load and the recombination load. Later, methods for assessing the segregation load under certain circumstances will be given. The extent of the recombination load is doubtful and at present cannot be assessed, so it will not be

discussed further. These loads would not exist in an asexual population; they may therefore be regarded as the price the species pays for the privilege of Mendelian inheritance.

3. *Incompatibility Load.* This load comes, not from any deficiency of a genotype itself, but from the fact that some genotypes have a reduced fitness with certain parents. For example, an *Rh*-positive embryo has a better chance of surviving when the mother is *Rh* positive than if she were *Rh* negative. So far, all examples of an incompatibility load are serologic, but future research may reveal other possibilities.

The mutation, segregation, and incompatibility loads can be measured, at least to an extent. It is convenient to distinguish between the *expressed load,* the load in the existing population (usually mating at random, or nearly so), and the *total load,* or inbred load, which would be found if the population were to be made completely homozygous.

Other loads can be considered, but are difficult or impossible to measure in the human population. Two of these are:

4. *Substitution Load.*[313, 447] In a changing environment some genes will be favored in the old environment but not in the new, and vice versa. The load is imposed by the fact that, in the new environment, the favored genes are rare and the harmful ones common. Haldane [313] demonstrated how to compute the total amount of differential mortality or fertility that is required to reverse the gene frequencies in order that the favorable allele is the common one. However, the requisite data for any such measurement in the human population do not exist at present.

5. *Dysmetric Load.* In an heterogeneous environment or in a society with division of labor, different genotypes are best suited to different niches. The optimum population is one in which the different genotypes occur in the right frequencies to fit most efficiently into the available niches. Haldane has suggested ways by which this might also be measured, but, so far, they have not been applied to the human population. This is an inclusive category and perhaps could best be studied were it broken up into smaller components. For example, the segregation load could be regarded as one part.

There are other possibilities. For example, if meiotic drive [746] or gametic selection exist in man, these could create a load, and so could conflicting aims between the individual and society or a phenotype that lowers the fitness of the population (for example, a criminal or a disease that calls for a great expenditure of medical or other economic resources). Only the mutation, segregation, and incompatibility loads will be discussed here.

The Mutation Load.—Consider first a population with average inbreeding coefficient, F, and with two alleles at some locus of interest. Multiple alleles present no complication, provided the heterozygote between two different mutant alleles has a fitness approximately the average of the two mutant homozygotes. This seems usually to be true, so is not an extravagant assumption.   Loci where alleles show complementarity should be regarded as separate loci for this purpose.   The method given here is that of Morton, Crow, and Muller.[575]

Genotype	AA	Aa	aa
Frequency	$p^2(1-F)+pF$	$2pq(1-F)$	$q^2(1-F)+qF$
Fitness	1	$1-hs$	$1-s$

The average fitness of the population is

$$\overline{W} = 1 - hs[2q(1-q)(1-F)] + s[q^2(1-F)+qF].$$

In the absence of mutation, the population would become entirely AA, with fitness 1.   Therefore the expressed mutation load is the difference, or

$$hs[2q(1-q)(1-F)] + s[q^2(1-F)+qF]. \tag{10}$$

The random mating mutation load $(F = 0)$ is

$$M_R = 2hsq(1-q) + sq^2. \tag{11}$$

The inbred mutation load $(F = 1)$

$$M_I = sq. \tag{12}$$

The equilibrium value of the "recessive" allele frequency, q, is now determined.   Notice that a newly arisen mutant can be eliminated from the population in any of three ways:

(1) by homozygosis because of inbreeding, with probability Fs,
(2) by homozygosis with a preëxisting allele, probability $sq(1-F)$ where q is the frequency of the preëxisting allele,
(3) by selection against the heterozygote, with probability $(1-F)(1-q)sh$.

Adding these, the total probability per generation of elimination of a new mutant is, if all high-order-product terms are neglected, $s(F+q+h) = sz$, where z is the sum of F, q, and h.

If a mutant gene has probability sz of being eliminated in any one generation, this means that it will persist on the average exactly $1/sz$ generations.   Hence the frequency of mutant alleles in the population at equilibrium will be u/sz, where u is the mutation rate from A to a.   Substituting this value for q in the formula for M_R and M_I, the mutation loads may be obtained.

The total load (that is, the inbred load) is the sum of qs over all loci.   Substitution of $q = u/sz$ gives

$$M_I = \sum \left(\frac{u}{zs} \cdot s \right) = \sum \frac{u}{z} = \left(\frac{1}{z} \right) \sum u. \tag{13}$$

The last term on the right is correct provided that z and u are uncorrelated.   This is not strictly true, for one component of z is q, which is dependent to some extent on u. However, the dominant term in z is usually h, which is probably independent of u. So, to a first approximation, the total load is the total mutation rate divided by the harmonic mean of z.

The expressed load may be obtained from equation (10). Substituting u/sz for q, the mutation load for a population with inbreeding F is

$$M_F = 2\Sigma qhs + \Sigma q(q + F)s, \tag{14}$$

when all second order terms are neglected. The sum is over all relevant loci. The first term considers heterozygous effects, the second homozygous.

If z is dominated by h (so that q and F can be neglected), $q = u/sz$, or in this case $q = u/sh$. Then

$$M_R = 2\Sigma qhs = 2\Sigma u \qquad (q + F = 0) \tag{15}$$

If h is negligible relative to $q + F$, then $z = q + F$

$$M_R = \Sigma q(q + F) = \Sigma u \tag{16}$$

In general, the expressed mutation load is the genomic mutation rate multiplied by a factor between 1 and 2, depending on whether the mutant is recessive or partially dominant. By a similar line of reasoning, the factor is $\frac{3}{2}$ for a sex-linked gene.[310]

If the trait being measured is survival rate, it is likely that h is considerably larger than either F or q. The latter two quantities are of the order of 10^{-3}, whereas, judging from *Drosophila* data, the average value of h in an equilibrium population is about .02 or .03.[358] Taking h as .025, the inbred mutation load is about twenty times as large as the random mating load.

Although it is not possible to measure the inbred load directly, it can be estimated from the reduced fitness in the children of consanguineous marriages. A regression equation can be written in the form

$$D = A + BF$$

where D is the reduction in fitness. For example, if death rates are being measured, D is the death rate of population of inbreeding F. More precisely, to take into consideration coincidences of independent causes of death in the same person,

$$S = e^{-(A+BF)}, \tag{17}$$

where S is the proportion of survivors.

On this formulation, A is the expressed load and $A + B$ is the total load. In actual data, A will also include some nongenetic deaths. Hence the total load is somewhat smaller than $A + B$, though larger than B.

Just as we can measure lethal effects as lethal equivalents, we can also measure detrimental equivalents. If s is the penetrance of the recessive homozygote and hs is the penetrance in heterozygotes, the same formulae can be used for other conditions as for those causing death. Following are some data from several sources summarized and analyzed by Morton: [568]

	A	B	B/A
Premature death (from late fetal to early	.1410	2.555	18.12
adult) France — two populations studied	.0893	1.482	16.60
Early American (1858) death	.1612	1.734	10.75
Congenital malformations (Japan)	.0102	.093	9.1
Childhood abnormalities (France)	.0443	2.196	49.6
(United States)	.1033	1.164	11.3
(Sweden)	.0821	2.020	24.6
Abnormalities, all sorts (early American)	.1067	5.792	54.3
Mental deficiency (normal parents)	.0129	.694	54.0

The French data suggest that the typical person carries from 1.5 to 2.5 lethal equivalents per gamete, or 3 to 5 per zygote. The data on childhood abnormalities from several sources suggest that, in addition to the lethal equivalents, the average person carries about 4 detrimental equivalents. The early American data are interesting in providing an estimate, admittedly crude, of what was true under a more primitive environment. The total load appears to be about twice as high when measured in terms of lethal and detrimental equivalents as expressed in that environment. The relationships given here can be used as an indirect estimate of the human mutation rates. These are summarized by Morton.[568]

The Segregation Load.—The most obvious cause of a segregation load is a locus where the heterozygotes are superior to the homozygotes. When all homozygotes are inferior to any heterozygote there is a stable equilibrium.[448] In fact, stable equilibria are possible under less restrictive conditions, as mentioned earlier.

From equation (5) the necessary conditions for equilibrium can be determined by setting

$$\Delta p_i = 0.$$

This leads to

$$W_i = \overline{W}.$$

If it is assumed that the optimum genotype, one of the heterozygotes, has fitness 1 and the fitnesses of the different genotypes are measured as deviations from this,

$$1 - s_i = W_i \quad \text{and} \quad 1 - \bar{s} = \overline{W}.$$

Thus the expressed segregation load is $\bar{s}$, and therefore equal to s_i.

In a randomly mating population the segregation load,

$$s_i = \frac{\sum_j p_i p_j s_{ij}}{p_i} = \sum_j p_j s_{ij} \tag{18}$$

where s_{ij} is the deviation from optimum fitness of genotype $A_i A_j$.

Therefore, the genetic load for all the alleles at a locus may be estimated by choosing any particular allele, A_i, and, for every genotype into which this allele enters, multiplying the reduction in fitness by the frequency of the other allele in the genotype. These products are then added. This remarkable principle makes possible an estimate of the total load, irrespective of the number of alleles at the locus, from knowledge of a single allele.

Morton [568] has applied this to estimate the segregation load for certain human diseases if these were maintained by some form of selective balance. A minimum estimate of the segregation load can be obtained by noting that the load must be greater than $p_i s_{ii}$, *a fortiori*, since the latter is one of several terms in equation (18). Morton has estimated the decreased fitness of the homozygotes for deafness, muscular dystrophy, and for low-grade mental defect. The expressed genetic load for these three conditions combined is estimated to be about .25. Morton therefore argues that there cannot be many such loci with such a large load, or the reproductive capacity of the population would not be sufficient to carry the load. Thus it is not likely that these traits are maintained by selective balance to any large extent.

The random mating segregation load, as just shown, is greater than $s_{ii} p_i$ for any allele A_i. However, the total (inbred) load is $\Sigma s_{ii} p_i$ where the sum is over all alleles. Therefore, the inbred load is less than k times the random mating load, where k is the number of alleles maintained in balanced polymorphism.

Morton, Crow, and Muller [575] and Crow [159] have used the high ratio of inbred to randomly mating load observed for mortality rates as an argument that most of the inbred load is mutational rather than segregational, since to assume the latter means that dozens of alleles must be maintained in balanced polymorphism at each relevant locus.

For all these reasons, it appears to me that loci maintained by selective balance make a much smaller contribution to the inbreeding effect than those maintained by recurrent mutation and therefore are a small fraction of all loci. However, this is not to say that the segregation load is necessarily smaller than the mutation load in a randomly mating population. The relative magnitude of these quantities has yet to be determined.

The Incompatibility Load.—The theory of the incompatibility load has been given by Crow and Morton.[163] The general principle is that a mother can produce antibodies only against antigens that she does not possess. Because antigens are ordinarily produced by dominant alleles, any increased death rate will be in heterozygotes. The potentially incompatible types may be summarized by writing the maternal genotypes and the allele contributed to the child by the father. Using the *ABO* blood group locus as an example, Crow and Morton have given the following table, where p, q, and r are the frequencies of the *A, B,* and *O* alleles.

Mother's genotype	Frequency (1)	Sperm genotype	Frequency (2)	(1) × (2)	Probability of death
OO	r^2	A	p	pr^2	d_A
BB	q^2	A	p	pq^2	d_A
BO	$2qr$	A	p	$2pqr$	d_A
OO	r^2	B	q	qr^2	d_B
AA	p^2	B	q	qp^2	d_B
AO	$2pr$	B	q	$2pqr$	d_B

The incompatibility load due to this locus is equal to

$$I = d_A(pr^2 + pq^2 + 2pqr) + d_B(qr^2 + qp^2 + 2pqr)$$
$$= d_A p(1 - p)^2 + d_B q(1 - q)^2. \tag{19}$$

More generally, if p_i represents the frequency of allele A_i and d_i is the decrease in fitness due to the antigen resulting from allele A_i, the incompatibility load can be written as

$$I = \Sigma d_i p_i (1 - p_i)^2 \tag{20}$$

where d_i may, of course, be zero for some alleles.

This assumes that d_i is the same irrespective of the mother's genotype (assuming she has no A_i allele) and of the other (non A_i) allele in the child, which is a reasonable assumption a priori, but not necessarily true. If it is not, a separate d for each maternal-fetal genotype combination has to be introduced.

The incompatibility load differs from the other two loads considered here in that it increases with the inbreeding of the mother, but decreases with the inbreeding of the child. The relationship of the load to inbreeding of both parent and child is given by Crow and Morton.[163] Data from Chung, Matsunaga, and Morton [136] indicate that the incompatibility load due to the ABO locus is .024 lethal equivalents expressed as prenatal deaths. This is a surprisingly large amount for a single locus, considering that the total embryonic death rate is considered to be 30 to 40 per cent.

The load due to the ABO locus becomes still larger when another factor is taken into consideration. The incompatibility effect by itself does not lead to a stable polymorphism. Chung and Morton have offered evidence of strong selection against group O children, in comparison with OA, in compatible matings. Thus it would appear that there is sufficient selection against OO homozygotes (and presumably also against AA and BB homozygotes) to maintain the polymorphism despite elimination of incompatible heterozygotes.

The combined load due to both incompatibility and segregation is estimated by Crow and Morton to be at least .066, of which .063 is estimated to occur in prenatal stages. Clearly there cannot be many loci with such a large effect. If the other major blood groups have comparable effects, a substantial fraction of all embryonic deaths may be attributed to maternal-fetal incompatibilities.

GENERAL REMARKS

In this article, attention has been concentrated on selected topics and methodologic points emphasized. Consequently, little attention has been given to discussion of more general aspects of selection.

It is especially important, if man is to understand his past, that information on selection in primitive and rigorous environments be obtained. The change of selection intensities with altered diet, the drastic decrease of infectious disease, and changed reproductive patterns merit detailed study. Studies of primitive societies will, of course, have to be done while such populations still exist. A summary of existing data and a thoughtful discussion of the direction future studies could take is given by Neel.[602]

Haldane [304] has argued that the major selective agent in human evolution in the last few thousand years has been infectious disease. See also Motulsky.[577] As agriculture and technology developed, human populations became at once much more dense and more protected from the dangers of nomadic life. An increased selective role of disease must surely have followed. Now that infectious disease is being eliminated as a major cause of death and premature death from all causes is low, selection through differential postnatal mortality becomes less and less important. That fertility differences are the major factor in recent human evolution has been effectively argued by Fisher.[230] The extent to which recent changes in reproductive patterns in civilized societies has changed the direction of human evolution and what the consequences will be are still open questions.

DISCUSSION

DR. BURDETTE: Dr. Stern will discuss Dr. Crow's paper.

DR. STERN: The mathematical treatment of selection which Dr. Crow outlined for us must be applied to empiric data. There is a great gap at present between the impressive accomplishments of the biomathematician in this field and those of the human geneticist working empirically. We have some measure of fitness of a few human genotypes, primarily rare dominant mutants, and, even here, we are still in an exploratory stage. What should we use as control population: relatives of propositi, the general population, or other, specific groups?

Still more difficult, and of great general significance, is the problem of selective fertility as applied to different socio-economic strata. How do we define these satisfactorily? How can we distinguish between over-all aspects of defined groups and different trends within any one group? What selective forces, if any, are involved in migrations, geographic, rural, urban, and in social mobility? How much are prenatal deaths selective, not only in blood-group incompatibilities for which good data are becoming available, but also for other genetically controlled traits? How selective is mortality before or during the reproductive period?

How complex the problems are in these fields is apparent from the Scottish Surveys of intelligence, performance, and body size of eleven-year-old school children.[792] In spite of careful collection of data and searching analysis of many interacting social variables, only a beginning has been made in an understanding of the dynamics of populations. How few and limited in scope are the studies on the genic content of two successive generations or of different age classes of a population! To complement the theoretical work on methodology concerning selection, we need increased efforts in the area of empiric methodology.

A.G. Steinberg, Ph.D.

POPULATION GENETICS: SPECIAL CASES

The formal genetics of man or of other organisms can be studied only by the analysis of family data. However, problems concerning the frequency of phenocopies, the number of loci affecting a given phenotype, mutation rates, mutation load, and segregation load may often be better analyzed by studying populations. Relatively inefficient methods to answer some of these questions were developed more than thirty years ago. Various investigations have refined them over the years, and some of these methods will be reviewed.

For many years geneticists have been determining allele frequencies at a single locus by estimating the number of affected individuals in a population on the assumptions that all individuals of the same phenotype have the same genotype, that the genotypes in the population being studied exist in a Hardy-Weinberg equilibrium, and that the sample studied has been randomly selected. These methods and their limitations have frequently been reviewed [828] and will be presented only briefly.

CONSANGUINITY

Dahlberg [167] has reviewed the classical techniques that employ marriages of cousins among parents of affected individuals for estimating the frequency of a rare, recessive gene. The most useful of these concerns marriages of first cousins. The equation is:

$$q = \frac{c(1-k)}{16k - 15c - ck},$$

where q = the frequency of the recessive allele, c = the frequency of first-cousin marriages in the general population, and k = the frequency of first-cousin marriages among the parents of affected individuals.

The use of the equation assumes that: [828]

(1) all occurrences of the character are genetically determined and due to the same gene,

(2) the gene frequency is uniform throughout the population being sampled,

(3) the genotypes occur in Hardy-Weinberg equilibrium, and

(4) the expected frequency of first-cousin marriages is essentially zero.

A source of information which has not been exploited concerns the frequency of homozygotes among cousins of homozygotes for recessive genes. It seems worth exploring this method as an adjunct to the methods developed by the Wisconsin group and, in the light of the rapid decrease in consanguineous matings in this country, perhaps as the only practical method for future studies.

Such information may be used: (1) to estimate the frequency of the allele in the population on the assumption that all affected individuals are homozygous for alleles at a single locus, (2) to check the validity of the assumption that only one locus is concerned with the character in question, and (3) to estimate the number of loci if more than one locus is concerned.

Estimate of gene frequency.—The probability that a first cousin of a homozygote for a recessive gene will inherit this gene from their common grandparent is $\frac{1}{4}$ (assuming heterozygosity among the parents); the probability that the unrelated parent of the cousin will transmit the gene to that cousin is q, the frequency of the gene in the population. Hence, the frequency of affected first cousins is q times $\frac{1}{4}$ or $q/4$. Four times this quantity equals the gene frequency, q. The standard error of this estimate is $[q(4 - q)/n]^{1/2}$, where n = the number of cousins.

Test of one locus versus more than one locus.—An estimate based on a count of affected individuals in the population assumes that all cases are due to genetic changes at a single locus. This assumption may be tested by determining q from the frequency of affected first cousins. Sporadic cases may be ignored if they are rare, or accounted for by using the methods devised by Morton.

If more than one locus is involved, the frequency of homozygotes in the population is approximately

$$\sum_{}^{k} q_i^2 = g,$$

where k equals the number of loci, q_i equals the frequency of the recessive allele at locus i, and where interactions for heterozygosis at different loci and simultaneous homozygosity at different loci are ignored.

The probability that a cousin will be affected if more than one locus is involved is

$$(\tfrac{1}{4}) \sum_{}^{k} q_i + (k - 1) \sum_{}^{k} q_i^2.$$

The derivation of this expression may be seen by considering an example with $k = 4$. The probability that a cousin of an individual homozygous for a given gene (say no. 1) is homozygous for at least one of the genes is

$$1 - [(1 - \tfrac{1}{4}q_1)(1 - q_2^2)(1 - q_3^2)(1 - q_4^2)],$$

where the expression in brackets is the probability that the cousin will not be homozygous for any of the genes. If this expression is expanded and terms involving the products of two or more of the terms $\tfrac{1}{4}q_1$, q_2^2, q_3^2, and q_4^2 are ignored (these will be very small since we are dealing with rare genes), the probability that a cousin will be homozygous *for at least one of the genes* is $\tfrac{1}{4}q_1 + q_2^2 + q_3^2 + q_4^2$. Proceeding in this way we obtain the following table for all four loci.

Locus	Probability of a cousin being homozygous
1	$\tfrac{1}{4}q_1 + \qquad q_2^2 + q_3^2 + q_4^2$
2	$\tfrac{1}{4}q_2 + q_1^2 + \qquad q_3^2 + q_4^2$
3	$\tfrac{1}{4}q_3 + q_1^2 + q_2^2 + \qquad q_4^2$
4	$\tfrac{1}{4}q_4 + q_1^2 + q_2^2 + q_3^2$
Total	$\tfrac{1}{4}\Sigma q_i + 3\Sigma q_i^2$

Since $k = 4$, three equals $k - 1$; hence the general equation for the probability of homozygosis of cousins (c) is $\tfrac{1}{4}\sum^{k} q_i + (k - 1)\sum^{k} q_i^2$. The mean of q, $(\bar{q}) = \Sigma q_i / k$, whence $\Sigma q_i = k\bar{q}$. The variance of q_i, $(\sigma^2) = \dfrac{\Sigma q_i^2}{k} - \bar{q}^2$ whence $\Sigma q_i^2 = k(\bar{q}^2 + \sigma^2)$.

These values for the Σq_i and Σq_i^2 may be substituted in the estimate of the frequency of homozygotes in the general population (g) and in the estimate of the frequency of affected cousins (c) thus:

$$g = \sum^{k} q_i^2 = k(\bar{q}^2 + \sigma^2) \quad \text{and}$$

$$c = \tfrac{1}{4}\sum^{k} q_i + (k - 1)\sum^{k} q_i^2$$
$$= \tfrac{1}{4}k\bar{q} + k(k - 1)(\bar{q}^2 + \sigma^2).$$

Since σ^2 is apt to be small relative to $\bar{q}^2$, these equations may be written as

$$g = k\bar{q}^2 \quad \text{and} \tag{1}$$

$$c = \tfrac{1}{4}k\bar{q} + k(k - 1)\bar{q}^2$$
$$= \tfrac{1}{4}k\bar{q}[1 + 4(k - 1)\bar{q}]. \tag{2}$$

The algebraic solution of these simultaneous equations yields

$$\bar{q} = g\,\frac{1 \pm [1 + 64(c + g)]^{\frac{1}{2}}}{8(c + g)} \quad \text{and} \tag{3}$$

$$k = \frac{32(c + g) + 1 \pm [1 + 64(c + g)]^{\frac{1}{2}}}{32g} \tag{4}$$

The minus root yields a negative estimate of $\bar{q}$ in equation (3) hence, the plus root is the appropriate one in this case. Using the plus root for $\bar{q}$ and either for k, the variances of $\bar{q}$ and k are:[511]

$$\sigma_{\bar{q}}^2 = \left[\frac{\bar{q}}{g} + \frac{4g^2 - 8(g+c)\bar{q}^2 + g\bar{q}}{[8(g+c)\bar{q} - g](g+c)}\right]^2 \sigma_g^2 + \left[\frac{4g^2 - 8(g+c)\bar{q}^2 + g\bar{q}}{[8(g+c)\bar{q} - g](g+c)}\right]^2 \sigma_c^2,$$

and

$$\sigma_k^2 = \frac{1}{g^2}\left[1 - k \pm \frac{1}{[1 + 64(g+c)]^{\frac{1}{2}}}\right]^2 \sigma_g^2 + \frac{1}{g^2}\left[1 \pm \frac{1}{[1 + 64(g+c)]^{\frac{1}{2}}}\right]^2 \sigma_c^2,$$

where

$$\sigma_g^2 = \frac{g(1-g)}{N}, \quad N = \text{number in the population sample},$$

and

$$\sigma_c^2 = \frac{c(1-c)}{n}, \quad n = \text{number of cousins}.$$

Recently Morton and his colleagues, Crow, Muller, and Chung, have developed methods based on consanguineous matings of various degrees which permit the estimation of the proportion of sporadic cases (phenocopies and those due to mutation and to other events of low probability), the number of loci at which mutations may give rise to the given phenotype, the average gene frequency, and the over-all mutation rate.

The method of "lethal equivalents" [575] is the starting point for the above estimates. Lethal equivalents were defined by Morton et al. as ". . . a group of mutant genes of such number that, if dispersed in different individuals, they would cause on the average of one death." This may be defined in another and perhaps clearer way: the lethal-equivalent value of an allele equals the value of the selection coefficient against the homozygote for that allele. An allele that causes death in all homozygotes is said to equal one lethal equivalent, one which causes death in 50 per cent of homozygotes is said to equal 0.5 of a lethal equivalent, et cetera.

Because much has been written about the derivation of these equations and because an expanded version of Morton, Crow, and Muller's derivation has been published,[575] it will not be reviewed in detail at this time. Suffice it to say that they arrived at the equation for P_D, the probability of death:

$$P_D = A + BF$$

where $A = \Sigma x + \Sigma q_i^2 s + 2\Sigma q_i(1 - q_i)sh$
$B = \Sigma q_i s - \Sigma q_i^2 s - 2\Sigma q_i(1 - q_i)sh$
x = environmental causes
q = gene frequency
s = selection coefficient against recessive homozygotes
h = a measure of dominance
F = coefficient of inbreeding (see following).

The value of F is computed for each degree of consanguinity observed in the preselected sample. The survival rate is observed for the offspring resulting from marriages of each degree of consanguinity. These values are used to estimate the values of A and B. Note that B is less than $\Sigma q_i s$, the mutational damage per gamete, and that

$$A + B = \Sigma q_i s + \Sigma x$$

is greater than $\Sigma q_i s$. Hence the number of lethal equivalents per gamete is somewhat less than $A + B$ and more than B. Note also that A equals the deaths resulting from random mating plus the deaths due to environmental causes.

This method has been extended to detrimental equivalents and to what Morton calls retrospective data. In the method previously discussed, the offspring of related parents were examined for defects. In such data the coefficient of inbreeding was known, but the proportion of affected offspring was not known. Often, and particularly for rare defects, affected individuals are ascertained, and then the frequency of consanguinity among their parents is determined; hence the term "retrospective" for such data.

The method that starts with parents of known consanguinity may conveniently be called the prospective method. Its application to detrimentals is straightforward and simply involves the replacement of lethals with detrimentals. The application of the method to retrospective data is somewhat different.

As before, the trait frequency for a given value of F may be written as

$$P \sim A + BF.$$

Let: c_i be the frequency of coefficient of inbreeding F_i in the general population; $\Sigma c_i = 1$.

     $a = \Sigma c_i F_i$ = mean coefficient of inbreeding in the general population.

     $\sigma^2 = \Sigma c_i F_i^2 - a^2$ = variance of F in the general population.

     I = trait incidence in the general population.

     $\overline{F}$ = mean inbreeding coefficient of the probands.

Then $I = \Sigma_i c_i(A + BF_i) = A\Sigma c_i + B\Sigma c_i F_i$

       $= A + Ba.$

Hence, $A = I - Ba$

     $\overline{F} = \Sigma c_i F_i(A + BF_i)/\Sigma c_i(A + BF_i),$

       $= (A\Sigma c_i F + B\Sigma c_i F_i^2)/I$, or

     $\overline{F} = [Aa + B(\sigma^2 + a^2)]/I.$

If $I - Ba$ is substituted for A in the last equation and the equation is solved for B, we have

$$B = I(\overline{F} - a)/\sigma^2$$

If I, the general population incidence, a, the mean coefficient of inbreeding, and σ^2 the variance of F, are estimated reliably (that is, have negligible variances),

$$\sigma_A = a\sigma_B$$
$$\sigma_B = I\sigma_{\overline{F}}/\sigma^2$$
$$\sigma_{(A+B)} = (1 - a)\sigma_B$$

Morton,[568] who originally derived these solutions, points out that they are good approximations of the maximum-likelihood solutions, which he also derived, but which shall not be presented at this time.

For completely penetrant recessive genes in the absence of sporadic cases ($h = 0$, $x = 0$)

$$A = \Sigma q_i^2$$
$$A + B = \Sigma q_i,$$

and an estimate of the number of loci may be derived as follows:

Let $\bar{q}$ = mean frequency per relevant locus and σ^2 = variance of $q = \Sigma q_i^2/n - \bar{q}^2$,
where n = the number of loci.
Then $n\sigma^2 = \Sigma q_i^2 - n\bar{q}^2$ and $\Sigma q_i^2 = n\bar{q}^2 + n\sigma^2$,
and $A = \Sigma q_i^2 = n(\bar{q}^2 + \sigma^2)$,
$A + B = \Sigma q_i = n\bar{q}$,
$$\frac{(A + B)^2}{A} = \frac{n^2\bar{q}^2}{n(\bar{q}^2 + \sigma^2)} = n\left(\frac{\bar{q}^2}{\bar{q}^2 + \sigma^2}\right).$$

This value is a minimum estimate of n, the number of loci, because $\dfrac{\bar{q}^2}{\bar{q}^2 + \sigma^2}$ is < 1.
It approaches n as σ^2 approaches zero. The standard error of n,

$$\sigma_n = 2\sigma_B/\bar{q}.$$
$$\frac{A}{A + B} = \frac{n(\bar{q}^2 + \sigma^2)}{n\bar{q}} = \bar{q} + \frac{\sigma^2}{\bar{q}}$$

Hence the mean frequency is slightly less than $A/(A + B)$ and approaches this value as σ^2 approaches zero. The standard error of $\bar{q} \sim \sigma_B/n$. Morton, in his paper, has discussed methods for determining the proportion of sporadic cases. Once these are determined, the above method may be applied.

SPORADIC CASES

Morton and his colleagues [142, 570] have developed two methods for estimating the proportion of all cases which is sporadic (x) . The first estimate does not depend upon an estimate of consanguinity and will not be discussed here. The second requires an estimate of various inbreeding coefficients. The inbreeding coefficient is the probability that both alleles at a locus are identical by descent; it is also the genetic correlation between uniting gametes. C. C. Li [502] presents an excellent, easily read discussion of the method of calculating F. The coefficients required are those of the general population (a), of the isolated cases (F_I), and of the familial cases (F_F). Then, if y is the proportion of isolated cases which is sporadic (note that y = sporadic cases/isolated cases, and that x = sporadic cases/all cases) ,

$$F_I = ya + (1 - y) F_F.$$

The first term on the right is the contribution of the sporadic cases to the inbreeding coefficient of all isolated cases; the second term is the contribution of the chance-isolated cases to the inbreeding coefficient. The sporadic cases should have the same rate of inbreeding as the general population, while the chance-isolated cases should have the same coefficient of inbreeding as the familial cases; hence the use of a and F_F.

Solving this equation for y,

$$y = \frac{F_F - F_I}{F_F - a}.$$

An estimate of x for complete or truncate ascertainment may be derived from the above equation as follows:

Let $n =$ the number of isolated probands.

             $N =$ the number of familial probands.

Then $c = (n/n + N)$ is an estimate of the proportion of isolated cases in the population of cases (isolated/all cases).

Recall that $y =$ sporadic/isolated.

Therefore $cy = \dfrac{\text{isolated}}{\text{all}} \cdot \dfrac{\text{sporadic}}{\text{isolated}} = \dfrac{\text{sporadic}}{\text{all}} = x,$

or $x = \dfrac{n}{n + N} \cdot \dfrac{F_F - F_I}{F_F - a}.$

MUTATION AND ADVANTAGE OF THE HETEROZYGOTE

The question of how lethal and deleterious genes are maintained in the population may now be examined. In general, recessive, lethal, and deleterious genes may be maintained in the population by mutation or by heterozygote advantage, or by a combination of these. The classical methods of estimating mutation frequency have ignored the possible contribution of heterozygote advantage. Recent studies, however, have indicated that in some instances such advantage may be of importance in maintaining gene frequency. It would be desirable to know how general such advantage is among all loci contributing to lethality or to reduction of fitness. Crow [159] has developed a method that permits an evaluation of the general importance of heterozygote advantage.

Consider a locus with 2 alleles in Hardy-Weinberg equilibrium:

Genotype	AA	AA'	$A'A'$	$\overline{W}$
Frequency	p^2	$2pq$	q^2	
Relative fitness	$1 - s$	1	$1 - t$	$1 - sp^2 - tq^2$

$$\overline{W} = \text{mean fitness.}$$

Sewall Wright has shown that the change in q per generation is

$$\Delta q = \frac{pq}{2\overline{W}} \cdot \frac{d\overline{W}}{dq}$$

$$= \frac{pq}{\overline{W}} [s(1 - q) - tq].$$

At equilibrium $\Delta q = 0$ and

$$\frac{pq}{W} [s(1 - q) - tq] = 0,$$

hence,

$$q = \frac{s}{s + t} \quad \text{and} \quad p = \frac{t}{s + t}.$$

Therefore at equilibrium the population is:

Genotype	AA	AA'	$A'A'$	$\overline{W}$
Frequency	$\dfrac{t^2}{(s + t)^2}$	$\dfrac{2st}{(s + t)^2}$	$\dfrac{s^2}{(s + t)^2}$	
Relative fitness	$1 - s$		$1 - t$	$1 - \dfrac{st^2}{(s + t)^2} - \dfrac{s^2t}{(s + t)^2}$

and the reduction in fitness is

$$\frac{st^2}{(s + t)^2} + \frac{s^2t}{(s + t)^2} = \frac{st}{s + t}.$$

At complete homozygosis the population is:

Genotype	AA	$A'A'$	$\overline{W}$
Frequency	$\dfrac{t}{s + t}$	$\dfrac{s}{s + t}$	
Relative fitness	$1 - s$	$1 - t$	$1 - \dfrac{2st}{s + t}$

The reduction in fitness at complete homozygosis is $2st/(s + t)$. Hence, inbreeding can, at most, double the reduction in fitness if only two alleles are present at a locus. If many alleles maintained by balanced polymorphism are present at a locus, the relative reduction in fitness will be greater than 2 as the result of inbreeding. Crow showed that the reduction in fitness due to segregation is k times as great in an homozygous population as it is in a population breeding at random. where k equals the number of heterotic alleles at the locus.

Recall that in the method of lethal or detrimental equivalents A is proportional to the genetic load due to random mating (it is somewhat larger) and B is proportional to the genetic load due to inbreeding (it is somewhat smaller). For heterotic loci B/A would be a minumum estimate of the number of alleles which must be maintained by heterozygous advantage if mutation does not contribute to their frequency. If the ratio is large, it would seem reasonable to assume that segregation is not involved and that mutation is the main source of genetic load. Current evidence indicates that the ratio is large. Hence, most genetic damage may be assumed due to mutation, and mutation may be assumed more important than heterozygotic advantage in maintaining deleterious genes in the population.

TWINS

The primary uses of twins in human genetics are to demonstrate a genetic component in the causation of a disease when such a component is not marked; to estimate penetrance for incompletely penetrant genes; and to estimate heritability for multifactorial (quantitative) characters.

Determination of Relative Frequency of Dizygotic and Monozygotic Twins.— The use of twins requires that their zygosity be determined. This can, in turn, be helped by a knowledge of the relative frequency of the two types of twins. Weinberg[916] developed (however, see Price[676]) the method used to estimate the relative frequency of twins. The assumptions inherent in the method are that members of monozygotic and of dizygotic pairs have the same prenatal mortality rate and that the relative prenatal death rates of males and females among twins is the same as that among single births. It is assumed that all data are taken at birth, therefore for the moment, postnatal differential death rates may be ignored.

The method is as follows:

Let p = frequency of newborn males, and
q = frequency of newborn females in the general population.

Then the distribution of families of size two, or of dizygotic (*DZ*) twins (which simply is a quick way of obtaining a family of size two) with regard to sex is $p^2 + 2pq + q^2$ (for families or twin sets of two boys, one boy, and no boys, respectively). If the discussion is limited to *DZ* twins, the above sum will equal all *DZ* twin sets, or one, since we are dealing with decimal fractions. The like-sexed, *DZ* twins cannot be distinguished (by sex alone) from monozygotic (*MZ*) twins. Hence, only the twins of unlike sex may be classified at once as *DZ*. Then $2pq/1$ = (all unlike-sex twins)/(all *DZ* twins), and all *DZ* twins = (unlike-sex twins)/$2pq$. Since p and q are known from the general population and the unlike-sex twins are determined by observation, the equation may be solved for the number of *DZ* twins. The difference between this number and the total of all twins is the number of *MZ* twins. The method may be applied to twins at various ages to obtain age-specific ratios of *MZ : DZ* twins. The sex ratio must be determined at the age of interest. As before, the assumptions are that the relative death rates of single-born males and females are the same as those of their twin-born counterparts.

*Determination of the Zygosity of Twins.—*Twins of unlike sex, of different eye color, skin color, hair color, or who differ for some character known to be genetically determined and to be completely penetrant, (blood groups, for example) are DZ and need be considered no further.

Twins who do not differ in any of the above ways may be MZ or DZ, and therefore a method for evaluating their zygosity is required. Basically the method consists of estimating (1) the probability that a pair of DZ twins would be as alike by chance as the pair in question, say $p(D)$; (2) the same probability for MZ twins, say $p(M)$; and (3) computing the fraction $p(M)/[p(M) + p(D)]$. The fraction is an estimate of the probability that the given set of twins is MZ. Robinow [79] used the expression $1/(1 + s)$ which may be derived from the above by dividing the numerator and the denominator by $p(M)$ thus:

$$\frac{p(M)/p(M)}{p(M)/p(M) + p(D)/p(M)} = 1/(1 + s) \text{ where } s = p(D)/p(M).$$

(As Robinow points out, Essen-Möller [206] developed the procedure for twins with unknown parentage. So far as I am aware, Robinow is the first to have published a procedure for twins of known parentage.) Smith and Penrose [817] used the same expression as Robinow and rederived his equations. They provided tables for $p(D)/p(M)$ for various characters using frequencies derived from various sources.

In principle, the value of $p(D)$ or $p(M)$ is derived as follows: compute for each of a series of *independent* characters the probability that DZ (or MZ) twins would be alike. A series of values $[p(D1), p(D2), p(D3)$, et cetera], one for each character, will result. Since the characters are independent, their probabilities are independent. The product of these probabilities equals the probability that the twins would be alike for all these characters, and this equals $p(D)$ [or $p(M)$ depending upon the probabilities used].

Two broad categories of characters may be used to evaluate $p(D)$ and $p(M)$. The first consists of those which are quantitative (continuously variable) and for which the mode of inheritance is not understood. The second comprises those characters, such as the blood groups, the genetic pattern of which is well understood and which are completely penetrant.

Quantitative characters, such as ridge counts in fingerprints, stature, et cetera, may be used if empiric data giving the frequency distribution of the differences between MZ and between DZ twins are available. As noted above, Smith and Penrose [817] have published such data, gathered from various sources.

Characters with a known pattern of inheritance supply most information when the genotypes of the parents are known; hence, when feasible, it is best to test the parents and siblings, as well as the twins. The data from characters with a known pattern of inheritance may be used if the parents' genotypes are not known, provided the frequencies of the genes in the general population are known. Essen-Möller,[206] Robinow,[729] and Smith and Penrose [817] described a method for using such frequencies to derive the probability that two siblings from the general population would be alike for a character of known pattern of inheritance.

Three approaches have been used to estimate $p(D)$ for characters with a known genetic pattern of inheritance when the genotypes of the twins' parents are known. The first (Rife[719]) asks, "Given these parents and these twins with the observed phenotypes, what is the probability that the twins, if fraternal, would have been alike in this way?" The second (Cotterman in Race and Sanger,[700] Smith and Penrose[817]) seeks an answer to the question: "Given parents with the observed characters and a child with the observed character, what is the probability that his fraternal twin will have the same phenotype?" The third (Robinow[729]) asks, "Given the parents with the observed phenotype, what is the probability that their fraternal twin offspring would be alike?"

The common practice in determining the zygosity of twins when parents and siblings are available is to test the entire family at once or to test the twins and, if they are alike, to test the siblings. Because both twins are observed simultaneously and because only the question about whether they are alike in some manner is asked before examining their relatives, the third question appears to be the most appropriate.

Table 8 presents a family composed of parents, a sibling, and a pair of twins.

Table 8

TWINS AND THEIR FAMILIES TESTED FOR VARIOUS BLOOD GROUPS

	Sex	ABO	MN	Rh	K	Fy^a *
Father		A	MN	R^1R^1	—	+
Mother		O	MN	R^2r	—	+
Sibling 1	F	A	MN	R^1R^2	—	—
Twin 1	M	O	MN	R^1r	—	+
Twin 2	M	O	MN	R^1r	—	+

* Anti-Fy^a serum only.

The twins show that the father is AO, and the sibling shows that both parents are heterozygous for Fy^a.

Table 9

PROBABILITY THAT FRATERNAL TWINS FROM THE FAMILY SHOWN IN TABLE 8 WILL BE ALIKE FOR EACH OF THE BLOOD GROUPS

Mating	Offspring	P of 2 alike
$AO \times OO$	$\frac{1}{2}A:\frac{1}{2}O$	$\frac{1}{4} + \frac{1}{4} = \frac{1}{2}$
$MN \times MN$	$\frac{1}{4}MM:\frac{1}{2}MN:\frac{1}{4}NN$	$\frac{1}{16} + \frac{1}{4} + \frac{1}{16} = \frac{6}{16}$
$R^1R^1 \times R^2r$	$\frac{1}{2}R^1R^2:\frac{1}{2}R^1r$	$\frac{1}{4} + \frac{1}{4} = \frac{1}{2}$
$kk \times kk$	all kk	
$Fy^aFy^b \times Fy^aFy^b$	$\frac{3}{4}Fy(a+):\frac{1}{4}Fy(a-)$	$\frac{9}{16} + \frac{1}{16} = \frac{10}{16}$

Table 9 shows, for each locus, the probability that dizygotic twins will be alike. The genotypes at each of these loci are independent of those at the others, of sex, and of the probability *a priori* that twins are DZ or MZ. All the probabilities may be combined to derive an estimate of the probability that the twins are MZ. The procedure is illustrated below:

Character	MZ	DZ	$DZ/MZ = s_i$
Zygosity (*a priori*)	.300	.700	2.330
Sex	1.000	.500	.500
ABO	1.000	.500	.500
MN	1.000	.375	.375
Rh	1.000	.500	.500
Fy	1.000	.625	.625
Product	.300	.0205	.0684

Using column MZ and DZ:

$$P(M) = \frac{p(M)}{p(M) + p(D)}$$

$$= \frac{.3000}{.3000 + .0205} = .936$$

Using $DZ/MZ = s_i$

$$P(M) = \frac{1}{1 + s} = \frac{1}{1.0684} = .936$$

If the sibling had not been $Fy(a-)$, the genotypes of the parents would not have been known. Since both parents are $Fy(a+)$, the mating (in a white family) could have been

$$Fy^aFy^a \times Fy^aFy^a, \; Fy^aFy^a \times Fy^aFy^b, \text{ or } Fy^aFy^b \times Fy^aFy^b.$$

The data for this locus can be used by calculating (a) the relative frequency with which each of these matings occurs in a panmictic population, and (b) for each of these matings the probability that DZ twins would be alike.

Let p = frequency of Fy^a,
q = frequency of Fy^b,

Then

	1.	2.	3.
Mating	Frequency *	Probability DZ twins alike †	Probability of occurrence = 1 × 2
$Fy^aFy^a \times Fy^aFy^a$	p^4	1	p^4
$Fy^aFy^a \times Fy^aFy^b$	$4p^3q$	1	$4p^3q$
$Fy^aFy^b \times Fy^aFy^b$	$4p^2q^2$	$\frac{5}{8}$	$(\frac{5}{2})p^2q^2$

* Each term should be divided by $p^4 + 4p^3q + 4p^2q^2 = p^2(1 + q)^2$.
† Only anti-Fy^a used .

The total probability that *DZ* twins from such parents will be alike

$$= p(D) = \frac{p^4 + 4p^3q + (\frac{5}{2})p^2q^2}{p^2(1+q)^2} = \frac{(2+4q-q^2)}{2(1+q)^2}$$

Race and Sanger [700] give, for England, $q = .5857$ and $q^2 = .3430$. Therefore for an English population,

$$p(D) = [2 + 4(.5857) - (.3430)^2]/2[1 + (.5857)]^2 = .7954$$

This value is greater than that obtained when the parents' genotypes were known, because, among the possible matings, the known mating $(Fy^aFy^b \times Fy^aFy^b)$ gave the lowest probability of concordance for fraternal twins. The new value (.795) would be used in place of .625 in calculating $p(D)$.

If, as is often the case, the parents are not available, it is still possible to compute the probability that the twins are *MZ*. The procedure is essentially the same (for each type of blood) as for the illustration given above for the *Fy* type of blood. The number of matings which may be considered is larger because the parental phenotypes are unknown. Thus $Fy(a+)$ twins may arise from $Fy^aFy^a \times Fy^bFy^b$ and $Fy^aFy^b \times Fy^bFy^b$ matings in addition to those listed above when the parents were both known to be $Fy(a+)$. The total frequency of the matings which could give rise to these twins is now

$$p^4 + 4p^3q + 4p^2q^2 + (2p^2q^2 + 4pq^3) = 1 - q^4.$$

(The terms in parenthesis are the new terms.)   The frequency of concordance among twins from these matings is

$$p^4 + 4p^3q + \tfrac{5}{2}p^2q^2 + (2p^2q^2 + 2pq^3) = \tfrac{1}{2}[2 - 3q^2 + 2q^3 - q^4]$$

(The terms in parenthesis are the new terms.)   Hence,

$$p(D) = \frac{2 - 3q^2 + 2q^3 - q^4}{2(1 - q^4)}.$$

With $q = .5857$, this expression equals $1.2551/1.7647 = .711$.

Investigators who collect samples of twins with at least one in each pair having a given phenotype often ask the question, "Is there an association between the phenotype and the zygosity of the twins?" Stern [839] treated this problem for twins selected by complete or truncate ascertainment. A more general treatment is given below:

Let p = proportion from a given mating type who are genetically liable to a specific phenotype;

   P = penetrance;

   π = probability of ascertainment;

   M = population frequency of monozygotic twins;

   D = population frequency of dizygotic twins.

Then the expected frequency of detected monozygotic twin pairs may be derived as follows:

Number liable	Number affected	Number ascertained	Frequency
2	2	2	$pP^2\pi^2$
2	2	1	$pP^2 2\pi(1-\pi)$
		Subtotal	$pP^2\pi(2-\pi)$
2	1	1	$p2P(1-P)\pi$
		Total	$pP\pi(2-P\pi)$

Expected frequency of detected MZ twins $= MpP\pi(2-P\pi)$.
The expected frequency of dizygotic twins may be derived as follows:

Number liable	Number affected	Number ascertained	Frequency
2	2	at least 1	$p^2P^2\pi(2-\pi)$
2	1	1	$p^2 2P(1-P)\pi$
1	1	1	$2p(1-p)P\pi$
		Total	$pP\pi(2-pP\pi)$

Expected frequency of detected DZ twins $= DpP\pi(2-pP\pi)$.

The ratio of ascertained MZ to DZ twins $= k = \dfrac{MpP\pi(2-P\pi)}{DpP\pi(2-pP\pi)} = \dfrac{M}{D}\cdot\dfrac{(2-P\pi)}{(2-pP\pi)}$.

If $\pi = 1$, $k = \dfrac{M}{D}\cdot\dfrac{2-P}{2-pP}$, the equation published by Stern.[839]   If $\pi \to 0$, $k = \dfrac{M}{D}$.

By a procedure similar to the preceding it can be shown that the expected frequency of concordance (C) for MZ twins

$$= P(2-\pi)/(2-P\pi), \quad \text{and} \quad P = \frac{2C}{2-\pi(1-C)},$$

where P and π have the same meanings as before. If $\pi = 1$ (complete or truncate selection)

$$P = \frac{2C}{1+C}, \text{ and if } \pi \to 0, P = C.$$

Hence, in single selection the ratio MZ/DZ equals that of the population, and penetrance (P) = concordance (C).

Only the more common special cases encountered in the genetics of populations have been reviewed, and these not completely. To do so would require a volume. Perhaps the pattern of thought involved in some of the methods of approach for solving problems of heredity in man which have been presented offer some insight into current solutions and will stimulate additional investigation into these interesting aspects of human genetics.

DISCUSSION

DR. BURDETTE: Dr. Eldon Gardner will discuss Dr. Steinberg's paper.

DR. GARDNER: The mathematical methods presented by Drs. Steinberg, Crow, Morton, and Lilienfeld indicate that human genetics as a discipline has now become mature. Objective data can be obtained and quantitative techniques can be applied. Statistical tools are being used effectively for resolving problems that in the past have at best been troublesome. Methods old and new are being brought together and applied with good success.

This improvement in methodology is fortunate because, in some respects, man is poor material for genetic study. Better methods compensate to some extent for an unfavorable material. Man's aversion to being the object of examination, small family size, long period between generations, difficulties in standardization, and lack of control in the environment, all conspire against the human geneticist. As has been demonstrated by the foregoing papers, improved methods have made it possible to learn a good deal from data based on small families covering only a few generations.

Data on human inheritance are where you find them. Human geneticists must take what information they can obtain and make the best of it. Different methods can be applied to different kinds of data. Obviously some problems require statistical treatment. Others can be resolved by the old-fashioned pedigree method that is most satisfactory for studying the inheritance of rare traits in large family groups. This method is dependent on information about related members of a family group. A certain amount of personal data is helpful in nearly every human study. Objective, statistical analyses become impersonal and tend to lose sight of the individual. Acquaintance with the people being studied still has an important place in some aspects of human genetics.

In one of our studies [267] on intestinal polyposis little statistical analysis was required. Much more important was good rapport with the family members and their excellent coöperation. Beginning with a proband, all fifty-one descendants from one progenitor covering three generations were located and examined. Through repeated clinical observations it was possible to identify a syndrome [266] that probably would have been missed if more objective and less personal methods of ascertainment and analysis had been followed. This study had the advantage that the investigator was dealing with a very rare trait dependent on a single, dominant gene.

It is illuminating to see how many approaches are now available for human studies and how powerful the statistical tools have become. Fortunately, procedures appropriate to the method of ascertainment are available to handle a wide variety of types of data.

DR. STEINBERG: I agree completely with what Dr. Gardner has said, and I am aware of the very fine study [266] he did on the family with multiple intestinal polyposis which led to the discovery of a new form of this disease. This method works well and efficiently when a rare character occurs in a large pedigree. If the pattern is clear by simple and straightforward tests, there is no need whatsoever to use more elaborate techniques. Some years ago, while working with medical colleagues, we encountered a sex-linked, recessive character,[2] lethal during early life in males. This was a single, large pedigree, and we used none of the elaborate, complicated, "quick and dirty" mathematics referred to earlier. These are not always necessary, and I do not think that those who use large pedigrees for study are necessarily old-fashioned. They may be using the best available data and techniques for the kind of study they are doing.

H. B. Newcombe, Ph.D.

POPULATION GENETICS:
POPULATION RECORDS

INTRODUCTION

A human geneticist may carry out his studies using pedigrees built up entirely as a result of his own effort or he may, in varying degrees, use information that has already been accumulated for other purposes. In an extreme case, he sometimes relies exclusively on data of the latter kind. The disadvantages of this approach are fairly obvious: access to the records may be difficult, the files may be excessively large, the medical information they contain may be unreliable, and they may fail to identify certain hereditary traits and environmental influences.

However, the more individual approach is, by its nature, restricted to studies of relatively small groups of people. In contrast, very large quantities of family information and of medical and social data are collected routinely in a modern society. In the past, only limited genetic use has been made of such records because of a number of serious practical difficulties, some technical and others organizational. But it is now apparent that with the presently available data-processing equipment and suitable handling procedures there is no serious obstacle to extracting pedigree information, together with medical and social data, on a fairly extensive scale from these routine sources.

It is not suggested that the extensive approach should take the place of the more conventional intensive studies. But, for the special purpose of acquiring an insight into the operation of selective forces in human populations, many of which may involve only slight differences in fertility and mortality in different lines of descent, no adequate alternative has been suggested. Whatever the present limitations of the approach, it seems the only way by which we are likely to secure the necessary histories of reproduction and death from very large numbers of families. Other studies of a more modest kind are possible by the same methods, and the quality of the records will undoubtedly improve in the future.

The present account will be concerned chiefly with the potential of genetically useful information that is already being recorded routinely for the population as a whole. The routine records, of course, differ from country to country, and I must necessarily confine myself to Canadian sources with which I am familiar. The total volume of population records is probably similar, however, in most of the more highly industrialized countries.

I shall not attempt to discuss the kinds of study which might be carried out, this having been done in some detail elsewhere,[616] and for details of the procedures by which genetic information may be extracted from large files of records of individual events the reader is likewise referred to material previously published.[620]

AUTOMATIC RECORD LINKAGE

For those who are unacquainted with these procedures, the present description of the routine records and their content of genetic information will be of interest only if the notion is accepted that this information may be extractable. Brief mention should, therefore, be made of the use of "automatic record linkage" as a method of deriving family data from existing population records. Family groups may be identified by "linking" individual birth records to the records of the marriages of the parents, using such common information as the names of the husband and wife, their birthplaces, and their ages.

Files of birth and marriage records in the form of magnetic tapes have been linked automatically by electronic computer with considerable rapidity and precision in the studies referred to. The accuracy depends almost entirely on the amount of identifying information common to the two records, but individual items of information do not need to be especially reliable in order to contribute significantly to the over-all accuracy. The speed achieved (about 30 births per minute with a marriage file of approximately 10^5 records, using a medium-speed computer) could probably be increased about tenfold with computing equipment presently available, and a further, sixfold increase in speed is claimed for a computer just recently put on the market. Future speeds will undoubtedly be greater still.

Most of the work in any such operation is associated with the preparation of the punch-card files, and a project involving total files from 10^5 to 10^6 records in punch-card form is not too unwieldy for one or two investigators to handle, provided that appropriate computing facilities are available to them. Records of

Table 10

CENTRALIZED POPULATION RECORDS,[1] BASED ON THE YEARS 1955–1957.*
(New records in any one year, or active files.)

Kind of record	Per 100 population per year [2]
Labor services	45.3
Welfare and social security	42.4
Income-tax returns	24.3
Population census (average over 10 years) [3]	20.0
Public-health services	17.7
Travel (customs declarations and overseas travel) [4]	5.5
Vital records	4.5
Veterans services	3.1
Defence forces (plus reserves, civil defence, and cadets)	3.0
Immigrant admissions and citizenship certificates granted	2.2
Civil service and crown corporation employees	2.0
Criminals and delinquents	.3
Total records	170.3

[1] *I.e.*, federal or provincial records centralized in Ottawa or in the provincial capitals.

[2] New records per year except where total file is indicated.

[3] *I.e.*, the 1951 and 1956 census.

[4] Includes records of Canadians traveling overseas and returning, but not immigrant admissions.

* The list omits Canadian passports issued (and in force), federal government bond sales to individuals, illness in the civil service, succession duties, and other records for which figures were not available.

deaths, illnesses, genetic handicaps, and of social particulars may be linked in a similar manner to the family groups derived from the records of births and marriages.

In discussing the various kinds of routine record it will be assumed that any technical difficulties in automatic record linkage are of a purely temporary nature. The organizational difficulties are likely to be much more important, but these will not be considered. Probably the sheer bulk of official and unofficial records will lead to a progressive increase in the use of centralized, data-processing facilities and greater integration of the records for reasons of economy alone. Such a trend would greatly reduce both the technical and the organizational difficulties associated with automatic linkage of the different kinds of population records.

The genetic and family data extracted by these procedures are, of course, as impersonal as the more familiar census statistics and the annual vital statistics. However, they relate to the family rather than to the individual as the important unit, and they describe a chronology of events within individual families rather than the events occurring over a brief period in time. Thus, they differ in two important respects from the more conventional sorts of population statistics.

Table 11

NONCENTRALIZED RECORDS FOR WHICH NUMBERS ARE AVAILABLE,*
BASED ON THE YEARS 1955–1957.

(New records in any one year, or active files.)

Kind of record †	Per 100 Population
Life-insurance policies in force	62.9
Personal-savings accounts (chartered banks)	55.6
Voters lists for federal elections	55.3
Residential telephones	20.6
Passenger-automobile registrations	19.8
School and university enrolment	18.2
Summary-conviction offences	13.7
Public-library borrowers	12.6
Credit-union membership	11.8
Industrial employees	8.4
Labor-union membership	8.4
Red Cross blood tests (donations, recipients, and clinical)	5.7
Loans from small loan companies	5.3
Life-insurance benefits per year	2.6
Fraternal-benefit societies — certificates	1.4
Post-office and provincial savings accounts	.8
Traffic-accident casualties (nonfatal)	.3
Total records	303.4

* *I.e.*, nonfederal records and nongovernment records; the distinction between these and those in table 10 is in part arbitrary.

† The list does not include drivers licenses, baptismal records, automobile insurance, non-federal voters lists, municipal property tax, credit cards, mail order and department-store credit plans, liquor permits, post office box rentals, bank loans, magazine subscriptions, prepaid medical-care plans, and a great many other records for which no nation-wide numbers are available.

Likewise omitted are employment records relating to a labor force making up about 35.5 per cent of the population, together with the associated pension and superannuation plans.

THE VOLUME OF POTENTIALLY USEFUL POPULATION RECORDS

The kinds of information required for genetic purposes have to do with the identification of family groups, the times and circumstances of the vital events of birth, death, and marriage (together with annulment, divorce, adoption, and legitimation), the nature of any illnesses or handicaps, and the social and economic characteristics of the family and its members. In addition, the very mundane fact that an individual is still alive is frequently of considerable importance and is not always easy to establish.

Some idea of the volume of records in which one or more of these items of information is contained may be gained from tables 10 and 11. The first relates to records kept as a product of the various functions of government and centralized either at the provincial or the national level. The second has to do with a somewhat less readily accessible group of records, many of which are not centralized. The distinction is necessarily arbitrary and neither list is exhaustive. The frequencies refer mainly to new records created during a year but, in the case of a few entries, active files of records have been quoted. Complete consistency with respect to one or the other method of expression was not possible, but the nature of most of the entries will be apparent on inspection and that of others will be clear from the breakdown in subsequent tables.

Among the various kinds of records for which data are available, about four new records per capita are created each year. This figure falls far short of the total number of records of all sorts which we create in the course of our various contacts with government (at its different levels), business organizations, educational institutions, and in our demands upon medical services, but it is at least sufficient to indicate that the total number relating to an individual throughout his lifetime is very large indeed. Whether we like it or not, we leave behind us a documentation of our lives which, if not complete, is at least impressive.

DISCRIMINATING POWER OF THE IDENTIFYING INFORMATION

Those who are familiar with the difficulties encountered in searching manually for individual records in large, accumulated files are likely to ask at this point whether these records contain sufficient identification to enable them to be linked. The question, of course, must be rephrased in terms of the degree of accuracy that can be achieved in linking two files having certain items of identifying information in common. Quantitative predictions of the accuracy can be made in advance when various combinations of items (such as surnames, Christian names or initials, ages, et cetera) are used. These predictions are based on a relatively simple formula developed originally for the purpose of the automatic-linkage study. The present use will be described in some detail because it may, in certain instances, permit one to reduce the amount of trial and error testing needed prior to starting a record-linkage study.

There is, for example, exceedingly little ambiguity in linkages of the vital records since these almost always contain identifying particulars of both a husband and a wife (either as groom and bride, as parents of a newborn infant, or as parents of a deceased individual; the deceased and the spouse of the deceased may also appear together on the death registrations). However, more ambiguities will occur where particulars of only a single individual are common to the two files.

The discriminating power of a specific agreement or disagreement may be derived from the formula:

$$\log_2 p_L - \log_2 p_F$$

where p_L and p_F are the frequencies with which that particular agreement (or disagreement) occurs respectively in linked pairs of records and in pairs which have been brought together by accident (see Newcombe, Kennedy, Axford, and James,[620] for the derivation of this formula). For example, agreement on the fact that an individual is of male sex will occur nearly 100 per cent of the time in linked pairs of records relating to male individuals but only half the time in pairs brought together by accident. In this case the formula yields a value of +1 for agreements with respect to male sex, indicating that the likelihood of a genuine linkage is increased twofold by the knowledge of this agreement (the +1 is the logarithm to the base 2 of this factor increase in the odds favoring a genuine linkage). Disagreement with respect to sex is rare in linked pairs of records (arising only as a result of clerical error), but common in those brought together by accident, and is thus associated with a large negative value. The numbers obtained from the formula have been termed binit values, and, being logarithms, the binit values from agreements (or disagreements) with respect to two or more items may be added together. The values p_L and p_F may be obtained from a relatively small number of linkages and accidental, name agreements arrived at by visual comparisons.

In the same manner, agreements with respect to different surnames, initials or Christian names, provinces or countries of birth, and agreements or partial agreements with respect to ages (together with disagreements in these various items) may all be assigned binit values indicative of the discriminating power of the particular information. Actual values have been derived from linked files of births and marriages, together with random comparisons of the two kinds of records, using the above formula. The sum of the binit values relating to two or more different items of information indicates the combined discriminating power of the agreements (or disagreements) in these items. It is convenient to remember that 10 binits represents an increase in the odds of about 1000-fold, and 20 binits an increase of 1,000,000-fold, in favor of a genuine linkage between a pair of records.

An alternative (although slightly less accurate) method of deriving the binit values is based simply on the frequencies of the particular identifying items using the expression:

$$- \log_2 p_R$$

where p_R is the frequency of the particular initial, birthplace, or other piece of information in a file of records.

The binit values for each category or kind or identifying information (such as first initials or birthplaces) in an actual file of records may be represented by a frequency distribution (some common initials and birthplaces having relatively low, and some rare ones relatively high, discriminating powers). The combined discriminating powers of two such kinds of identifying information may likewise be represented by a frequency distribution, the shape of which may be calculated from the two separate distributions. (A somewhat tedious operation is involved in which each frequency in one distribution is multiplied by every frequency in the other, the binit value assigned to the product being in each case the sum of the two original binit values.)

A test of the reliability of the calculation is possible where the combined discriminating powers can be derived both directly from a file of records and also from the calculation. This has been done for the surnames (in phonetically coded form) represented in a file of approximately 100,000 marriage records (see table 12). Agreement between the observed and the calculated distributions of

Table 12

CALCULATION OF THE DISCRIMINATING POWER OF TWO PIECES OF IDENTIFYING INFORMATION IN BINIT UNITS*

Binits	Single surnames (bride)		Surname pairs (groom and bride)		
	Obs. number	Per cent	Obs. number	Obs. per cent	Calc. per cent
5	0	.0			
6	1,299	1.1			
7	3,850	3.4			
8	18,272	16.1			
9	22,320	19.6	0	.00	.00
10	21,701	19.1	100	.09	.09
11	17,422	15.3	0	.00	.00
12	13,763	12.1	47	.04	.04
13	7,944	7.0	223	.43	.30
14	4,193	3.7	2,221	2.0	1.9
15	1,588	1.4	6,692	5.9	5.9
16	746	.7	15,100	13.2	13.2
17	548	.5	89,163	78.5	78.5
18	0	.0			
Total	113,646		113,646		

*The table uses surnames of grooms and maiden surnames of brides in phonetically coded form. It is based on a file of 113,646 records of marriages contracted in the Province of British Columbia over the ten-year period 1946–1955. "Obs. no." indicates number of records in a file of 113,646 marriages, and "obs. per cent" and "calc. per cent" refer to percentages of the total file. The "binit" values are based on the second of the two formulas referred to in the text.

binit values for grooms' surnames and brides' maiden surnames in combination is surprisingly close, indicating that the calculation is reliable and may be extended to more than two categories of information.

Since many records contain the surname, first given name, second initial, and age of single individual, the calculation has been carried out for these categories in various combinations, starting with data on the distributions of binit values derived from vital records. The manner in which the discriminating powers increase with increasing numbers of these items is indicated by the frequency distributions in table 13.

Table 13

CALCULATION OF THE DISCRIMINATING POWER OF A NUMBER OF ITEMS
OF IDENTIFYING INFORMATION *

(in terms of the binit values associated with different per cents of a file or records)

Binits (range)	(a) Surname	(b) First given name	(c) Second initial	(d) Age	Combined (ab)	Combined (abc)	Combined (abcd)
−10−−6		.2					
−5−−1		3.4	4.3	2.1			
0− 4	.8	14.8	62.6	38.9	.1	.1	
5− 9	20.6	66.1	33.1	58.9	3.0	.8	.3
10− 14	72.3	15.5		.1	13.2	6.0	1.8
15− 19	6.3				52.0	21.5	8.3
20− 24					29.1	49.8	26.8
25− 29					2.6	20.2	43.5
30− 34						1.6	17.6
35− 39							1.7

*Values for (a), (b), (c), and (d) were drawn from the British Columbia vital records. The values for combinations of identifying information were obtained by calculation from those for single items.

Such distributions may be used to derive the likelihood of error in linking individual records to a main file of a given size, by plotting them as accumulated percentages against increasing binit values. The percentages will then represent an increasing proportion of false linkages as attempts are made to link records to main files of increasing size (see figure 1). Thus, if all the above four categories of information are present on a group of records (each of which relates to an individual represented in the main file), the proportion of false linkages arising through ambiguity will be 2 per cent with a main file of 2^{14} (that is, of 8,192) and 8 per cent with a main file of 2^{18} (that is, of 263,144). Any admixture of records relating to individuals not represented in the main file would, of course, increase the errors (for example, a 50 per cent admixture would double them). Further identifying information common to both sets of records would reduce the errors: (1) an indication of the sex of the individuals, for example, would approximately halve the errors, (2) address or telephone number provides virtually positive linkage for all individuals who have not moved to a new residence, (3) year of birth has greater discriminating power than has age, while date of birth (day and month) further reduces the number of wrong linkages by a factor of approximately $\frac{1}{365}$ (equivalent to between 8 and 9 binits), (4) province or country of birth and city or town each add substantially to the accuracy, and (5) even social particulars such as occupation, race, or religion, which may be stated differently on two successive records, nevertheless possess considerable discriminating power. When information about husband, wife, and children is already linked through a common address as in the census, or in some other way, the linkage with other family records becomes exceedingly precise.

Fig. 1. Size of file. Percentage of false linkages.

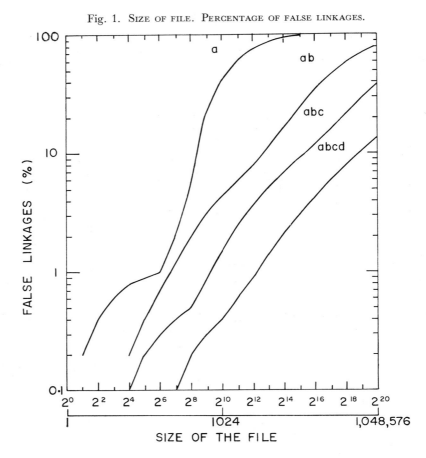

The application of the binit values and the computing procedures are both extremely flexible and lend themselves to a variety of refinements that permit nearly maximum use to be made even of very unreliable identifying particulars. In most cases, a reasonable degree of accuracy may be expected, adequate for many kinds of statistical study, and in some cases the accuracy may be very high.

SOURCES OF SPECIFIC KINDS OF INFORMATION

Of the various kinds of information inherent in the routine population records, family information is basic to virtually any genetic study. Registrations of marriages, births, and deaths identify the members of a family unambiguously. Since the marriage registrations contain details of the parents of both bride and groom, successive generations may be linked, provided that the families remain within the area under study. The larger the area, the more complete will be the genealogic information.

Identification of family groups occurs also in other records (see table 14). The census links all members of a family who reside together, and a still more

Table 14

RECORDS IDENTIFYING FAMILY RELATIONSHIPS

Kind of record	Per 100 population per year[1]	Relatives identified			Punch cards
		Parents	Spouse	Offspring	
Family allowance	33.4	+			
Income tax	24.3		+[2]	+[2]	
Population census (av. over 10 yrs.)	20.0		+[3]	+[3]	+
Red Cross, blood donations	3.1		+[4]		
Birth registrations	2.9	+			+
Physicians notices of births (B.C.)	2.9	+			+
Immigrant admissions	1.7		+	+	+
Death registrations	0.8	+	+		+
Marriage registrations	0.8	+	+		+
Handicapped children registered (B.C.)	0.06	+[5]	—	—	+
Total	89.96				

[1] Based on the national population except when British Columbia is indicated.
[2] Where classed as dependents.
[3] Where resident at same address.
[4] Where also a donor.
[5] Approximately 2 per cent of births.

complete picture of the family groups may be derived by linking the records from successive census years and by linking census to vital records. The 1951 census for England and Wales included a question concerning the date of marriage (which is absent from the Canadian census form), and this information further improves the accuracy of such linkages. Family-allowance-account cards (records of government payments for all children under the age of 16) and income-tax returns identify any offspring during their years of dependence and are capable of being linked with one another and with the vital records and the census. Blood-donation cards identify husbands and wives when both are donors (through their common address, with a small admixture of ambiguities involving in-laws who share the same residence). The other records listed in the table identify limited numbers of offspring with one or both parents. Many of the above records already exist in the form of punch cards, but these have not been produced with a view to such linkages.

Table 15

RECORDS CONTAINING SOCIAL PARTICULARS

Kind of record	Per 100 population per year[1]	Occupation	Income	Religion	Race	Citizenship	Education	Punch cards
Income tax	24.3	+	+					
Unemployment insurance	23.2[2]	+	+					+
Population census (av. over 10 yrs)	20.0	+	+	+	+	+	+	+
Hospital admissions								
general	15.6	+[3]		+[4]				+
federal	.5	+						
provincial	.3	+		+	+	+	+[5]	+
Customs declarations	> 5.5					+		
Birth registrations	2.9				+	+		+
Immigrant admissions	1.7					+		+
Marriage registrations	.8			+	+	+		+
Death registrations	.8				+	+		+
Nursing (V.O.N.) discharges	.4	+						+
Sickness in the civil service	.27	+						+
Indictable offences	.20	+		+	+	+		+
Juvenile delinquents	.05	+		+		+		+
Penitentiary admissions	.02[6]	+	+	+	+	+		+
Total	96.54							

[1] New records per year except where total file is indicated; based on the national population.
[2] Total active file of persons insured; the numbers of individuals newly insured was not available from the published material and is probably of less interest.
[3] N.S., N.B., Ont., and B.C. only.
[4] Nfld., N.S., N.B., Ont., and B.C. only.
[5] On forms for mental hospitals but not for tuberculosis sanitoria.
[6] Approximately twice this number are inmates at any one time.

Social particulars are most fully stated in the census, but, as may be seen from table 15, many other records contain overlapping information and much of it is already on punch cards.

Of the various routine sources of medical information (see table 16), few are of as much interest as the registers of handicapped children. The broad categories of disability represented in the British Columbia Register is shown in table 17, and a number of genetically interesting conditions appearing in it are listed in table 18. Registers of this kind are of special importance when (as in the case in British Columbia) the value of their statistical products as aids to understanding the causes of disease is recognized. In planning such registers, it is important that the research uses shall not be wholly subordinated to the more immediate applications to programs of rehabilitation.

The British Columbia Registry of Handicapped Children is expected to extend its activities shortly to include handicaps of adults. It is not presently kept as a genetic or family register, but information from the Registry may be linked

Table 16

RECORDS CONTAINING MEDICAL INFORMATION *

Kind of record		Per 100 population per year [1]	Punch cards
Admissions to reporting, public hospitals		15.60	+
Red Cross, blood typings			
Donations to bank	3.10		
Matched blood	.20		
Recipients	1.02	5.73	
Clinical *Rh* tests	.81		
Other tests	.60		
Physicians notices of births (B.C.)		2.86	+
Death registrations		.83	+
Reported cases of notifiable disease		.82	
Admissions to federal hospitals		.54	
Visiting nursing services, cases		.40	+
Records of malignant neoplasms (B.C.)		.27 [2]	
Veterans, casualty rehabilitation — total registration		.26	
Nonfatal motor-vehicle accidents — persons		.25	
Admissions to mental institutions		.22	+
Allowances for disabled persons — recipients		.19	
Admissions to tuberculosis sanatoria		.13	+
Blindness census — totally blind (1951)		.10	
Deafness census — totally deaf (1951)		.10	
Handicapped children's register (B.C.)		.06 [3]	+
Allowances for blind persons — recipients		.05	
Total records		28.41	

[1] New records per year except where total file is indicated.

[2] Records per year in British Columbia = 4000; total records in the file of live cases = 25,000.

[3] About 2 per cent of the children born in 1952 had appeared in this register by 1958; total records in the file of live cases = 11,000.

* The list does not include the various records arising from medical care of North American Indians and Eskimos, a considerable number of provincial records, and those of voluntary societies dealing with health problems.

to the family groupings currently being derived from the vital records for that province by means of automatic record linkage.

Diagnoses from hospital admissions and discharges will provide a further extensive source of medical information. As a result of Canada's Hospital Insurance and Diagnostic Services Act it is expected that soon each province will have a mechanically punched card for all hospital admissions which will show not only the accounting data for that admission but diagnostic information as well. Most of the provinces already have such cards (see table 19) but no indication of the frequencies of diseases is yet available from this source for inclusion here. Also omitted from the present account, but for a different reason, are the frequencies of the various causes of death; these are readily available elsewhere and their uses and limitations as a source of medical information are widely known.

Table 17

CONDITIONS REPRESENTED IN THE BRITISH COLUMBIA REGISTER
OF HANDICAPPED CHILDREN †

Nature of the disability	Percentage of total cases registered *
Nervous system and sense organs	30.9
Mental defects and personality disorders	20.7
Bones and organs of movement	12.6
Circulatory system	10.0
Infectious and parasitic diseases	6.9
Digestive system	4.9
Allergic and endocrine diseases	3.8
Genito-urinary system	2.2
Neoplasms	2.2
Other	5.8
	100.0

† Half the cases are affected by the following individual disabilities: mental retardation (16.3%), strabismus (8.3%), congenital malformations of the heart (6.8%), cerebral palsy (5.5%), impairment of hearing (4.5%), clubfoot (4.5%), poliomyelitis (4.3%).
* Based on 10,867 living cases age 0–20 years in 1958, registered during the period 1952–1958 from a population of approximately 1.4 million at the end of this period.

The very large quantities of blood-group data acquired each year in the course of running blood banks, transfusion services, and clinical blood tests are of particular interest. There are numerous studies that could be carried out if this information (relating chiefly to the *ABO* and *Rh* groups) were linked to the reproductive histories of the individuals. In addition, the blood types of brides and grooms could readily be determined at the time of marriage, and in one province (Manitoba) this was, in fact, done over a number of years. The reproductive histories of couples, if both were typed at the time of marriage, would permit studies of any correlations with stillbirths and early infant deaths and of the selective forces which operate to maintain the frequencies of the various blood-group alleles. To those interested in investigations that combine precise identification of hereditary traits with precision in the measurement of slight, selective differentials through following large numbers of families, the blood characters offer exceptional opportunities when combined with family data derived from the vital records.

Disease categories reported by visiting nursing services are listed in table 20; case discharges are already recorded routinely on punch cards for most of the Canadian provinces. It is likely that these nurses come in contact with many individuals who are never hospitalized and whose conditions might be difficult to ascertain in any other way.

Records of notifiable diseases (see table 21) are of little genetic interest but might be useful in studies of effects upon the fetus when the infections involve

Table 18

CERTAIN CONDITIONS OF GENETIC INTEREST APPEARING IN THE
BRITISH COLUMBIA REGISTER OF HANDICAPPED CHILDREN

Condition	Per million births *
Strabismus	2,300
Club foot	2,290
Cardiac malformations	1,700
Cleft palate and hare lip	1,700
Cerebral palsy	1,400
Mongolism	1,000
Spina bifida and meningocele	940
Hydrocephalus	590
Epilepsy	500
Digestive-tract anomalies	420
Genito-urinary-tract anomalies	350
Osteochondrosis	250
Dislocated hip, congenita	240
Cataract, congenital	170
Diabetes	130
Deafness (not including deaf-mutism)	100
Skin anomalies	70
Fibrocystic disease	60
Achondroplasia	54
Muscular dystrophy	50
Amputations, congenital	42
Albinism	13
Deaf-mutism	12

* The bulk of the above entries were calculated from registrations of children age 5–9 years in 1958, from approximately 150,000 births over 5 years.

pregnant women. Unfortunately the present method of reporting does not lend itself to this kind of study, since the notification forms (at least in British Columbia) do not identify the affected individuals by name. Some of the diseases are believed to be under-reported, but this in itself would not present any serious barrier to such studies as long as the cases that are reported are correctly diagnosed.

The results of a special sickness-survey in 1950–51 (see table 22) should be mentioned. This study was designed to assess the prevalence of illness through the visits of interviewers to a sample of 10,000 households. The methods of the study are of little use for genetic purposes, since only a small fraction (about 0.2 per cent) of the population are contacted and there is no provision for following the individuals and families over prolonged periods.

The tables omit records arising from the medical care of North American Indians and Eskimos together with a considerable number of provincial health records, about which information was not available at the time of writing. In most provinces malignant neoplasms are reported, and in British Columbia additional records arise from school medical inspections. Further, each health unit and sub-

Table 19

"Linkability" of Certain Population Records

Kinds of record	Forms							Punch cards						
	Name	Address	Age	Birthplace	Marital status	Occupation	Family[1]	Name (or no.)	Address (or code)	Age	Birthplace	Marital status	Occupation	Family[1]
Population census (1951, 1956, 1961)	+	+	+	+	+	+	+	+	+	+	+	+	+	+
Blindness census (1951)	+	+	+		+									
Immigrant admissions	+	+	+	+	+	+	+	+	+	+	+			
Citizenship certificates	+	+	+	+	+	+	+	+	+	+	+	+	+	+
Passports	+	+	+	+	+	+	+							
Customs declarations	+	+		+										
Birth registrations	+	+	+	+	+[2]	+[2]	+	+	+	+	+	+[2]	+[2]	
Death registrations	+	+	+	+	+	+	+	+	+	+	+	+	+	
Marriage registrations	+	+	+	+	+	+	+	+	+	+	+	+	+	
Physicians notice of birth (B.C.)	+	+	+	+	+[2]			+	+	+	+	+[2]		
Handicapped Children's Register (B.C.)	+	+	+			+[2]	+	+	+	+				
Cancer register (B.C.)	+	+	+	+			+	+	+	+				+
Visiting nurses (V.O.N.), discharge	+	+	+		+	+		+		+		+	+	
Sickness in the civil service	+	+	+		+	+		+	+	+		+	+	
Red Cross, blood donors	+	+					+[3]							
Sickness survey (1950–51)	+	+	+		+	+	+	+	+	+		+	+	+
Hospital admissions general (Nfld.)	+	+	+		+				+	+		+		
(N.S.)	+	+	+	+	+	+		+			+	+		
(N.B.)	+	+	+		+	+								
(Ont.)	+	+	+	+	+	+		+	+	+		+		
(Man.)	+	+	+		+		+	+		+		+		
(Sask.)	+	+	+		+			+	+	+		+		
(B.C.)	+	+	+		+	+		+	+	+	+			
Admissions to tuberculosis sanatoria	+	+	+		+	+		+			+	+	+	
Admissions to mental institutions	+	+	+	+	+	+	+	+			+		+	+
Income-tax returns	+	+			+	+	+							
Family-allowance applications	+	+	+				+							
Unemployment insurance	+	+	+		+	+		+			+		+	+
Indictable offences	+		+	+	+	+		+			+	+	+	
Juvenile delinquency	+		+	+		+	+	+			+	+		
Penitentiary inmates	+	+	+	+	+	+		+			+	+	+	+

[1] Identification of any two or more members of a family.

[2] Relates to parents of the individual.

[3] But only if two or more members of a family are donors.

Table 20

DISEASES RECORDED BY VISITING NURSING SERVICES
(Victorian Order of Nurses)

Conditions	Visits per 10^6 population per year	Per cent of all visits *	Cases per 10^6 population in one year
Diabetes	7,560	15.9	164
Heart	7,170	15.0	340
Central nervous system and sense organs	5,790	12.2	209
Anemias and blood diseases	4,840	10.2	218
Malignancies	3,500	7.4	201
Arthritis and rheumatism	2,450	5.2	87
Other	16,190	34.1	1,731
Total	47,500	100.0	2,950

* A total of 19,415 cases with 6 diseases accounted for 66 per cent of all visits in 1956, exclusive of maternity and newborn care.

Table 21

RECORDS OF NOTIFIABLE DISEASES OF POSSIBLE INTEREST IN STUDIES OF
TERATOGENESIS AND ENVIRONMENTAL EFFECTS ON THE FETUS*

Notifiable disease	Cases per 10^6 population per year
Measles	3,480
German measles	3,380
Chicken pox	2,540
Mumps	1,950
Venereal disease	1,040
Epidemic influenza	750
Scarlet fever	730
Tuberculosis	580
Whooping cough	530
Other notifiable diseases	140
Total	15,120

*Based on data for 1956. Frequencies of some of these conditions differ substantially from year to year. An extreme example is influenza, which has varied in incidence by a factor of 100 within the period 1954 to 1958.

office in this province maintains a folder containing a family record, immunization records, records of roentgenograms of the chest, and of certain medical treatments.

Numerous voluntary societies, organized both at the provincial and at the national levels, deal with health problems and some of these maintain their own registers of diseases. At the national level there are societies with specific interests in blindness, deafness, mental health, cancer, arthritis and rheumatism, crippled chil-

Table 22

CANADIAN SICKNESS SURVEY 1950–51 *

Nature of condition	Per cent of total illness
Respiratory system (including common cold)	17.3
Symptoms, senility and ill-defined	16.0
Circulatory system	12.9
Bones and organs of movement	10.6
Digestive system (including digestive upset)	8.8
Nervous system and sense organs	6.4
Allergic, endocrine, metabolic, and nutritional	5.6
Genito-urinary system	5.0
Skin and cellular tissue	4.7
Infectious and parasitic	4.0
Other	8.7
Total	100.0

* Based on approximately 33,000 persons from 10,000 households across Canada in 1950–51, representing 0.2 per cent of the population. The total illness on the day the survey began was estimated to be 97.7 per 1,000 population.

Table 23

RECORDS ARISING FROM DEMANDS UPON PUBLIC WELFARE
AND SOCIAL SECURITY SERVICES

Kind of record	Per 100 * population per year	Punch card
Family allowances—children	33.40	
Unemployment insurance—persons insured	23.20	+
Employment offices—applications for employment	18.60	
Old-age security—pensioners	4.81	
Workmen's compensation—accidents and compensations	3.45	+
Government annuities—contracts in force	2.29	
Veterans' pensions—in force	1.16	
Returned-soldiers' insurance—policies in force	.66	
Mothers' allowances—children	.63	
Old-age assistance—recipients	.54	
Unemployment assistance—recipients	.53	
Veterans' "awaiting returns allowance"—total	.38	
War-veterans' "allowances"—in force	.33	
Veterans' casualty rehabilitation—total registration	.26	
Allowances for disabled persons—recipients	.19	
Veterans' insurance—policies in force	.18	
Veterans' social services—annual requests	.08	
Veterans' assistance fund—annual recipients	.05	
Allowances for blind persons—recipients	.05	
Total records	90.79	

* Except where total file is indicated.

Table 24

RECORDS ARISING FROM CRIMINAL ACTIVITIES AND DELINQUENCY

Kind of record	Per 100 population per year [1]	Punch card
Summary conviction offences	13.70 [2]	
Indictable offences, adult	.20	+
Juvenile delinquency	.05	+
Penitentiaries—admissions	.02 [3]	+
Total records	13.97	

[1] New records per year.
[2] No federal records kept.
[3] Approximately twice this number are in custody at any one time.

dren and adults, paraplegia, heart conditions, multiple sclerosis, and muscular dystrophy. There are also hemophilia societies in the provinces of Quebec, Ontario, Manitoba, and Alberta (but with no national organization) having registers that are believed to include about half to three quarters of all cases of hemophilia in their areas.

Certain other characteristics of individuals, not associated with illness, are sometimes of interest in genetic studies of families and populations. A recurrent need for social assistance, for example, might be detected from records of demands upon government programs of assistance, welfare, and social security (see table 23). For the sake of completeness an attempt has been made to include in the table all such programs for which information is presently available.

Recurrent involvement in criminal activities or delinquency would be even more readily ascertained from routine records (see table 24), since the more serious offenses are fully documented and the records exist in punch-card form.

LINKABILITY OF THE DIFFERENT KINDS OF RECORDS

Mechanical methods of data processing are already used routinely to derive statistics from many of the records referred to. The handling procedures, however, have not been designed with a view to integrating these, either through linkage or in other ways, and the task would be too difficult with electro-mechanical accounting machines. As a result there is exceedingly little uniformity in the various official forms and in the punch cards derived from them. In particular, the manner in which individuals are designated or identified (by names, age or birth date, birthplace, et cetera) shows little uniformity in the various records.

In spite of this, many of the forms and some of the punch cards contain sufficient, overlapping information about the individual to permit record linkage with a fair degree of accuracy (table 19). Only slight modifications in the forms and cards would be needed to permit highly accurate, record linkage to be carried out. Integration of such enormous quantities of records will become increasingly possible as more use is made of large capacity computers and a considerable degree of integration may well be introduced for reasons of economy alone.

THE BRITISH COLUMBIA POPULATION STUDY

Individual geneticists wishing to carry out modest studies using routine sources of information may perhaps be deterred by the discussions of such large numbers of records. To counter any such feeling on their part, it should be emphasized that a study involving somewhat more than 100,000 records per year is presently being carried out at Chalk River in collaboration with the Dominion Bureau of Statistics and is really a modest undertaking. The chief requirement for the individual research worker is the use of a computer capable of handling magnetic tape files, together with the necessary programming. The faster the computer, the simpler the operation becomes. By far the greatest part of the effort lies in the initial programming and in the preparation (punching, phonetic coding, and sorting) of the card files.

The British Columbia study is based on records of 114,000 marriages contracted in that province over the ten-year period 1946 to 1955. To these are being linked the birth records of children from the marriages who are born within the province over the twelve-year period 1946 to 1957 and the deaths of such children during the thirteen-year period 1946 to 1958, together with records of hereditary defects as derived from the register of handicapped children. Of the 35,000 births in 1955, just over half were from marriages contracted in the ten-year period. Linkage of the 1954 and 1955 births to the appropriate marriage records has now been completed, that of the 1956 births is in progress, and punch-card files for the 1957 births are ready for linkage. Files of death records and of records of handicapped children will be prepared and linked later. A full description of the plan for this investigation has been published elsewhere.[619]

The study has two main objectives. It will attempt to detect differences of fertility and mortality in families carrying hereditary defects, and it will be used to demonstrate the technical feasibility of extracting genetic information from routine records of individual events. In addition, a number of by-products are expected in the form of twin studies, child-spacing studies in relation to stillbirth and infant death, and studies of fertility in relation to social characteristics.

The statistical products are presently limited to those which can be derived from one year of births (1955) linked to the preceding ten years of marriages. However, studies of birth order in relation to duration of marriage have been used to show the manner in which fertility in the mothers of children born in 1955 is associated with age at the time of marriage, the birthplaces of the two parents, and with urban as compared with rural residence. The results will be described elsewhere.[619] These products have served chiefly to illustrate the rapidity with which statistics that would be virtually unobtainable by conventional methods can be derived from linked records using electronic computers.

ACKNOWLEDGMENTS AND SOURCES

I would like to thank Mr. F. F. Harris of the Dominion Bureau of Statistics and Mr. J. H. Doughty of the Division of Vital Statistics for the Province of British Columbia for information relating to many of the official forms and punch cards, Dr. B. P. L. Moore of the Red Cross Blood Transfusion Service for statistics relating to blood typings, and Dr. F. Clarke Fraser of McGill University for information concerning the hemophilia societies in Canada.

The tables were compiled in part from information obtained through the above individuals and in part from published statistics relating to activities of various federal government departments and of certain provincial governments. The following published sources have been used but have not been specifically referred to in the tables:

> Annual Report of Notifiable Diseases, 1958
> B.C. Hospital Insurance Service, 1955
> Canada Yearbook, 1957–58
> Canadian Sickness Survey, 1950–51
> Census of Canada, 1951
> Census of Canada, 1956
> Crippled Children's Registry (B.C.), 1955
> Home Nursing Services, 1957
> Hospital Morbidity Study (Ontario), 1954
> Illness in the Civil Service, 1958
> Physicians' Notice of Birth Statistics (B.C.), 1953
> Registry for Handicapped Children (B.C.), 1958

I am also indebted to Mr. P. O. W. Rhynas for carrying out the calculations from which the accuracies of linkages to files of different sizes are derived.

The automatic record linkage studies referred to in the text were carried out at the Computing Center of Atomic Energy of Canada Limited using a Datatron 205 computer with two magnetic tape units and associated I.B.M. punch-card equipment. The linkage of one year of births (35,000 records) to the marriage file requires eleven hours of computing time for phonetically coding the surnames, seven hours for writing the tape, twenty hours for the linkage operation, and three hours for punching the linkage cards (the "readout" from the linkage tapes). The purchase price of a computer of this size is in the vicinity of $175,000, and the rental of the associated I.B.M. equipment is about $1,500 per month. Use of A.E.C.L's computer for any one project is generally limited to about twenty hours per month. The I.B.M. equipment required for this particular study includes a reproducer (number 514), a tabulator (number 407 or other model), and a sorter (number 82), together with the standard key-punch and verifier used in the preparation of the card files.

DISCUSSION

DR. BURDETTE: Dr. Newcombe's paper will be discussed by Dr. Woolf.

DR. WOOLF: Dr. Newcombe should be congratulated for venturing into such an ambitious project as this. Human geneticists are handicapped frequently because of the lack of good population records, necessitating the usual laborious methods employed in obtaining data on frequencies of consanguineous matings, gene frequencies, mutation rates, ethnic origins, and so on. Then, unfortunately, because of inadequate medical and vital-statistics records for most of our population, much of these data are of questionable validity. Dr. Newcombe is proposing an unique method of extracting information from a population. It will be of interest to follow his investigations and determine the practicability of his method of solving problems confronting human geneticists.

Every human geneticist dreams about an ideal population in which genealogic records are complete and medical records for all individuals are accurate and available. In such a population a geneticist could start with a propositus and very quickly obtain information on relatives. Estimates of parameters could be obtained expeditiously. Of interest on this subject are the records that are available for the Mormon population in the western part of the United States. Here is a source material of population records that can be used to great advantage by a human geneticist working on this specific population. Members of the Church of Jesus Christ of Latter-day Saints are encouraged to be active in genealogical research. The reason for this is that one of the basic teachings of this church deals with vicarious baptisms and other ordinances for the dead.

The Latter-day Saints Genealogical Society has sponsored an extensive program of microfilming and obtaining records of genealogical value from many parts of the world, such as wills, land records, marriage licenses, vital statistics, orphans-court records, estate settlements, minutes of early churches, and bible and tombstone records. From these records and other sources, the patrons of the society compile pedigree and family-group records that are of practical interest to the human geneticist. The family-group records cover three generations and contain the names of the offspring and parents of a married couple, dates and places of birth for all of these individuals, dates of death, names of spouses, plus other information of an ecclesiastic nature. These are filed alphabetically by the surname of the husband.

These records have already been of great value in the research carried out at the Laboratory of Human Genetics, University of Utah [937, 938, 939, 940]. It is a general practice to visit the archives before any Utah kindred is studied. If the names of the kindred members can be obtained, then much time and effort can be saved. The main difficulty so far encountered is that so many family-group records have been submitted by the patrons that some have been bound in bundles and are not accessible. A microfilming program is now in progress for these records, and this should permit additional studies on these records in accord with the imaginative spirit of Dr. Newcombe.

A visit to the archives of the society is a most interesting experience. Dr. Curt Stern was pleasantly surprised on a recent tour to find that the family group record of John Lambert (wife: Mary) of Sapiston, Suff. England, is on file. The son, Edward, born March 29, 1719, was the first "porcupine man." Edward and his descendants constitute the kindred which at one time was used as the classical example of holandric inheritance.

DR. NEWCOMBE: I should like to add a brief comment and close with a quotation. There is, of course, reason to discuss whether the accuracy and completeness of the diagnostic and other information routinely recorded is sufficient for genetic purposes. Clearly, for some types of study, it is not, even when used in conjunction with very extensive, pedigree data; while other kinds of investigation may be possible only through the employment of such routine sources of population data. The potential usefulness will be easier to assess as greater familiarity is gained with the procedures for extracting family information on an extensive scale. In the meantime, some reservations on the part of geneticists are perhaps understandable.

However, similar reservations apparently exist with regard to the use of the routine sources of population data by sociologists and human ecologists and have prompted a comment by the editors and part authors of a recent book on demography [336] that may be relevant in the present context. The editors write, ". . . one is tempted to conclude that the problems of obtaining adequate population data for their [the sociologists' and ecologists'] studies pale into insignificance beside the problem of getting the practitioners of those disciplines to recognize the scientific potentialities of the great quantities of data that do exist." This somewhat wry comment reflects a feeling which I have encountered elsewhere on the part of individuals who maintain such records. The researcher is often unaware that information he needs concerning human populations exists in an accessible form, and that, as a result of this lack of appreciation, the records themselves are not yet fully used for research purposes.

Edward Novitski, Ph.D.

COMPUTER PROGRAMMING

Innumerable books and manuals have dealt with the principles of operation and programming of automatic digital computers,[586] and a very concise account of the applications of computing machinery in the biologic sciences has been presented by Ledley.[485] This discussion will be devoted to those aspects of computer operation and programming relevant to the understanding of this by biologists with no great interest in the mechanical details involved, but with some interest in the possible use of these machines for their own purposes. In particular, it is hoped that this presentation will offer information on the following points: What kinds of problems may be handled effectively by a computer? What steps must be taken to get a problem in shape for computation? How much technical training is necessary before attempting to program a problem? What are the relative expenses in time and money of computer operation compared to ordinary calculation? In what general fields has high-speed computation already proved profitable?

PRINCIPLES OF COMPUTER OPERATION

The answers to these questions depend to a great extent on understanding how a digital computer works, not in detail, but in general terms. The following analogy may be adopted: an operator of a calculating machine has been employed who is not very bright but who will follow instructions exactly, will never make a mistake (or perhaps one in an entire lifetime), but cannot do anything for which precise instructions have not previously been given. Imagine further that it will be desirable to have some data analyzed or processed for a period of three months during which time you will be incommunicado. Before leaving you must write out a set of instructions that will tell exactly what to do and furthermore you must include in those instructions specific steps to be taken in any eventuality. Because some unforeseen error in your instructions may cause the operator to stop working after only a week or two has gone by, you will probably want to test, in some simple way, your written program to make sure that your instructions are correct. What you will have done, then, before leaving is exactly the equivalent of writing out a program for a computer.

A closer look at the operations involved in the program so devised can tell more about what a computer does and does not do. It can perform the four elementary arithmetic operations of addition, subtraction, multiplication, and division. Differentiation or integration cannot be handled, although the numerical answers can be obtained by substitution of approximation procedures that involve only the operations that also can be performed on a hand calculator. If a program requires the addition of two numbers, which two numbers are to be added and what is to be done with the total must be specified. The pencil and paper that an assistant would use is replaced in the computer by memory locations. The computer would be asked, by a special language that will vary from one machine to the next, to clear its arithmetic registers, put into the proper register for addition a number located in some memory position, say 123, add to it a number located in some other position, say 132, and then store the total in another memory position, say 500. Thus we are not concerned with the specific value of the total at this point, but we know exactly where to find it when we need it for further work.

The instructions that are given to the machine are stored in its memory in sequence and are carried out in sequence, until one of the instructions tells the machine not to go to the next memory location for an instruction, but to a specified one. These transfers are of two types, conditional and unconditional. The latter is uninteresting, amounting to nothing more than a jump of some memory locations, usually necessitated by the fact that the machine has momentarily been working on some special orders, has finished, and must return to the main program. The conditional transfer is more fascinating; it is one of the two powerful operations of a computer. It depends on a test of the characteristics of a number; let us consider a simple one available in many machines, that of whether a number is positive or negative. Suppose that our instructions tell the machine to take the square root of some intermediate result that has been stored in its memory by a previous computation. It is clearly necessary to determine whether this number is positive or negative, for if it is negative, this operation cannot be carried out. Prior to the operations that will give the square root, we therefore ask the machine to test the number. We give it an order directing it to go on to the next orders in sequence, which will give the square root, if the number is positive, but to jump out of the sequence to a different set of orders, if negative. These latter orders might be, for instance, to type out the offending number along with some other number which can later be used to pinpoint the part of the problem that gave rise to the difficulty, and then to transfer back into the main part of the program and continue with its operations. In this way we can be sure that the large number of operations required for the solution of the problem will not be interrupted by the occasional occurrence of an impossible request.

Individual orders are stored in the machine in the form of numbers, and one of the great features of flexibility in programming stems from the fact that these orders may be manipulated mathematically as numbers. Let us suppose that part of our problem demands as an intermediate result the sum of a hundred numbers located in successive memory positions starting with 256. We could ask the machine to clear its registers, add to it the contents of 256, add 257, add 258, and continue until one hundred additions were made. Fortunately we do not have to repeat this kind of order one hundred times because the machine can be coded to do it automatically. We ask the machine to "add 256," bring this order itself into the arithmetic unit just as any number, add 1 to it and replace the result, which now says "add 257," back into the memory location previously occupied by "add 256." Now we send the operation of the machine back to the instruction we just changed. We have now coded what is called a loop, and if we set a counter to 100 so that the operations will go through the loop that many times, the machine will add 100 successive numbers starting with the one stored in memory position 256, because each time around this loop of instructions, the machine will read consecutive memory locations to be added.

In this way, long series of operations may be programmed simply. This feature of the machines, along with the possibility of testing numbers to determine alternative courses of action, makes efficient programming possible. In fact, these qualities make the digital computer the powerful instrument it is, rather than its high speed, which is at first sight its most striking feature. If we had to give a single instruction for each and every operation of the machine, from the beginning to the end of the program, the high speed at which these gadgets work would be meaningless, for the time and work involved in writing out the program would make the procedure prohibitively expensive.

The operations described above are fundamental to computer operation, whether they be of large or medium size. By large size is meant computers of the class of the Remington Rand Univac or the IBM 700 series; examples of medium-size computers are the IBM 650, Bendix G-15, and Burroughs 205. The larger machines, of course, embody many additional features, but these are basically refinements or conveniences which can be enormously helpful but are dispensable. The IBM 704, for instance, has 22 different instructions that call for a transfer in the order of operations depending upon certain characteristics of a number. All these instructions can be simulated in a smaller machine, with some sacrifice of programming space, by sets of orders which convert them to a test of whether a number is positive or negative.

With these thoughts in mind, we can now ask what must be the nature of a problem which will lend itself to digital-computer analysis. Some that suggest themselves immediately are:

1. The problem must be one which is susceptible to numerical treatment. This does not mean necessarily that only numbers must be involved, for some nonnumerical problems are handled by machines. But in these cases, a translation must be made to a numerical format. For instance, language translation is regularly programmed for computers. This is accomplished by the expedient of representing letters by digits. Comparison of one word with another is then a simple matter of subtraction of one number from another, followed by a test to see if the result is zero.

2. The steps in the arithmetic treatment must be rigidly defined. The machine obviously will not carry out any operation that is not specifically asked for in the program. Consequently, one cannot expect miraculous results, but only those a person could obtain given sufficient time and energy. Note that even when there may be a choice in the sequence of operations depending on a test of some characteristic of a number, this must specifically be anticipated and be incorporated into the program at the outset.

3. The program must involve repeated computations using the same instructions; otherwise, the expenses of programming the problem exceed the benefits of high-speed, accurate computation. I have seen, in one recent publication, a simple average of some twenty numbers, followed by the parenthetical comment that this average was presented with the compliments of a computing device. It would seem that the time spent in setting the numbers to be averaged on punched cards far exceeded the time that would have been necessary had an ordinary adding machine been used.

Let us take the case described earlier of computing the square roots of N numbers. If N is large enough, then it becomes worthwhile to write a generalized program that will extract the square root of any number. Alternatively, since a program once written is good indefinitely, and if many different problems call for the square root of a number, it becomes profitable to write such a program.

In summary, the question as to whether a problem is worth considering for programming depends on a positive answer to these two points: first, is the problem one which I could show an unimaginative assistant how to do on an ordinary hand calculator; and second, is the amount of work involved prohibitive for such hand calculation?

PROGRAMMING

When the investigator has decided that some data might profitably be handled by a computer, his next question is concerned with his relationship to the program. In many instances, perhaps in most, this problem is most easily handled by taking the data and the problem to the nearest statistician or mathematician with the request that he arrange to have an appropriate program written for whatever machine is most readily available at the nearest computing center.

This is, however, the cowardly way out. Very often biologists are curious enough about these machines to want to try, at least once, to see what a computer can do. Sometimes there is no professional programmer available locally, or his fees may be adjusted to the more opulent sciences and cannot be afforded by the biologist. In such instances programming becomes a personal problem. If the investigator feels that he could write out the arithmetical instructions to the assistant described in the introduction, he has every right to believe that he can program a problem himself.

It is obviously necessary to program for a specific machine. Usually there is no choice. It is the one located on the campus or at some nearby educational institution. Sometimes governmental agencies (highway departments, for instance) have computers that they make available during their unused periods to scientific workers. Salesmen for companies that manufacture these machines sometimes are willing to arrange for time to use a machine; the competition between companies is now extremely keen and each can be depended on to promote its own product to the utmost. In any case, a computer is generally available to any worker who needs one, usually at a modest charge per hour of operation.

At this point, one must obtain a manual for programming the machine to be used. Two other necessities are: (1) a numerical analyst or mathematician who can give expert advice on the method of handling the problem and (2) an experienced programmer, who can give specific information on problems as they arise. It is virtually impossible to learn to code by studying a manual alone, and courses that purport to have as their purpose teaching programming methods can have little impact. By far the most effective system is to learn by programming a specific problem that makes mathematical sense to the learner. In this way, the advantages of machine programming, the tremendous flexibility allowed by testing and arithmetic manipulation of instructions, become apparent.

During this discussion, no mention has been made of the specific components that make up a machine, how information is fed into a machine, or how answers are extracted. These, of course, vary from one model to the next; but they are, in a sense, as trivial as asking about the color of a hand calculator before proceeding with an addition. The fact remains, however, that there will inevitably be a mental block to proceeding without some of this information, so it may be very desirable at some point to watch the machine being programmed in operation. I can recall that during my learning period I was bothered by the problem of how I would go about getting the computer to start its operation or, to put it more simply, where the on-off switch was located. What I did not know was that the actual mechanical operation of the machine was in the hands of a professional operator who handled such petty details and who, in fact, seemed to know far more about my problem, when difficulties arose, than I did myself.

As an illustration of how a complicated problem may be handled by a mathematic ignoramus,[626] with complete success, I should like to reproduce an equation whose solution was necessary as part of an analysis of data on laboratory populations.

$$\frac{\delta \log L}{\delta q} = -\frac{A+C}{1-q} + \frac{(A-C-D)(2k_1-1)}{1-q+2qk_1} + \frac{2B+D}{q}$$
$$-\frac{2(A+B)(qk_2+q-2qk_1+k_1-1)}{(1-q)^2+2q(1-q)k_1+q^2k_2} = 0$$

Here, *A, B, C, D,* are observed numbers, k_1 and k_2 are assigned constants, and q is a gene frequency to be solved for, a maximum-likelihood fit being achieved when this equation equals zero. From the genetics of the problem, we know that q must be between 0 and 1, and furthermore that there is only one solution in that range. It is seen from inspection that as q approaches 1, the value of the equation will be highly negative because of the influence of the first term, and that as q approaches 0, it becomes highly positive because of the third term. Furthermore, a little algebraic manipulation will show that there is no direct algebraic solution to this equation. These two conditions lead us to conclude that, if an estimate of q is too small, the value of the equation will be positive, and, if it is too large, the value will be negative. Therefore, some system of successive approximation is indicated by which successive guesses become more and more accurate, depending on whether the preceding estimate is too small or too large.

In addition to the memory locations needed for storage of the results of the individual arithmetic operations necessary to give first the values of the individual terms, and then the value of the entire expression, three were reserved respectively for (1) q, (2) Δq, the absolute amount by which q would change in each iteration, and (3) a counter telling the number of iterations to be done. Initially both q and Δq were set to $\frac{1}{2}$, and the counter was set to 14. The steps in the program are now approximately as follows: Step one, divide Δq by 2 and replace Δq by $\frac{1}{2}\Delta q$. Step two, with the current value of q, evaluate the expression. Test the sign of the result, if positive, subtract Δq, if negative, add Δq. Subtract 1 from the counter, and replace the old number with the new, which is one less. Test this number; if positive, transfer the sequence of operations back to step one; if negative (which will occur when 1 has been subtracted from 14 a total of fifteen times) transfer back to the main program.

Since Δq becomes successive powers of $\frac{1}{2}$ at each successive iteration, and is added or subtracted depending on whether the previous estimate was too small or too large, at the conclusion of these calculations the machine will have stored in its memory a solution to the equation correct to $(\frac{1}{2})^{14}$.

ELIMINATING ERRORS

One of the aspects of programming that invariably puzzles the beginner is how it is possible to make sure that the machine has been properly coded to give the correct answers to a given problem. In practice this is far simpler than one would imagine. The majority of errors (or "bugs," as they are colloquially termed) involves faulty instructions rather than numerical errors, and the faulty program simply refuses to work at all. If, during the "debugging" process, the machine is asked to give in its output all intermediate values to a longer calculation, the incorrect instructions can generally be determined quickly. Finally, when the program appears to be running properly, a set of sample data can be fed into the machine; if the results are identical with those that have been previously worked out on a hand calculator, the program can be assumed to be correct. In this procedure, it is wise to select unusual or extreme cases for test in order to ascertain if the program will proceed successfully even under exceptional conditions. The problems of "debugging" are among the more fascinating facets of programming, more stimulating than a Chinese puzzle (until the novelty of the game wears off).

Elimination of errors may cost an appreciable part (from a quarter to a third) of the total problem preparation. Special systems have been devised to diminish the time involved. Examples are interpretative routines that trace a problem step by step through its entire course and the memory dump, which produces in the output the complete contents of the memory locations at any given time.

CHOICE OF COMPUTERS

Generally, one has to take advantage of whatever machine is available. If given any choice, however, the investigator should usually select one of the larger types of scientific computers. These have speeds of tens of microseconds per individual operation, rather than tens of milliseconds characteristic of most medium-cost machines (although many companies are now putting on the market machines of small capacity but of high speed). These monstrous gadgets are more expensive to run per hour, but their greater speed may make them less expensive per arithmetic step. The fact that all scientific computers use a system of binary numbering, both for arithmetic and for orders, sometimes makes coding unpleasant on small machines but is not obvious on the larger ones. Here the machine automatically converts to binary form any number presented to it in decimal and at the conclusion of the operations converts the answer back to decimal. There may be decimal-point difficulties, as there usually are on any ordinary calculator; the large computers have built into their hardware means of taking care of this problem routinely. (It should be added that the more refined features mentioned above are all found on some individual small- and medium-size machines and that programs are usually available which simulate them.) Their larger mnemonic capacity makes possible working on more elaborate and extensive problems. Their great advantage, however, is found in the simplified programming techniques.

There are three stages of development of programming methods. In the most primitive, previously mentioned, the machine must be told precisely in its own language a numerical code that the programmer must memorize. At the second level is a mnemonic equivalent, with a one-to-one relationship between each instruction given to the machine and each instruction needed by the machine to carry out its mission. Thus, the operation, multiply, may be represented in the original program by "mpy," a special translating program converts these letters to the number that the machine recognizes as its multiplication order. The program may also assign memory space, as it is needed, so that the original program need not refer to specific, numbered, mnemonic locations, but to some simple, meaningful combination of letters. A total, for instance, might be called "tot," the translating program assigns an available mnemonic space to that symbol, and whenever that symbol is referred to subsequently, the proper number is substituted.

The highest level of sophistication is found in the formula-translation programs for the largest size computers (Math-matic for the Univac and Fortran for the IBM 704 and 709). Programs written for these computers are in the English language, or essentially so. Equations in virtually unaltered form from that of standard practice may be translated by the computer itself into the series of operations which give the desired result. A single order in original form may take the machine, on the average, five or six of its own operations to complete. Finally, the translation program itself accomplishes a certain amount of "debugging," pointing out certain types of obvious errors in the original program which must be corrected before it will translate the original into its own machine language. In one recent experience with such a translation procedure, the original program contained 176 statements, which were converted into 1,075 machine instructions, but not until we were asked to correct two errors that the translation program uncovered. There was, in addition, one other error of a numerical sort that the computer could not recognize, but which was immediately obvious after examination of the intermediate values obtained by the machine after a sample problem was run. The total time that elapsed between the start of the problem and obtaining the results from the machine was one week, and this included a considerable amount of time to decide upon the proper mathematical procedure for this particular problem. The total running time of the machine (an IBM 704) was three hours.

Such an efficiency in time spent along with an almost complete absence of "debugging" could never be matched by simple machine programming, which, for this case, might well have taken four to six months. Thus the greater cost of operation of the machine per hour (several hundred dollars) was far offset by the saving in time required for preparation of the program.

In calculating the cost of completing a program, it should be kept in mind that once finished, it can be stored away on punched cards, punched tape, or magnetic tape indefinitely until a similar set of data needs analyzing, whereupon, with a minimum of effort, it is possible to repeat the procedure. In this sense, a completed program represents a permanent investment in future biologic analyses.

AVAILABILITY OF SUBROUTINES

If a problem requires the value of, say, the sine of a number already computed, it will not be necessary to write a simple routine to make that calculation. This is a common need, and a subroutine will already have been written by someone somewhere to take care of the need. In fact, in the large machines, it is sufficient only to mention in a program that the sine is needed, and the machine will search its magnetic tapes until it finds the subroutine that gives the sine and insert it into the program. Similarly, most mathematic functions are readily available on prepackaged routines which may be inserted into a larger program with very little effort. In this way, programming effort expended at one installation can be made available at others. It has been estimated that upwards of 90 per cent of all programs may consist of such "canned" routines skillfully tied together.

This does not help the biologist very much, however. For one thing, his problems are usually not of the type that involve uncomplicated, straightforward, mathematic computations. In the second place, many of the functions he might be interested in have not yet been programmed; for instance, I am not aware of a single case of a subroutine having been written which will give the appropriate probability value based on the t, F, or chi-square distributions. In the third place, there as yet exists no organization of biologists which might work towards the end of making such subroutines generally available. On the basis of these points, it seems sensible to suggest that whenever biologists are involved in computer programming, they should, if possible, program for a commonly-used, commercial machine in order to take advantage of available subroutines and to write their programs in units of subroutines, so that others may eventually take advantage of this work.

There are four general categories into which the uses of computers in biology may be put:

1. *Data processing.*—Data collected in biology (in those fields where data can be collected) tend to be multidimensional. Common statistical techniques, such as analysis of variance or calculation of coefficients of correlation, are time-

consuming and cumbersome when done by hand. Programs for these operations are either generally available or can be written with a minimum of effort. However, in many cases, the data are so extensive that even machine analysis is impeded because of the enormous effort that would be required to transfer the data to some input medium for a computer. It would seem sensible at this stage in the development of the science for any investigator who is about to embark on a project involving the collection of masses of data to consider most seriously the desirability of starting out with some kind of punched-card or punched-tape system of recording data, in the event that computer analysis later becomes necessary or desirable. In this connection, it should be noted that one need not be too concerned that the specific medium for recording data be precisely that to be later accepted by a specific computer, for instruments are now available to convert any one system into any other. In this application, there may be less justification for preferring a high-speed machine to a medium-size one on two counts. In the first place, the programs involved are relatively simple so that there may be no great advantage to the complex-formula-translation programs available on the larger machines. In the second place, the bottleneck may be in the speed with which the machine can accept data; if data are read from cards at the rate of 250 per minute, a high-speed machine could perform more than 5,000 operations in the interval between each card. If all that the program asks is a simple running total and cumulative sum of squares amounting to less than a dozen operations, clearly the machine is not being used to capacity.

2. *Equation solving.*—Some fields of biology lend themselves to precise mathematical formulations. Genetics is one of these fields, and the sample problem given earlier is based on the precise mathematical expectations implied in Mendel's laws. In such instances we may be much more interested in the numerical answer to a given problem than in any general solution, and this, of course, is precisely what the computer can give. The complexity of the equations, the necessity for seeking out all solutions if more than one exists, and the desirability of including in the program the possibility of varying all parameters, either one at a time, or in combinations, indicate a high-speed machine. As pointed out earlier, there exist at the present time no generalized programs for calculating probabilities from distributions of various sorts, nor any general way of handling maximum-likelihood problems or discriminant functions, all of which would be extremely useful to the biologist. In this connection, as an aside, I should like to point out a problem now facing us which will grow more acute in the next few years. So far, published data have been considered public domain. Anyone with a computer may feel free to write a program to analyze them more extensively, or in a different way, and may publish the results under his own name. In this case, the program was his direct contribution. As generalized programs become more available, however, such an entrepreneur might take a prepared program developed after many months of work by one person, have it work on data col-

lected by still another person after years of work, allow the machine to spend a few hours teasing information out of the data, and come out with an essentially new insight into the data or a set of new conclusions. In this case, who has made what contributions, and how should they be acknowledged in print? Should the title of the computer take first place in the series of authors?

3. *Simulation of biological systems.*—As an example of this application, there is the growing usefulness of programming the computer to behave like a genetic population.[34, 259, 501] A population may be defined as an array of genotypes and phenotypes; a new generation is formed by selecting gametes at random and combining them to form new individuals. Such selection coefficients or other modifications in population behavior as are pertinent to the system under investigation can be inserted into the scheme computationally at the point corresponding to its biologic effectiveness. The powers of iteration of the machine are here put to good advantage, for the problem may involve finding out what happens after an enormous number of generations, and this can be handled easily. Such programs usually ask for an output after each generation and must incorporate some simple test for stopping the machine after the population has reached a steady state.

Such schemes very often involve Monte Carlo methods in which a random number is generated by the machine, and either of two alternatives is decided on by the magnitude of the number. In populations of very small size, for instance, the stochastic element of gene-frequency determination can be best simulated if a random number between zero and one is manufactured by the machine. If it falls between zero and the gene frequency, a gamete is designated as carrying one allele, if between the gene frequency and one, the other allele. In this way gene frequencies will fluctuate from one generation to the next precisely as one would expect them to behave in nature.

4. *Retrieval of Information.*—Professor Newcombe (this volume) has discussed the problems of recording and collating data on vital statistics, for which purposes computers of special types are being developed. The problem here is that of storing enormous amounts of information and then of retrieving a single item with maximum speed. For this purpose, high-speed memories tend to be too limited in capacity, whereas the enormous capacity of magnetic tape is offset by the necessity of searching serially until the needed item is found.

The accumulation of published literature and its inaccessibility to the average worker are serious problems at the present time, and it is customary in discussing this matter to point out the value of computers in the eventual solution of this problem. A description of this brave, new world conjures a picture of some enormous computer, located in the geographic center of the United States, into which pour all the latest research results. Output leads are directed to every laboratory in the country, where investigators sit in special thinking (or perhaps absorbing) chairs, with wires fastened directly to their cerebra and perhaps other parts of the body.   In this way, will Science advance to the greater glory of mankind? If so, it will advance without me.   I like to identify myself with those living in the day of sealing wax and string and, if necessary, will retire from the frenzied, hectic whirlwind of scientific work to the relaxed calm, peace, and quiet of big business.

DISCUSSION

DR. BURDETTE: Mrs. Schultz will discuss Dr. Novitski's paper.

MRS. SCHULTZ: I have been asked to substitute for Dr. Victor McKusick and to describe briefly the computer program being set up at Johns Hopkins to estimate linkage in man.   It is probably a good example to begin with, as it illustrates the two facets of the characteristics of a computer.   Most people, when they think of a computer, think of its ability to solve arithmetic problems.   However, an even more important part of the computer's potential is its ability to solve logical problems.   The program I will describe is really composed of two parts: the first is a logical problem, that is, the elimination round, which determines all possible genotypes for each person in the pedigree, and the second, the probability round or linkage calculation, is mostly an arithmetic problem.

The method used by the computer is essentially that described by C. A. B. Smith in 1953.[801]   In the same paper, he suggested that a digital computer be used to handle the not-so-difficult, but tedious, calculations involved.   Simpson, at the Rothamsted Experimental Station in England, wrote a program for the Elliott 401, which estimated linkage for two generations.   He was somewhat limited by the size and capabilities of his machine; and for multigeneration families or for more complicated situations, such as consanguinity, a larger and faster computer is needed.

The program I will describe was set up by Dr. Renwick and me at Johns Hopkins and has been written for the IBM 709, which has 32,768 storage locations and is quite a fast machine.   The time taken to calculate the information on a pedigree depends on the complexity of the pedigrees, the number of persons in the pedigree, the amount of information given on each, and so on; but most pedigrees take only a few minutes of machine time.

The recombination fraction, X, gives an estimate of the distance between two loci.   So, essentially, the machine calculates the probability of the family given various values of X and assumes the value of X that gives the maximum probability for the family to be the correct one.

Before going into any details, a few terms should be defined.   By main parent is meant that parent directly connected to the pedigree and by mate that parent who is connected only by marriage.   Order of genes is very important to the machine as this is its means of telling whether there has been a crossover. For example, OA and AO are considered as two distinct genotypes, the first meaning that the child obtained the O gene from the main parent and the A gene from the mate and AO meaning that the A gene came from the main parent and the O gene from the mate.

The first step, of course, is to read in the information on the pedigree. This includes the values of X that is to be tested, and the frequencies of genotypes with which the genotypic possibilities of the mate are weighted. Also for each person in the pedigree, the parents, the generation he represents, his marital status, as well as his phenotype, must all be identified. In a problem such as this, where there is a large amount of input, it is always wise to place as many internal checks on the input as possible, since the greatest probability of error will come from reading in faulty information.

The elimination round determines the possible genotypes for each person in the same manner as it would be done by hand. For example, from the phenotype of the main parent a list of possible genotypes is set up and the same is done for the mate. Then these lists are compared with the genotypes of their children, and any genotypes that could not have produced such children are eliminated.

In the probability round, the genotypes of the main parent and mate are compared to those of the child to see if there has been crossover. In this manner, an expression for the probability of the pedigree is deduced in the form of a polynomial of sequential products and sums in terms of X.

Do-it-yourself programming is becoming easier. Quite a number of systems have been written which make it possible for almost anyone to program. Consequently, I hope that geneticists as well as the medical profession will be using digital computers much more frequently in the future.

GENETICS OF DISEASE

J. A. Fraser Roberts, M.D., D.Sc.

INHERITED DISEASES

I approach a conference on methods in human genetics with considerable dif-
fidence, for I am a user of statistical methods and not a mathematician. What I
wait for hopefully and greet with pleasure is the devising of methods of analysis
by others, so that I can apply them to my own data. I thought therefore that perhaps the main contribution I should try to make to this discussion is to consider
some points about obtaining data, so as to take advantage of opportunities and to
minimize or evade difficulties. We should probably all agree that the elaboration
of methods of analysis tends to run ahead of the provision of suitable material on
which to exercise them. "First catch your hare," as the words are usually mis-
quoted; the cooking can come later.

HOSPITAL POPULATIONS

Every center at which work on human genetics is based offers some opportunities and suffers from some disadvantages. Our own Unit is at The Hospital for
Sick Children at Great Ormond Street in London. The great advantage is the
high proportion of patients suffering from relatively rare conditions which particularly interest us. On the other hand, for a number of reasons, London is not a
good place for studying population genetics, as for example on the lines pursued
by Stevenson and his colleagues in Northern Ireland. It would in practice be very
difficult to attempt a total ascertainment in an area where so many different hospitals, authorities, and organizations have to be approached individually. A considerable proportion of our work is based on material provided by our own hospital, but a further proportion involves extension to other hospitals, either in
London or in various other parts of the country. Of course, in our country, in

common with many others in Western Europe, we enjoy the advantage of limited distance. It is by no means impracticable to follow up families throughout the whole country.

We are all familiar with the difficulties of insuring that ascertainment of index cases is carried out so that the subsequent analysis can be carried out efficiently. Here it seems to me that one way of evading this formidable difficulty is to make the analysis on subsequent children. Of course, a lot of information is sacrificed, but the sample should be a good one and freer from bias than samples obtained in most other ways. Naturally, if parity and age of parents are important factors, one is in difficulties, but if these effects are absent or relatively small, and this should be known from other studies, then I think we get in this way one of the safest bodies of material on which to base estimates. I will quote an example.

It is known, largely as the result of the excellent surveys carried out by McKeown and his colleagues at Birmingham, that, when a couple have had a child with anencephaly or spina bifida, the risk of recurrence in subsequent children is about 1 in 25. But what is the risk when a couple have had two such children? It is important to discover this both for theoretical and practical reasons. It was decided to obtain a sample of such couples, a task which involved the coöperation of many maternity hospitals in different parts of the country. First of all, the records must be available in such a form that all cases of these conditions over some predetermined period can be readily identified and the hospital notes made available. We then examine these, noting the necessarily rare instances in which a previous child has been affected. The all-important point is, of course, that ascertainment must be through the second affected child. The mothers are then located, a family history obtained, and any subsequent children examined. Up to the present, eight have been affected out of fifty-three subsequent children, or about 1 in 7. The 5-per-cent confidence limits are wide, ranging from 1 in 5 to 1 in 18, but at least it can be said that the risk is significantly increased following the birth of two such children as against the risk following the birth of one. It has been a most formidable task, and I rather quail when I think of the number of man- and woman-hours expended; but it was something that had to be done and we were well placed for doing it. It is to be hoped that others may add to these numbers, though it is unlikely that the estimate will be very appreciably improved until another fifty subsequent children are collected. The detailed family histories, which have not yet been analyzed, should also prove useful.

CHILDREN OF PATIENTS WITH PYLORIC STENOSIS

In surveys on conditions that come to light and are treated in early life, it is usually fairly easy to obtain adequate numbers of siblings of index cases; but it is much more difficult to secure an adequate sample of children. So it was decided to carry out some investigations starting specifically with index cases who were the parents of children. The first condition I will mention is congenital pyloric stenosis. This work has been carried out by my colleague, Dr. Cedric Carter, and I am very grateful to him for allowing me to quote his results. It was known, principally from the work of McKeown and his colleagues, that the incidence of pyloric stenosis in siblings of affected children is about 1 in 12 for boys and 1 in 50 for girls. The over-all incidence in the general population is about 1 in 300 births, or roughly 1 in 200 male births and 1 in 800 female births. It has only become possible fairly recently, however, to observe the children of affected persons, for prior to Rammstedt's operation, introduced to Great Britain in 1920, severely affected children in hospitals rarely survived. A list was therefore prepared of

Table 25

CHILDREN OF PATIENTS WITH CONGENITAL PYLORIC STENOSIS

Parents	Children				
	Sex	Normal	Affected	Total	Per cent affected
Males (160)	Boys	132	10	142	7.0
	Girls	134	2	136	1.5
Females (30)	Boys	22	7	29	24.1
	Girls	22	1	23	4.3
Total		310	20	330	6.1

all children treated by Rammstedt's operation at The Hospital during the years 1920 to 1939. They numbered 941. Then came the task of tracing them. First, individual letters were sent to the original addresses. The usual procedures for attempting to trace removals were then followed, but naturally the haul was not very large, for the addresses were anything up to forty years old. The next step was, with the help of the Registrar General, to verify exact dates of birth and full names. At this point the National Health Service offers special opportunities for tracing individuals. The Ministry of Pensions and National Insurance cannot, of course, disclose addresses, but they were prepared to help by filling them in on our envelopes and posting the letters. Married women were often difficult to trace owing to change of name, but recourse to the marriage registers, with the help of the Registrar General, produced a further set of new surnames and addresses, and thus an additional quota was traced. To summarize the results of a rather complicated but quite practicable procedure, no fewer than 673 out of the 941 patients operated on between 1920 and 1939, or 72 per cent of the total, have so far been duly traced. Those patients having children have been visited. Table 25 (Dr. Cedric Carter's data) shows in brief summary the results up to date.

At one time it was thought that the genetic component in the determination of pyloric stenosis might be a recessive gene, but, of course, a gene expressed in only some of the homozygotes. It is known that a number of nongenetic factors are concerned, and there is, of course, the notably higher incidence in boys than girls. All this could be regarded as partial penetrance, the most important additional factor being sex. But clearly the gene, if it is a single gene, is dominant and not recessive. Furthermore, the remarkable difference between the offspring of affected boys and girls respectively shows that affected girls are much more genetically predisposed than are affected boys, for this is duly reflected in their children. In fact, a single-gene hypothesis becomes impossible, unless a model can be visualized involving additional or modifying genes. And a very complex model it would have to be. Personally, I think such an explanation almost untenable. A multifactorial hypothesis, with a threshhold effect, what Grüneberg calls quasi-continuous variation, also seems difficult to devise. The proportional difference between the offspring of affected men and affected women seems too large. In a word, girls are four times less susceptible than boys, and those that are affected have four times as many affected children. On a multifactorial hypothesis, this difference should be diluted in the offspring. Of course, the confidence limits are wide enough to make this possible in a larger sample.

The affected women, who are mostly young and will have many more children, are a specially useful group. It should be possible to study the development of the condition from birth onward and to throw further light on possible minor degrees of muscle hypertrophy as well as on the environmental factors responsible for the development of the condition. There will be no difficulty about keeping in touch. Already the knowledge of the high degree of risk to the offspring of these women has led to early diagnosis and prompt treatment in more than one instance.

CHILDREN OF PATIENTS WITH HARELIP

We have carried out a similar study on harelip and cleft palate, that is, starting with index cases who are those treated by operation at The Hospital between 1920 and 1939. We were encouraged to undertake this study because Fogh-Andersen's [235] fine survey, though providing ample data on siblings of index cases, included relatively few children of index cases. Nearly a thousand subjects have been traced, and, of those with children, 211 have been seen so far, together with their 423 children. Table 26 shows the results for harelip with or without cleft palate. The numbers for central cleft palate are rather small as yet. Incidentally, and not at all surprisingly, our findings entirely confirm Fogh-Andersen's in showing that the two conditions are genetically distinct.

Once again the same pattern emerges. Harelip is commoner in boys than girls, the ratio of the sexes being about two to one or rather more. The affected girls have substantially more affected boys than have the affected boys.

It seems to me that we shall have to revise our thinking about conditions that clearly have some hereditary basis, but which yield proportions of affected among first-degree relatives appreciably lower than one in two or one in four. It is fatally easy to postulate a single gene, and then fit the figures by assuming partial penetrance, the unknown factors, genetic or nongenetic, or both, being left conveniently vague. The sex differences shown by pyloric stenosis and harelip contradict a hypothesis of this kind and more complicated explanations will have to be sought. And, of course, there has always been the basic difficulty of explaining why a harmful gene, lowering fitness substantially or drastically, could be as common as must be assumed. I hope I am not biased, but for many years I have

Table 26

CHILDREN OF PATIENTS WITH HARELIP (WITH OR WITHOUT CLEFT PALATE)

Parents	Children				
	Sex	Normal	Affected	Total	Per cent affected
Males (99)	Boys	99	2	101	2.0
	Girls	94	2	96	2.1
Females (52)	Boys	54	6	60	10.0
	Girls	46	0	46	0.0
Total		293	10	303	3.3

distrusted single-gene explanations, when the firm ground of theoretical Mendelian ratios has been left behind, as the basis of genetic resemblances in common conditions.

I mentioned earlier that London is not a very good center for studying the frequencies of genes producing abnormal conditions. I am once again indebted to my colleague Dr. Cedric Carter for allowing me to quote a piece of work he has recently started. It is a simple and correspondingly ingenious idea. For a selected range of conditions he includes patients at The Hospital who are domiciled in a defined area of Southeastern England, including London. This area has a population of some 17,000,000 and provides the great bulk of the patients. The birthplaces of the parents and grandparents are then ascertained. Numbers are small as yet and, with one exception, no appreciable differences have emerged; but the one exception is very striking and shows the potential usefulness of the method. The exception is phenylketonuria. Fifty-three per cent of the patients with this condition, though domiciled in Southeastern England, have one or more grandparents who were born in Ireland, or Southwestern Scotland, which includes a large, Irish, immigrant population. The corresponding percentage of patients suffering from the other conditions studied is only 10 per cent. The percentage of those with one or more grandparents on both sides, paternal and maternal, who were born in Ireland or Southwestern Scotland, is 16 as against only 1 per cent for the other conditions. So here is a very simple method, particularly suitable for hospital work, for obtaining data on relative frequencies of conditions in different geographical areas, without the immense labor of carrying out complete regional surveys. I should imagine it would be a very suitable method for use in some parts of the United States.

So far I have talked about investigations on a single hospital or more than one hospital. The advantages of starting with hospital patients are very great. A grateful patient or a grateful parent is an excellent person with whom to begin. In the work on pyloric stenosis and on harelip and cleft palate, the general attitude has been one of pleasure that The Hospital should be interested in seeing what has happened to its patients and their children after so many years. Here are some figures for the study on harelip and cleft palate already quoted. Out of 214 patients so far approached, and this so many years after operation, there have been only two refusals to coöperate and one partial refusal.

DATA ON TWINS

Hospital patients also may be very useful in obtaining samples for human genetic study outside the field of disease. For a short while we have been conducting a study starting with twins. The Hospital agreed to add a question to those routinely put by the Records Staff to parents of patients new to The Hospital, both inpatients and outpatients. The question is: "Is the child one of twins?" Actually we hoped to ascertain, in addition, the potentially important group of children who had a twin who died, and it seems likely that we are doing so. The number of pairs obtained in this way is about 350 a year. On these subjects and their siblings who are aged 5 to 15 years we make some 15 anthropometric measurements and 2 mental measurements. For older siblings and parents we make 9 anthropometric measurements. The purpose of this work is to obtain basic data on resemblances between relatives in continuously distributed characteristics. So far, observations have been confined to the twins and their siblings and parents. Later we hope to add grandparents, uncles, aunts, and cousins. Now the striking thing is this: out of some 280 families, nearly all seen at their homes and not in hospital, there has not been a single refusal to coöperate. It remains to be seen, however, how far gratitude, and the general knowledge that interest in twins is natural and harmless, will spread to grandparents and uncles and aunts. Of course a hospital sample can be criticized on the ground that it may not be representative of the general population and also because a twin comes to notice because it is ill. But these seem rather minor objections to set against completeness, and, in any event, the great majority of subjects are outpatients, often suffering from trivial or ephemeral disorders.

BLOOD GROUPS AND RELIGION

Another point about hospital patients is that questions can be asked without giving offense which would be difficult to include with samples differently chosen.

To quote one example, possible stratifications in the population are of much interest, and one simple measure is religion. This is not an easy question to ask subjects forming many kinds of sample. Blood donors, for example, might well resent it, or might perhaps think that giving blood was a more hazardous business than they had innocently supposed. But so far from the question being resented by hospital patients, it is asked as a routine and the information is already there in the notes. A survey has recently been carried out on *ABO* and *Rhesus* blood groupings in relation to mental diseases. Twenty-thousand consecutive admis-

Table 27

BLOOD GROUPS AND RELIGION,
MENTAL HOSPITAL PATIENTS, SOUTHERN ENGLAND

	No.	Percentage Frequencies			
		O	*A*	*B*	*AB*
Church of England	8,908	45.4	43.9	7.7	3.1
Other Protestant	1,452	44.6	42.9	9.3	3.2
Roman Catholic	1,158	50.2	36.6	10.4	2.8
Jewish	186	36.0	38.2	20.0	5.9
Other and not known	556	46.0	41.7	9.4	2.9
Total	12,260				

sions to thirty-seven mental hospitals have been grouped, with negative results: within the limits imposed by these numbers there do not appear to be any associations.

The stratifications are revealed in table 27, which shows results for Southern England, a region over which there is little variation in *ABO* blood group frequencies.

The Roman Catholics are quite notably higher in Group *O* than the remainder. This is undoubtedly a reflection of their predominantly Irish origin, though the point cannot be established directly as it was decided that patients' names should not appear on the forms. The sample of Jews is very small, but *B* is notably high, as would be expected if a high proportion of these patients are of fairly recent Middle- and Eastern-European origin. This finding, coupled with some others, is of interest. It shows that there are stratifications in the British population, and quite notable ones, but they have not produced any false, or rather secondary, associations between blood groups and disease, an explanation of the findings which has appealed to some. For the benefit of those not very familiar with some of the institutions in Britain, it should be mentioned that the very high proportion of patients shown as Church of England reflects a tendency of those with somewhat vague and shadowy religious affiliations to describe themselves as members of the Established Church. It does not necessarily imply any very active participation.

DIFFERENCES IN AGE AND SEX

With simply inherited conditions, variable age of manifestation can be allowed for in some appropriate way, provided good data are available on a sufficient scale. With diseases, often common diseases, when there is a genetic element in causation, the task is equally straightforward in principle. Provided that figures are available for incidence in the general population, separately for the sexes, and by successive age groups, then the incidence in relatives of subjects can be compared with expectation. A notable example, to quote just one, is provided by Doll and Buch's [182] admirable work on peptic ulceration. They were able to show that the incidence among brothers of index cases is about two or two and a half times greater than expectation in a comparable sample from the general population. The requirement is, of course, studies on ample numbers so as to establish these incidences, a point which Professor Penrose would have mentioned had he been at this symposium.

But this refers to the presence or absence of all-or-none disease conditions. If we suspect, however, that some of or all the underlying genetic element may be continuously distributed, we may need to use the methods appropriate to continuous variation. Thus in congenital dislocation of the hip, the width of the acetabular angle is of importance, and this is different in the two sexes and also changes rapidly as children pass through the relevant ages.[117] In diabetes mellitus the frequency distribution of glucose tolerance may be relevant to our analysis. In benign essential hypertension, it is thought by one school with which I have been associated—and I find the argument convincing—that hypertension represents the positive tail of the frequency distribution of arterial pressures, a distribution that is fairly normal if plotted on a logarithmic scale. In such instances, an obvious way of adjusting measurements so as to equalize for differences in age and sex is to fit regression lines and score individuals according to the deviation from the fitted line. However, variance is also increasing, so a given deviation at one age has a very different meaning from the same deviation at another. Psychologists often work in terms of standard scores, that is, scores that are units of the standard deviation; but they have advantages not easily realized in other fields. Scales of intelligence are so designed that sex bias is eliminated and sometimes also designed to yield equal variances at different ages. Moreover, because the span of age under consideration is often small, changes of variance are of little moment. All this is very different with, say, arterial pressures, for variance increases something like sixfold between childhood and old age. If different groups are being compared, say, parents, siblings, and children of index cases, their mean ages will be very different; and a given deviation from expectation, as specified by the regression line, will imply something very different in the different groups of relatives.

I first encountered this problem twenty-five years ago in connection with some group-intelligence-test results on a sample of children showing a rather wide scatter of age. It struck me that a rational procedure would be to establish standard deviations at successive ages by fitting the regression of variance of score on age. All that was then required was to multiply a given deviation of score from the norm by the ratio of the estimated standard deviation at that age to some fixed standard deviation; but, needless to say, when it came to testing the fit of the regression line, linear in that instance, I was completely at sea. I took the problem to Professor Fisher. He said he had not met it before; I think that perhaps he was too polite to say that he would not have thought of approaching it in that way; but, characteristically, he at once gave me the solution. Later, with the kind help of my former colleague, Dr. Peter Armitage, the treatment was extended to the fitting of polynomials of higher order. The method has been used for several measurements, including arterial pressures.[320, 726, 727, 728]

The variance in each array is weighted by the number in the array. The mean square, derived by dividing the weighted sum of squares of deviations from the regression line by $k - r$, where k is the number of arrays and $r - 1$ the order of the polynomial is:

$$\Sigma n_p (y_p - Y_p)^2 / (k - r)$$

The theoretical sampling variance of an estimate y of a true variance, Y, based on a sample of n, is $2Y^2/n$. The equation therefore estimates approximately the mean value of $2Y^2$, taken over the k arrays, namely

$$2\Sigma Y_p^2 / k.$$

The approximation, which should lead to little discrepancy, is due to the fact that the observed variances are weighted only according to the number of observations and not according to estimates of their true values. Thus the mean square due to linear regression is compared with twice the square of the mean variance, with degrees of freedom 1 and ∞; the extra mean square due to the quadratic term with $2/k$ times the sum of squares of the expected variances given by the linear function, et cetera; the remainder with $2/k$ times the sum of squares of the expected variances given by the highest polynomial fitted, with degrees of freedom $(k - r)$ and ∞.

Again using arterial pressures as an example, it so happens that the variances for both systolic and diastolic pressures are almost equal in the two sexes at about the age of sixty years. Hence, adjustment to age 60 equalizes for sex as well. The multipliers for systolic pressures in women range from 2.48 for the age group 15–19 to 0.94 for the age group 70–74. With men, the corresponding highest and lowest multipliers are 2.42 at ages 25–29 and 0.81 at ages 80–84. If one likes, the observed deviation, multiplied by the multiplier, can be added algebraically to the expected score at age 60. This gives a figure with a direct physical meaning, that is, the equivalent pressure as at age 60. Incidentally, I now doubt whether full adjustment tables are necessary. One usually has a batch of cards to score, and it is simpler in practice to set up the multiplier on the machine and multiply successively the deviations for subjects of that particular age group.

I have noticed a certain hesitation, perhaps even distaste, amongst statisticians when confronted by these polynomials. It is perhaps a rather clumsy approach, but the adjustments seem to work. For example, adjustment in this way for changing variance removes the bulk of the troublesome lack of independence of Terman-Merrill I.Q. and chronological age, at least up to the age of 12 years. It is true that the fitted regression lines tend to be somewhat unstable from sample to sample, but it is difficult to think of alternative methods that would not lead to an equal instability. Thus, logarithmic transformations make it necessary to use arbitrary constants, which also vary from sample to sample.

In general, the increase of variance with age is not linear, so that polynomials of higher order have to be fitted, usually cubics. It is usually found that a linear function, while it accounts for the great bulk of the association, gives impossibly low, or even negative, values for the variance at the youngest ages. It might be asked whether it would not be better to work in terms of standard deviations instead of variances. Increase of standard deviation certainly tends to be more nearly linear, but it turns out with arterial pressures at least that higher-order polynomials are still required. One point is that the curves are sometimes truly inflected. There is a tendency for variance to be higher in adolescence, then to fall, then to rise continuously, but again to show a fall in old age. It seems to me that the variances are more sensitive than standard deviations, though the mathematicians can tell us whether this is due to the weighting of the observed variances according to the number of observations instead of according to estimates of the true values, which I take it would be a rather troublesome procedure. To sum up, one would naturally welcome an improved and simpler method for allowing for change of variance with age, but, in the meantime, the polynomials do seem to provide a reasonably adequate method.

PUBLICATION OF ISOLATED PEDIGREES AND NEGATIVE RESULTS

The last points I should like to mention concern publication. There does seem to be a case for encouraging the reporting of isolated family histories or of small groups of families, series that are too small to make much contribution by themselves, but which, when added to others, would provide adequate data for analysis. The risk of biased selection for recording is, of course, always with us; but it should now be possible to minimize or perhaps eliminate this hazard. The other question is more important. This is the tendency to publish results if they are positive and not to publish if they are negative. Statisticians are, of course, well aware of this danger and of how it biases tests of significance. In my own work this problem has come to notice very forcibly in connection with studies on associations between blood groups and disease. I know of several studies, mostly carried out incidentally, which have given negative results and so have not been published. The subject of human genetics lends itself to brief reporting, at least in regard to these two points, namely, the handful of family histories and negative results in the search for associations. It would be a useful service if journals could devote a small part of their space to these humble but nevertheless useful tasks.

DISCUSSION

DR. BURDETTE: Dr. Herndon will discuss Dr. Roberts' paper.

DR. HERNDON: The general topic of inherited disease probably includes at least two-thirds of the practice of clinical medicine. As the entire subject cannot be covered, I will confine my remarks to comment on a few of the points touched upon by Dr. Roberts. I was particularly interested in the data reported from the study of pyloric stenosis done by Dr. Roberts and Dr. Carter. From the point of view of methodology, this study differs from the usual approach in which a series of patients, usually infants and children, is collected and studies are made of the siblings and parents of the index cases. The approach they used is to collect a group of adult individuals who survived pyloric stenosis in infancy, and then to study the children of these affected individuals. This approach of ascertainment by means of an affected parent of the sibship to be studied, rather than by ascertainment of a member of the sibship to be studied, has obvious statistical advantages, and can be expected to provide evidence of a different type. As Dr. Roberts has pointed out, these results will require some revision in our concept of genetic factors concerned in etiology of pyloric stenosis, and provide empiric risk figures of predictive value.

From the technical viewpoint, the method of study used by Drs. Roberts and Carter requires that one locate a group of patients who were hospitalized 20, 30, or 40 years previously. In American populations, I would suspect that this method would be quite useful in certain population groups, but would not be very effective in other groups. The degree of effectiveness of his method would likely be inversely related to the degree of mobility of the population. In a metropolitan area such as New York City and most of our Eastern seaboard, it would be quite difficult to locate adults on the basis of hospital records 25 years old. In some of our older and more stable rural populations, this method would have a high degree of effectiveness. For example, in North Carolina we have carried out certain studies in which the index cases were obtained from old hospital records, and we have had a high degree of success in locating these patients.[343] I feel certain that the yield from this method would also be quite high in the Utah population. In areas where families move at frequent intervals, a large proportion of index cases might never be located. It would seem advisable to obtain information on the degree of mobility of a specific population, and an estimate of the expected yield, before applying this method to proposed studies.

One of the charts presented by Dr. Roberts classified individuals by religion. This table showed a heavy preponderance of members of the Church of England. Dr. Roberts remarked that this category was probably somewhat overloaded because the Church of England was the popular church in which to claim membership in the population studied. Several investigators have referred to the effect of religion on mate selection, and the effectiveness of religion in producing stratification in population samples. This factor should certainly be taken into account in studies of gene frequency and in studies of statistical association of various characteristics with disease processes.

Dr. Roberts alluded to uncertainties in religious classification due to the tendency of those not staunchly connected with a particular religious group to claim membership in the popular group in the community. We have encountered this problem in North Carolina also, where it is said that the Baptists are outnumbered only by the English sparrows. Our group works at the North Carolina Baptist Hospital, and we are quite aware that a significant number of our patients are Baptist only upon entering the front door of the clinic. We attempt to compensate for this factor of uncertainty in classification by adding two additional questions to our classification sheets. In addition to asking the patients to name a religious preference by denomination, we later ask them for the name of the specific church that they attend and for the name of the minister. Those who can specify neither the church nor the minister are not considered to be members in good standing. In addition to studying the effects of religious stratification within populations occupying a restricted geographic area, it is of interest to compare religious groups in different geographic areas. In areas where one church group is preponderant, one may expect to find isolate effects in members of the smaller religious groups. Of interest also is a comparison of different preponderant groups from different areas. For example, I would be interested to see comparisons of the frequencies of certain genes among members of the Church of England in England, among Baptists in North Carolina, and among Mormons in Utah.

At the risk of restating the obvious, I wish to emphasize a concept that I believe was implicit in much of Dr. Roberts' discussion. If one considers the present relationship of the discipline of medical genetics to basic problems in pathogenesis of disease, it becomes apparent that genetics now stands in a position rather analogous to that occupied by the discipline of bacteriology a half-century ago. A century ago, the names of diseases were based upon symptom complexes, and knowledge of etiologic agents was meager. Fifty years ago the bacteriologists were still concerned with identifying specific bacteria and in reclassifying infectious diseases upon an etiologic basis. The microbiologists demonstrated that certain symptom-complex categories, such as pneumonia and meningitis, could be caused by different varieties of bacteria, and that an etiologic classification presented therapeutic and prognostic advantages. In a similar manner, they also demonstrated that symptom complexes that appeared to be very different clinically could still represent the effects of a single micro-organism, such as tuberculous meningitis, pulmonary tuberculosis, and tuberculosis of the bone. Geneticists are now concerned with etiologic classification of inherited diseases, and are undergoing a similar process of subdividing certain categories of symptom complexes and combining others. Many examples may be quoted. One is the separation of phenylketonuria from the general group of undifferentiated, mental deficiency. It is now recognized as a specific metabolic defect occurring in those homozygous for a specific, autosomal, recessive gene. A more recent example is the demonstration that certain cases of hemolytic anemia occurring after ingestion of drugs of the primaquine group occur only in persons whose erythrocytes have a decreased concentration of glutathione, which is secondary to a deficiency of the enzyme, glucose-6-phosphate dehydrogenase.[115] Studies by Childs and associates [131] have shown that the enzymatic deficiency is gene-controlled, and other studies have demonstrated significant differences in genic frequencies between major ethnic groups. Numerous examples could also be quoted of apparently different symptom-complex categories that can now be shown to represent variations in expressivity of a single gene. The clinical variance observed in Von Recklinghausen's neurofibromatosis and in osteogenesis imperfecta are classic examples. It seems that a rapid evolution in our concept of inherited disease is in progress, with emphasis upon the identification and study of specific etiologic agents or specific genes.

Dr. Roberts has pointed out to us certain areas in which our knowledge of inherited disease is rapidly expanding, and certain techniques of study that have proved to be effective. He has also demonstrated that the discipline of medical genetics has come of age, and that its subject matter is still growing at a rapid pace.

DR. BURDETTE: Dr. Buckwalter, I know that you have continued your interest in the work initiated by Professor Aird and his group at Hammersmith. Would you care to say anything about the present status of this work?

DR. BUCKWALTER: To Dr. Roberts I am indebted for a good share of my understanding of human genetics. My interest in the *ABO*-blood-group-disease association was awakened while working with Professor Ian Aird at Hammersmith Hospital as Dr. Burdette suggests. In figures 2 and 3 are summarized the results of studies done by our group in Iowa, St. Louis, and with South African populations. The frequencies of *ABO* blood groups within each category of disease and population were compared with the frequencies observed in controls (vertical columns) for each population, using the chi-square method. The findings did not indicate differences in frequencies of blood type between the patients and their controls of statistical significance in some instances. However, when the results are

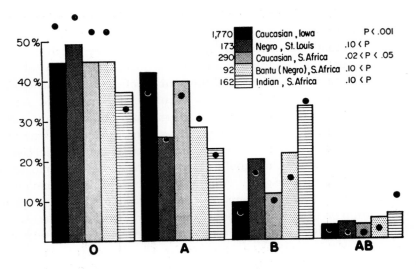

Fig. 2.   DISTRIBUTION OF BLOOD GROUPS IN PATIENTS WITH PEPTIC ULCER.

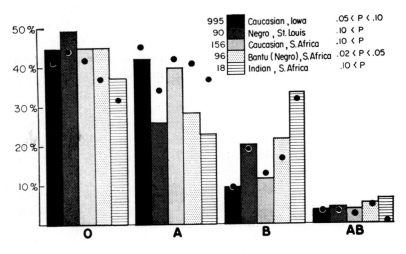

Fig. 3.   DISTRIBUTION OF BLOOD GROUPS IN PATIENTS WITH GASTRIC CARCINOMA.

pooled, they indicate associations of statistical significance characterized by an increased incidence of gastric carcinoma in persons of type *A* and of peptic ulcer in persons of type *O*.[106] Less convincing evidence of associations in other disorders has been found.[106, 107] The indication of associations in racially and ethnically differing populations provides support for a causal rather than noncausal or specious explanation for the indicated associations.

DR. STEINBERG: I support Dr. Roberts' plea for publication of negative results. Although I also agree that it is useful to have pedigrees available, I am not convinced that the expense of publishing them is always warranted. A central registry for storage with published lists of available pedigrees would make it possible for those interested to obtain them from the clearing house.

DR. BURDETTE: Dr. Roberts, do you have any concluding remarks?

DR. ROBERTS: Dr. Burdette, I have nothing appreciable to add, except to thank Dr. Nash Herndon very much for his kind remarks. Dr. Buckwalter's work is very familiar to me. Some of these racial differences that he has been finding, particularly in his South African material,[105] will, I am sure, be of the greatest interest to everyone.

I am very glad to hear Dr. Steinberg's favorable response, particularly as regards publication of negative results. A central depository would be a help, though I do wonder how much it would really be used. Such a depository would, one hopes, operate in conjunction with a journal. As regards good genetic material, insufficient in amount, but all that the observer has, one would like to see it put on record in some form or other. Then when enough material has been accumulated analyses can be made.

C. Nash Herndon, M.D.

EMPIRIC RISKS

The term, empiric, often has a somewhat less-than-complimentary connotation. As a noun, the term was applied to an ancient sect of physicians who disregarded all theoretic study and based their knowledge and practice upon experience alone. Secondary and apparently derived meanings include: (a) "one who deviates from the rules of science and regular practice," and (b) "one who relies upon practical experience; hence a quack, a charlatan." [614] As the last of several meanings of the adjective form, empiric is defined in Webster's International Dictionary as "pertaining to, or founded upon, experiment or experience; as empirical knowledge." While the latter meaning may be the one least favored by the editors of dictionaries, it is certainly the meaning intended by geneticists when this word is used in scientific communication.

In genetic usage, the phrase, empiric risk, may be defined as the probability of occurrence of a specified event based upon prior experience and observation rather than upon prediction by a general theory. The specified event with which we are usually concerned is the clinically manifest appearance of a particular disease or defect. When a prediction is made of the likelihood or frequency of occurence of a disease under certain circumstances that is based primarily upon previous observation under similar circumstances, this prediction simply states an empiric risk. In this frame of reference, a life-insurance table is simply an empiric-risk statement of the probability that one may die of any cause at a specified age, based upon previous mortality experience of the population. The determination of empiric risks may thus be regarded as a specialized type of epidemiology.

APPLICATIONS OF EMPIRIC RISKS

Empiric-risk figures may be useful in a variety of ways. Their determination is essential to certain types of genetic research. The first step in segregation and linkage analysis is to make empiric observations of the proportion of affected and normal individuals within family units, and these observations are then tested for goodness of fit to a series of genetic hypotheses. If an adequate hypothesis is found, then predictions concerning the family and population dynamics of the identified gene can be based on general genetic theory. If empiric observations fail to fit a satisfactory genetic hypothesis in any respect, then one must suspect either the action of some modifying influence not previously identified or the appropriateness of the original sampling technique; or one may even suspect the existence of a previously unknown mechanism. A series of empiric observations may deviate from expectancy based on a reasonable genetic hypothesis in a consistent manner and, in so doing, may point the way to further investigation to determine the cause of the deviation. The determination of empiric risk is therefore an essential step in the formulation of an adequate genetic hypothesis for any inherited disease or defect and may often serve as an indicator of environmental interaction, heterogeneity of material, or other types of modification. Properly interpreted, empiric observations often provide the guidepost leading to new and productive areas of research concerning the pathogenesis of disease.

Empiric-risk figures for specific diseases also have clinical usefulness. The clinical applications are clearly related to the general concepts of preventive medicine that are concerned with primary prevention of disease, early diagnosis of disease, and limitation of disability.

At the level of primary prevention of disease, empiric-risk figures may be applied in two ways. The most common use is in connection with counseling in families that have produced one or more children with severe malformation or fatal disease. Parents who have produced a malformed or severely ill child immediately demand a prognosis concerning possible future children. It is clearly the responsibility of the attending physician to provide realistic risk information or to refer the couple to a suitable source of such information. In the present state of our knowledge, empiric observations provide the only realistic figures for risk of recurrence of a long list of malformations and pediatric diseases. Space will not permit even an outline of the general principles of counseling in such situations, but two observations seem especially pertinent. First, the empiric-risk figures now available usually prove to be reassuring to the parents of a malformed child, as their fears are usually much worse than realistic risks. Second, the physician should never attempt to dictate an acceptable risk level for a specific family. There are many personal, psychologic, moral, and religious factors that enter into a decision to have, or not to have, more children, and a level of risk acceptable to one family may be totally unacceptable to another. Each mentally competent couple should make its own decision in the light of its own unique set of circumstances.

A second type of application of empiric-risk figures at the level of primary prevention of disease is only beginning to be developed. This relates particularly to disease of late onset and to disease that becomes evident clinically only upon interaction of a genetic and an environmental component. The clearest examples at present are afforded by diseases for which adequate genetic hypotheses exist. For example, galactosemic children are homozygous for a specific recessive gene; but the clinical disability resulting from retarded growth and mental development occurs only if there is a continuing supply of galactose. This disease is preventable by environmental manipulation, specifically by elimination of milk sugar from the diet. Other examples of genetically determined susceptibility to specific environmental insult are known, and there are probably many others not yet adequately defined. Porphyria hepatica is probably conditioned by a dominant gene with reduced penetrance, but heterozygotes cannot always be identified by clinical methods. However, heterozygotes are particularly sensitive to drugs of the barbiturate group, and a fatal paralytic crisis can be induced by a single intravenous dose of a barbiturate.[175] It is probable that other unidentified precipitating factors exist also. Even without a clinically reliable method of heterozygote identification, an empiric risk for development of symptoms of porphyria hepatica can be assigned to any blood relative of a known patient, and those with significant risk can be protected against at least one precipitating agent, namely barbiturate drugs. The general concept of prevention in a wide variety of genetic diseases depends upon two factors: (a) identification of individuals with an increased risk for development of a specific disease and (b) identification of environmental stimuli that may precipitate clinical disease in a susceptible individual. While identification of susceptibles by specific laboratory test would be the most satisfactory method, it seems likely that empiric-risk evaluation may remain the only practical method for a number of conditions. Empiric-risk figures and other epidemiologic data may also be helpful in identification of environmental stimuli that are harmless to most of the population but highly dangerous to those with specific susceptibilities.

From the viewpoint of preventive medicine, empiric-risk figures can also be useful at the levels of early diagnosis of disease and of limitation of disability. There are a number of diseases of the middle and later years of life that are insidious in onset and which may cause permanent disability before the victim seeks medical attention. One example is provided by chronic, simple glaucoma. A familial aggregation of cases of glaucoma is usually noted, and, in certain families, there is reason to suspect the action of an autosomal dominant gene. Other families fail to conform to expectancies based on this hypothesis, and it seems likely that all cases so diagnosed are not etiologically homogeneous. Any relative of a patient with glaucoma has a measurable risk of incurring permanent damage to the optic nerve during the fifth or sixth decade, this risk being somewhat less than the coefficient of relationship of the individual to a known case. If each individual with increased risk for development of glaucoma would visit an ophthalmologist regularly for examination and tonometry, a large proportion of the blindness and visual defect due to glaucoma could be prevented. Pernicious anemia provides another example of a chronic disease of insidious onset, in which the exact genetic mechanism is not yet clear. Permanent damage to the spinal cord may occur before the patient is aware of his difficulty and seeks therapy. The risk for development of pernicious anemia among siblings and children of known patients is not known exactly, but is certainly not less than 10 per cent and may be higher than 30 per cent. Periodic health examinations, including hematologic studies and determination of gastric acidity, would permit early diagnosis and prompt treatment of these affected relatives and would prevent the development of disabling neurologic complications.

It seems clear that empiric-risk figures have two major areas of usefulness in human genetics. First, they constitute an essential tool for certain types of genetic research. Second, they have clinical applications that are particularly pertinent to preventive medicine and public health.

TYPES OF EMPIRIC RISKS

With respect to inherited disease, empiric-risk figures may have continuing usefulness in several types of genetic situations. First, there are many inherited diseases for which an adequate hypothesis of genetic mechanism exists, but the observed proportion of clinically affected individuals deviates from the expected proportion in some consistent manner. A common type of deviation is in the direction of a significant reduction in the proportion of affected individuals within sibships. Pedigree studies of certain dominant genes often demonstrate that a clinically normal child of an affected person may transmit the disease to his descendants, indicating that he was a carrier (or transmitter or conductor) of the gene even though clinically normal. This phenomenon is referred to as reduced penetrance. Penetrance has been defined as the frequency with which a characteristic expression of a gene or genotype is manifested among those who possess the gene or genes in question.[822] Other usages have extended the concept to include the frequency with which any detectable effect is exhibited. For certain conditions it is thus possible to define two levels of penetrance, one at the level of clinical disability or illness and a second at the level of laboratory abnormality in an apparently healthy carrier. Estimates of penetrance rates at both levels may have clinical usefulness. Neither level can be predicted for a specific disease by general theory. Penetrance figures must be obtained for each disease by empiric observation and pedigree study. For example, observation and analysis of large groups of pedigrees of the autosomal dominant type of peroneal atrophy have demonstrated that about 90 per cent of those heterozygous for this gene exhibit the physical signs of the disease. Except by the progeny test of ability to produce affected offspring, no method of identifying the carriers who are clinically normal is available. The risk that any normal child or sibling of a patient with dominant peroneal atrophy may be capable of transmitting this disease to his children would be obtained by multiplying the coefficient of his relationship to the patient (in this case, 0.5) by the complement of the penetrance rate (in this case, 0.1), leading to a risk estimate of 5 per cent.

The deviation of observed proportions of affected individuals from theoretic expectancies may exhibit a consistent relationship to sex. It is not uncommon to find that penetrance of a particular gene is more reduced in females than in males, leading to some shift in the sex ratio. More pronounced deviations with excess of affected males may be indicative of genetic heterogeneity, such as an unsuspected mixture of cases due to autosomal recessive and sex-linked recessive genes. At times such deviations in sex ratio can be adequately explained by physiologic or anatomic differences in the sexes.

An example of this situation is afforded by hemochromatosis, in which the ratio of affected males to affected females is about 10 to 1, and the age of onset of symptoms averages about 48 years in males and 52 years in females. The empiric risk of development of clinical hemochromatosis for brothers and sons of patients is not known with accuracy, but it is probably in the range of 10 per cent. An adequate genetic hypothesis could not be formulated until studies of plasma-iron levels of patients were carried out.[176, 222] It now appears that those heterozygous for the gene for hemochromatosis absorb iron from the intestinal tract in greater quantity than is normal. This leads regularly in males to increased levels of plasma-iron and to eventual deposition of hemosiderin in liver, pancreas, and other organs. Females, however, have an additional avenue of iron excretion via the menses and are usually able to compensate for the overload of iron. About half the female patients with hemochromatosis give a history of absent or scanty menses. The data of Debré and associates [176] on levels of plasma-iron in children of patients with hemochromatosis suggest that the penetrance level of the postulated, dominant gene (if defined by levels of serum iron in excess of 200 micrograms) may approach 100 per cent in males over age 15 years, but presence of the gene is still not detected with certainty in females with normal menses. It should now be possible to prevent the hepatic and pancreatic damage in hemochromatosis through identification of susceptibles by determinations of serum iron and restricting intake of iron or increasing loss of iron via phlebotomy. The mere fact that the phenomenon of reduced penetrance of a certain gene exists should give rise to the hope that a rational method of prevention or therapy may be found.

A second and major type of empiric risk is associated with diseases that recur within family groups more often than can be accounted for by chance alone, but which fail to conform to any simple genetic pattern. An adequate genetic hypothesis for the disease in question is therefore lacking. It seems probable that certain of the diseases falling into this category are not etiologically homogeneous entities. Diverse causes may produce clinical pictures that are identical or so closely similar that they are erroneously classified as one disease.

An example of this situation is afforded by microcephaly. A number of pedigrees exist that strongly suggest that the affected children are homozygous for a specific recessive gene. It has also been shown [583] that exposure of the fetus to therapeutic doses of radiation may result in microcephaly. Microcephaly is also one of the types of malformation that may follow maternal rubella during the first trimester of pregnancy. Microcephaly has been observed in mice subjected to anoxia during early embryonic development,[395] and it is possible that anoxia may be an effective etiologic agent in man also. It is also possible that other unidentified etiologic agents may exist. Thus, at least three and possibly more etiologic types of microcephaly exist, but we are usually unable to distinguish these by clinical examination of the patient. Any series of cases collected for study is likely to represent a mixture of cases of various causation in unknown proportion. The probability of recurrence of microcephaly among the siblings of those cases caused by one or more recessive genes would be 0.25, while the same probability for siblings of those cases caused by environmental insult to the fetus might approach zero. The observed empiric risk of recurrence of microcephaly derived from such a series of cases would then be a function of the proportions of cases of different causes among the propositi. Risk figures derived from different series could be expected to vary considerably. They would have little usefulness aside from affording some estimate of the proportion of cases of genetic and environmental cause among the original propositi. It is thus apparent that the question of etiologic homogeneity of material is of considerable importance in interpretation of empiric-risk figures.

Empiric-risk data can be more useful, but must still be interpreted with caution, in conditions under which there is no direct evidence of etiologic heterogeneity. In certain conditions a specific genotype may determine a state of susceptibility, and specific precipitating environmental factors may be required for development of clinical disease. Congenital clefts of the lip and palate may provide an appropriate example. It seems well established that cleft palate alone and harelip with or without cleft palate are clinically and genetically distinct conditions, and that other special types of cleft palate associated with other defects, such as mandibulo-facial dysostosis or cysts of the lower lip, also exist. Fraser [262] has investigated etiologic factors associated with both major types of palatal clefts extensively and has concluded that they are influenced by multiple genetic and environmental factors. Empiric-risk figures for the appearance of affected children in families with various types of family history have been derived for both clinical types. For families in which both parents are normal and one child has harelip with or without cleft palate, the risk for each additional child lies in the range of 4 to 7 per cent; in the same family situation but with the affected child having cleft palate without harelip, the risk is in the range of 2 to 5 per cent. If one parent is affected and there are no affected children, the risk for each child, if the parent has harelip, is about 2 per cent, while the risk, if the parent has cleft palate alone, is about 7 per cent. For families in which one parent and one child are affected, the risk for harelip is about 11 per cent, while the risk for cleft palate alone is about 17 per cent. Fraser [261] has pointed out that these figures are averages and are not as precise as one might wish, and that the true risk for a specific family may differ widely from these figures. If applied with some caution, these and similar figures do have some usefulness in counseling.

A third major group of empiric risks are those that express statistically significant associations of diseases with nonspecific traits or with other diseases. For example, it is a well-known observation that diabetes mellitus and hypertensive cardiovascular disease appear in obese individuals with a higher frequency than in the general population. In these instances, obesity may serve as a precipitating factor for development of clinical symptoms in a susceptible individual. However, a statistical association certainly does not imply a cause-and-effect relationship. Obesity could simply be a by-product of a more specific dietary factor of etiologic significance. Another well established association is the increasing incidence of mongolism associated with advancing maternal age. A child born to a mother 35 years of age has a greater probability of being a mongol than a child born to a mother 20 years of age. However, the younger mother has a greater risk of bearing a second mongol, even if both younger and older mothers have the same number of children born subsequently.[284] It is not clear just how these associations fit with the recent discovery that mongols have an extra chromosome of the acrocentric group.[418]

Other nonspecific associations that have recently attracted attention include the association of blood type *A* with gastric carcinoma and the association of blood type *O* with peptic ulcer.[104, 724] Associations of this type are intrinsically interesting and provide leads for new investigation into the pathogenesis of disease. The accumulation of knowledge of such associations may also be directly helpful in identification of individuals with increased risk for development of specific diseases.

A fourth and final type of empiric risk is represented by the concordance rates that may be derived from twin studies. The concordance rate is the frequency with which both members of a twin pair exhibit the trait under consideration. If this rate is 100 per cent for monozygotic twin pairs (and significantly lower for dizygotic pairs), we have evidence that the characteristic considered is determined predominantly or exclusively by the genotype and that environmental variation within the range experienced by the twins studied is insufficient to modify the expression of the trait. In most twin studies, concordance rates of less than 100 per cent are observed. The reduction from perfect concordance in monozygotic pairs when compared with dizygotic rates offers a rough measure of the effectiveness of environmental variation in influencing expression of the trait. Concordance rates are related to penetrance rates, but the two should not be considered as always equivalent. The environmental experiences of twin pairs are usually more similar than the experiences of two unrelated persons and may give rise to significant differences in these rates. Discordant monozygotic twin pairs are interesting in themselves and offer a unique investigative opportunity. They represent two individuals of identical genetic constitution who have reacted differently even in rather similar environments. Opportunity is thus offered for intensive study of environmental differences that may be significant in the pathogenesis of genetic disease.

It would seem superfluous to comment upon methods of collection of empiric-risk data. The requirements concerning sampling techniques and analysis are no more and no less than those required for any other type of genetic study and are included elsewhere in this volume. The major problem of methodology with respect to empiric risks is one of interpretation of data. At times it is difficult to decide what is an empiric risk and what is not. When analysis of a series of observations is complete, we are often faced with two temptations. The first temptation is that of over-generalization of conclusions. Most observations of genetic significance are collected under certain restrictions and requirements from a restricted population. These restrictions are sometimes minimized in an effort to arrive at a conclusion of general applicability. Errors can and have been committed by attempting to apply risk data of restricted origin to general situations or in attempting to translate them from a special population base to a general population. The second common temptation arises when observed data fail to fit simple genetic hypotheses adequately. It is quite tempting to formulate secondary hypotheses and modifications until the data manage to conform to some general genetic theory. Sometimes such secondary hypotheses have served as a point of departure for further investigations that have been informative and productive. Other hypotheses have been only ingenious speculations and have often created more problems than they have solved. Neel [601] has reviewed some of the pitfalls that may be encountered in attempting to formulate a sound genetic hypothesis. The danger of assigning a causal relationship to statistical associations upon inadequate ground should also be mentioned. It is sometimes better to leave an observation in an empiric state and use it with caution and under specific restriction than to attempt to erect a finished hypothesis upon an inadequate base.

SUMMARY

In summary, an attempt has been made to classify certain broad types of genetic situations in which estimates of empiric risks may be made. The determination and evaluation of an empiric risk is often an essential step in investigation of the pathogenesis of a disease and can be considered as a special type of epidemiologic study. Empiric risks are thus useful as guideposts to further research needs and also have important clinical applications in the field of preventive medicine and public health.

DISCUSSION

DR. BURDETTE: Dr. Anderson will open the discussion of Dr. Herndon's paper.

DR. ANDERSON: Often geneticists have a feeling of ambivalence toward empiric risks and are not fully agreed that they can be used or are needed. A decision on this point certainly enters into the choice of research problems and into experimental design. Some workers believe that empiric risks are not needed at all. Dr. Roberts [725] has pointed out that the proportion of people who need such risks is indeed small, but they need it very much. The decision to have another child, for example, is an important one. It will be based upon assumed empiric risks if someone does not provide more accurate information. A major contribution of Dr. Herndon's paper is his clear statement of the important uses for empiric risks. They are often used in consultations concerning children to be placed for adoption. In a series of 165 such consultations at the Dight Institute, there were fifteen questions about epilepsy, fourteen about mental deficiency, and eleven about schizophrenia. Even though the genetic evaluation of these conditions is not yet clear, the available estimates of risks can help in reaching a decision.

A second objection to empiric risks is that they are too imprecise and do not provide us with enough relevant information. Admittedly, some of the more urgent questions are those for which genetic analysis has not been completed. In such cases, empiric risks include the chance of repetition of environmental factors. Parents of an affected child are not always concerned about whether environmental or genetic factors have been responsible for the condition. Further research should make it possible, however, to make a distinction between genetic and environmental components and thus provide more precise estimates of risks. The earlier discussions today also have indicated clearly the need for adequate methods of analysis. We need to use the concept of sporadic cases, identify carriers whenever possible, and study the effect of factors such as age of the mother, sex of the child, and the pattern of symptoms. Although the empiric risks currently available may be imprecise, it is certainly possible to make them more precise and more useful.

A third objection is that it is unwise for those uninformed about genetics (including some physicians) to use empiric risks. Relatively small numbers of qualified genetic counselors are available to consider individual cases, and tabulating all information that should be considered about each situation ordinarily encountered is usually thought not to be feasible. A somewhat similar problem exists in the field of psychology. Dr. Paul Meehl [534] has compared the use of psychometric data and other types of observational information as they have been combined by mechanical and by personal means. "Clinicians often hold the view that no equation or table could possibly duplicate the rich experiences of the sensitive workers." Meehl evaluated twenty studies involving a comparison of clinical and actuarial methods. In all but one of these the predictions made by use of tables were approximately equal or superior to those made by a clinician. In the field of human genetics, empiric risks should be stated in such a way that they can be used adequately in all situations where they are needed.

Attention is now directed briefly to methods of calculating empiric risks. Earlier today Dr. Roberts stressed the importance of subsequent risks. Haenszel [302] has presented related ideas included in the "concept of case order." In practice, it is surprisingly difficult to decide upon the appropriate method of procedure. The effect of different methods may be illustrated by a simple set of calculations which Dr. Sheldon Reed [709] carried out for the data of Munro on phenylketonuria (for which the theoretical subsequent risk would be 25 per cent). If whole sibships are taken, excluding the probands, 21 per cent were similarly affected. The apparent reduction may indicate the effect of "reproductive stoppage." If whole sibships are taken with each affected person as a proband, the risk becomes 24.6 per cent. In a third method, the risk subsequent to the first affected in each family is 29 per cent. This high value suggests a biased inclusion of multiplex families. In a fourth (and apparently most suitable) approach, Dr. Reed tabulated the siblings subsequent to each affected child and obtained a risk of 25.4 per cent.

Another problem in the calculation of empiric risks is introduced by variability in age at onset. This may be illustrated by data from the Dight Institute study of mammary cancer.[12] It is quite obvious when one studies a series of sisters of patients with cancer of the breast that these women have incomplete life experience. In our study, 4.4 per cent of the sisters had already developed breast cancer. This number is not a risk figure of any kind, since it clearly depends upon the age distribution of the sisters at the time of the study. Some of the living sisters still have a chance of developing the disease. By using a modified life-table method, it is possible to estimate the number who will have developed mammary cancer by the time the sisters have all died or reached age 85. This is 7.4 per cent for sisters of the patients with mammary cancer, as compared with slightly over 4 per cent for the sisters of the patients' husbands, a control group. The morbid risk, assuming that all would live to age 85, is 10 per cent for the sisters of the patients. In conditions with considerable variation in age at onset, the point to be stressed is that some type of adjustment may be in order to derive a meaningful risk value.

In summary, our use of empiric risks should profit from consideration of the following points:

1. Management of related problems encountered in other fields is instructive. Dr. Herndon has mentioned some examples, and survival rates for conditions such as cancer might be added.

2. Dr. Herndon's list of medical conditions for which empiric risks are needed can be expanded profitably.

3. The degree of precision really required, particularly for counseling, should be determined. Dr. Schull[784] has suggested that the "decision-making area" may be tested. A risk of 5 per cent probably does not affect a family's decision to have more children. A risk of 25 per cent or more for a very serious condition probably does make a difference. It should be possible to study the psychological effects of different risk levels and thus to determine the degree of precision necessary.

4. Finally, it should be possible to agree upon adequate but simple procedures for calculating empiric-risk values.

DR. PATAU: Dr. Herndon wondered how the empiric risks established for mongolism relate to the recent discovery of trisomy as the cause of this disease. In a general way, these risks are compatible with the cytologic situation since trisomy comes about by nondisjunction, most likely during the first meiotic division. Mongoloids can be viewed as an heterogeneous group. First, these are the cases caused by accidental nondisjunction, which almost certainly will occasionally occur at random. The sharp increase of the empiric risk with maternal age evidently means that, in aging women, the probability of this kind of meiotic accident is greatly increased. Presumably this holds true not only for the chromosome responsible for mongolism, but for other chromosomes as well. The implied expectation that the incidence of trisomic syndromes other than mongolism should also increase with maternal age has already been confirmed.

Second, there undoubtedly are cases in which nondisjunction is the result of a special mechanism rather than of accidental malfunction. Two classes of such mechanisms can be distinguished. In both cases, they are likely to lead to a familial incidence of abnormal chromosomal complements, irrespective of the age of the mother. The first class comprises genes that interfere in some way with the orderly course of meiosis and thus produce nondisjunction. It can be expected that their effect is not restricted to any particular chromosome with the result that in members of one and the same family syndromes due to different abnormal-chromosome constitutions may occur rather than mongolism alone. We know of a case in which a sibling of a *D* trisomic displays Turner's syndrome. Of course microsymptoms would not be observed in healthy members of such families. The other class of mechanisms that can cause familial incidence of mongolism, or other trisomic syndromes, includes different kinds of structural heterozygosity, such as inversions or reciprocal translocations, which may cause frequent nondisjunction of a specific chromosome in carriers with an essentially normal genic complement. In consequence, only the given type of trisomy, for example, mongolism, can be expected to recur in the offspring of a carrier, except for deficient chromosomal complements, which most likely would be lethal. It would not be surprising if the carrier showed microsymptoms due to position effect or genic mutation accompanying breakage. Our group, together with Dr. Irene Uchida, has investigated a case of familial incidence of mongolism that suggests a situation of this kind.

DR. HERNDON: I share Dr. Patau's enthusiasm for the new work on chromosomes in relation to disease and find his remarks of interest. Dr. Anderson's able discussion both posed and answered a rhetorical question, "Do we need empiric-risk figures?" I agree with his conclusion that we do need empiric-risk figures for specific diseases and malformations and that we must make use of these in clinical medicine. There is now widespread interest in medical genetics among clinicians in practically all fields of medical practice. The professional geneticist can no longer remain detached from clinical demands to provide consultation service in routine medical problems. When the diagnosis is clear and the genetic hypothesis concerning etiology of the disease is supported adequately, genetic counseling and clinical consultation are relatively easy. However, many clinical problems are not so easily solved. Where only empiric-risk figures exist, we must be prepared to present these fairly and explain their meaning and limitations. A major risk is that of giving the impression that we know more than we do. The medical geneticist offering clinical consultation must be prepared to define the limits of knowledge in his field and to point out the many areas of uncertainty that still exist. It is becoming apparent that the geneticist can make a valuable contribution to clinical medicine. In many diagnostic centers, the geneticist has become established as a valuable member of the medical team, and this trend can be expected to continue.

Rody P. Cox, M.D., and Colin MacLeod, M.D.

RELATION *between* GENETIC ABNORMALITIES *in* MAN *and* SUSCEPTIBILITY *to* INFECTIOUS DISEASE

Susceptibility of man to infectious diseases encompasses a subtle combination of factors, some environmental and others inherited. Although hereditary factors are universally recognized as being important in the susceptibility of the individual to infection, the specific characters responsible for this predisposition are unknown in many instances. Separation of genetic from environmental factors in familial aggregations of infectious disease is difficult and is responsible in large part for the paucity of specific information. Familial tendency to certain infectious diseases may be due to prolonged or intimate contact with the microorganisms and exposure to high dosages; however, hereditary susceptibility contributes in many instances.

In addition to the difficulties of separating the environmental from genetic factors, the ill-defined and nebulous qualities of resistance and susceptibility resist objective analysis. Occasionally, recognizable genetic lesions are responsible for or associated with susceptibility of resistance to a specific infection of man. Until recently, few genetic markers have been available and little attempt has been made to correlate genetic traits with susceptibility or resistance to infection.

Sickle-cell anemia [387] is an exception. The substitution of a valine residue for a glutamic acid in the globin portion of hemoglobin is the molecular basis of this disease. Patients with sickle-cell anemia are unusually susceptible to *Salmonella* infections with osteomyelitis as a frequent occurrence. The reason for this susceptibility is not known, although it seems probable that there is impairment of a factor or factors concerned specifically with resistance to systemic *Salmonella* infection.[49] On the other hand, evidence suggests [7, 607] that individuals with sickle-cell trait may be partially resistant to falciparum malaria or to its cerebral manifestations.[706] The frequency of the sickle-cell gene in certain parts of Africa has been explained on the basis of a selective advantage conferred by this mutant gene.

In the past, decreases in the severity and in the clinical manifestations of a number of infectious diseases have been explained as being due to a selection of a more resistant population. The natural history of many infectious diseases indicates that when introduced into a virgin population they are widely prevalent and killing diseases. Gradually their rapid killing power abates, and the clinical symptoms become milder. Smallpox and syphilis,[294] when introduced into Europe, followed this pattern. A number of diseases, even those thought of as characteristically mild or chronic, such as measles or tuberculosis,[109] have been widespread and lethal, when introduced into a previously inexperienced population.

The factors responsible for the mitigation of the severity of such diseases are unknown. The selection of genetically resistant individuals is but one possibility. Maternal antibodies may reduce the severity of infection in newborns and partially protect them until active immunity is acquired. With a rising herd immunity there is a decrease in susceptibles with fewer contacts and smaller infecting dosages. Environmental factors contribute by improved sanitation and better nutrition. Because of the multiplicity of factors involved, it is hazardous to conclude, in any given instance affecting human populations, that the severity of the disease declined because of the elimination by the disease of genetic susceptibles.

In other infectious diseases there has occurred cyclic change in morbidity, mortality, and prevalence. These fluctuations have been interpreted as being caused by enhanced virulence of the microörganism or by increase in its communicability, by antigenic alteration of the parasite conferring the ability to spread among a partially immune population, or because of diminished resistance of the host. The importance of any one of these factors in a particular epidemic is difficult to assess. The epidemic events represent a fluctuating equilibrium between the virulence of the parasite and the resistance of the host. To prove a rise in virulence or a fall in resistance is difficult.[932] One can but describe a disease that is more prevalent, more fatal, or more communicable at one time than another.

An example is scarlet fever in England in the last years of the eighteenth century. Scarlet fever was much more deadly than it had been a quarter of a century before or was again a quarter of a century later.[294] The prevalence of the disease was similar in both epochs. The reasons for the change in severity are not known. However, a change in the genetic susceptibility of the host population is improbable. It is more likely that, at times of great severity, a highly virulent strain of Group *A* streptococcus was widespread.

Diphtheria is another example of a disease that has shown variation in prevalence and severity. Burnet [109] has explained these fluctuations as possible variations in bacterial virulence associated with changes in the amount of diphtheria-toxin production, and similar ideas have been expressed by others. This explantion would appear to be an oversimplification.

We have recently experienced an epidemic of influenza associated with a major antigenic alteration of the virus. The Asian or A2 strain, possessed of a "new" antigenic mosaic, was able to spread widely, causing a high incidence of infection and disease the world over.

VIRULENCE OF MICROÖRGANISMS

Because the emphasis in this volume is on human genetics, only a limited discussion of microbial virulence is indicated. Pathogenicity refers to the capacity of microörganisms to cause disease. Virulence is the term used to indicate the degree of pathogenicity of a type or strain of microörganism.[520]

One of the best examples of genetic factors controlling the virulence of microorganisms is the relationship of the amount of specific capsular polysaccharide (SSS) synthesized by pneumococci to their virulence for mice. The capsular polysaccharide contributes to the virulence of pneumococci by its antiphagocytic effects. MacLeod and Krauss [521] conclusively demonstrated with three different types of pneumococci that murine virulence depends on the *amount* of capsular polysaccharide (SSS) synthesized by them. Strains which produced no SSS are avirulent; those that synthesized small or intermediate quantities of SSS had slight to intermediate virulence for mice; while the fully encapsulated strains were highly virulent. The genetic control of the virulence factor (SSS) was directly demonstrated by the transformation of rough, nonvirulent pneumococci to various degrees of encapsulation. The virulence of the transformed strains and the amount of polysaccharides produced by them approximated that of the strains from which the transforming extracts were prepared. This evidence demonstrates that the important factor concerned in virulence of pneumococci for mice is the genetic control of capsular-polysaccharide synthesis.

Our knowledge of virulence factors in other microbial species is less complete. Toxigenicity of diphtheria bacilli depends on the presence of a lysogenic bacteriophage, and, moreover, the amount of toxin produced is inversely proportional to the iron content of the nutrient medium. The virulence of streptococci is related to the presence and amount of type-specific *M* protein on the bacterial surface which inhibits phagocytosis. For many important groups of organisms, such as the staphylococcus, the virulence factor(s) are either not well defined or are unknown.

MICROBIAL VARIATIONS AFFECTING PATHOGENIC PROPERTIES OTHER THAN VIRULENCE

Genetic variations other than those directly affecting virulence may contribute to disease patterns. When a major antigenic alteration occurs in an organism, as was recently observed when the A2 (Asian) variant of influenza virus emerged, the lack of immunologic resistance placed most of the population in jeopardy. The infectious agent spread rapidly throughout the entire world because specific immunity to this virulent antigenic variant was present only among the oldest age groups, who, it appears, experienced infection by an antigenically similar variant during the epidemic of 1889.

Changes in microbial metabolism and extracellular products may contribute to the clinical expression of disease without altering virulence. These changes may be responsible for the variation in symptoms and signs of an infectious disease from one epidemic to another. Variation in communicability, which is the capacity to spread from one host to another, will affect prevalences and rate of dissemination. To assess the contributions of each of these factors to an epidemic is difficult. It seems probable that the emergence of a parasitic strain of high virulence is an occasional event, rather than a normal process, in the fluctuations that occur during the course of infectious disease.

SUSCEPTIBILITY OF THE HOST

Experimental infections in laboratory animals have demonstrated the importance of the host's genetic "soil" in determining susceptibility to infection.[773] A few inbred strains of laboratory mice show specific resistance or susceptibility to a particular strain of virus or bacterium. Genetic analysis of resistant, inbred strains of mice indicates that only one of these strains, the Asiatic mouse, has a single gene difference that determines its resistance [278] to a particular bacillus. In other susceptible inbred-lines of mice, more than one allele is involved in the diminished resistance to specific infections.[278]

Genetic factors are of importance in some infections of man. Multiple mechanisms may be responsible for human susceptibility to a particular organism, the clinical expression of infection, or infection at a particular anatomic site. In hypogammaglobulinemic children,[281] the susceptibility to most bacterial infections and to viral hepatitis is due to a defective immune mechanism. The failure to produce sufficient quantities of antibody globulin results in repeated infections often with the same species of microörganism. Children with cystic fibrosis of the pancreas have an abnormal respiratory-mucus secretion [180] that impairs clearing capacity of the lungs and favors recurrent pulmonary infections. The hereditary defect is confined to the exocrine glands and in most instances is characterized clinically by symptoms of pancreatic insufficiency and chronic pulmonary disease.

More obscure reasons for susceptibility to infection are associated with the predisposition of the dark-pigmented races to disseminated coccidioidomycosis. This racial predilection is strikingly demonstrated in the Filipino and also in the American Negro.[811] Dissemination is one hundred and eighty times [275] more frequent in Filipinos than in Caucasians.

It should be profitable to examine the mechanisms responsible for susceptibility to infection in man in terms of specific heritable diseases. The mode of inheritance, in the following text and tables, is usually expressed according to classical, Mendelian concepts of dominance and recessiveness. Frequently, however, dominance and recessiveness are incomplete and penetrance variable. An example is sickle-cell trait, which is classed as recessive, although in heterozygotes nearly half the hemoglobin molecules have the *S* defect.

GENETIC FACTORS PREDISPOSING TO INFECTION WITH A SPECIFIC MICROÖRGANISM OR TO A PARTICULAR CLINICAL EXPRESSION OF THE DISEASE

Hereditary predisposition to a specific infection (table 28) as seen with *Salmonella* infection in sickle-cell anemia, and the predilection to disseminated coccidioidomycosis in the dark-skinned races, have been mentioned above. The underlying reasons for this susceptibility are unknown. It is probable that resistance to these infections is impaired in a specific manner.

Leprosy.—Familial susceptibility to leprosy has been recognized since 1848. This disease, which is caused by a mycobacterium, is naturally transmitted only by

Table 28

GENETIC FACTORS PREDISPOSING TO INFECTION WITH A SPECIFIC MICROÖRGANISM OR TO A PARTICULAR CLINICAL EXPRESSION OF THE DISEASE

Genetic character	Infectious disease or clinical expression	Mode of inheritance	Evidence for genetic determination of character	References
Sickle-cell anemia	Salmonellosis, particularly osteomyelitis	Autosomal recessive; heterozygotes show trait in variable degree. (Negroes)	Convincing	7, 49, 387, 607, 706
Dark pigmented skin	Disseminated coccidioidomycosis	Dominant with variable expression.	Convincing	275, 811
	Leprosy	Unknown; probably multiple genes.	Questionable	26
	Tuberculosis	Unknown; probably multiple genes.	Questionable	433
	Paralytic poliomyelitis	Unknown; recessive (?)	Suggestive	25, 185, 344, 611, 744

prolonged contact with an infected person. However, the failure of the disease to follow the usual pattern of communicability suggests an inherited predisposition. The evidence for genetic susceptibility is based on studies [26] of isolated centers of endemic leprosy such as New Brunswick (Canada) and Louisiana. In these circumscribed areas of prevalence, the majority of cases were confined to a few families. The rarity of conjugal leprosy and the infrequent infection of non-related intimate contacts strengthens the hereditary hypothesis. The evidence is not decisive, but a tendency to an inherent susceptibility to leprosy seems possible.

Tuberculosis.—Tuberculosis resembles leprosy in that it is a contagious disease of chronic nature, caused by a mycobacterium, and hereditary predisposition has been proposed. The primary importance of familial contacts with the opportunity for a large infecting dose complicates studies of familial incidence. Kallman and Reisner's [433] review of the incidence of tuberculosis in identical and fraternal twins and their siblings, parents, and spouses furnishes important data. Sixty per cent of monozygotic twins without known exposures were concordant for tuberculosis and the extent of the disease was similar. Only 13 per cent of dizygotic twins without known exposure were concordant for tuberculosis, and the severity of the disease in fraternal twins was frequently dissimilar. Rates of concordance in parents, spouses, and other siblings were lower than in dizygotic twins but higher than in the general population. The evidence indicates susceptibility in proportion to genetic relationship. The objections to twin studies are the closeness of environmental contacts and sampling errors. Kallman and Reisner's investigative design minimizes these pitfalls but does not eliminate them. In any discussion of susceptibility to tuberculosis, the data that show the morbidity and mortality of adult males to be about twice as high as in females of the same age groups should be mentioned.

Poliomyelitis.—Poliomyelitis virus is a common, enteric pathogen. Central-nervous-system involvement occurs in only a small fraction of infected persons. Clinicians writing the earliest descriptions of poliomyelitis were impressed by the frequency of paralysis in robust, healthy children. Anthropologic investigations [25, 185] showed that paralysis occurred more commonly in plump, well-developed children who had definite facial and dental characteristics. The growth curve of these susceptible children had a tendency toward earlier growth and development. Studies on twins, siblings, and parents indicated a greater concordance of *paralytic* poliomyelitis in monozygtic twins than in dizygotic twins.[344] The fraternal twins were similar to other siblings in the incidence of paralysis. Neel [611] suggests that a recessive gene with about 35 per cent penetrance when homozygous might be consistent with the twin data in paralytic poliomyelitis.

In the three foregoing examples of leprosy, tuberculosis, and poliomyelitis, analysis is beset with the complex equilibrium between host and parasite which determine clinical disease. The importance of a common environment and nutrition coupled with intimate and prolonged contact between infected and susceptible members of a family make the analysis of inherited predisposition for a particular infection difficult. An analagous situation is seen in mammary cancer in mice. The interaction of the host's genetic susceptibility, a tumor virus transmitted by nursing, and the influence of hormonal effects upon the mammary glands are all causative factors responsible for disease.[64]

Tuberculosis and leprosy are infectious diseases of unknown incubation times, and infections contracted in childhood may not manifest themselves clinically until late in life. The importance of age, infecting dosage, and the lack of knowledge concerning the pathogenesis of leprosy make any definite conclusion hazardous. The similarity of favorable or unfavorable environmental circumstances may explain in part the familial and racial aggregation of these infectious diseases. Environmental influences may affect the incidence of infection in a manner suggesting that the experience is characteristic of the family or isolate when in fact it is the particular environment that is crucial.[869] The use of twin studies suffers some of the same limitations. It would appear, even when due allowance is made for the above limitations, that the evidence still suggests a multifactorial genetic mechanism of susceptibility for tuberculosis and leprosy.

The studies on poliomyelitis are of a different type in that paralytic infection is a rarity in the manifestations of disease. The occurrence of paralysis in the robust, vigorous child may not be an association between diminished resistance and anthropomorphic characteristics, but rather that strenuous physical activity after the onset of the major illness predisposes to paralytic disease.[744] However, the concordance of paralysis in identical twins at some time during their life strongly suggest the existence of a genetic susceptibility.

HEREDITARY DISEASES ASSOCIATED WITH AN IMPAIRED CELLULAR OR ANTIBODY RESPONSE TO INFECTION

Defense against infection encompasses a combination of factors. Two of the more important are the primordial cellular and the specific antibody responses to infection. A genetic impairment of these defense mechanisms usually results in a lowering of resistance to many diverse species of pathogens. See table 29.

A. *Impaired Cellular Response*

Familial Neutropenia.—Familial neutropenia is a form of chronic neutropenia accompanied by recurrent episodes of bacterial infection. The number of cases reported is small,[79,124] and the syndrome is distinguished from other chronic forms of agranulocytosis by its familial occurrence. Certain pedigrees show a dominant type of inheritance.

The disease, *Chediak-Higashi Syndrome,* [127, 183, 347] is characterized by abnormal cytoplasmic inclusions in the granulocytes with progressive granulocytopenia and increased susceptibility to pyogenic infections with a fatal termination during

Table 29

HEREDITARY DISEASES WITH AN IMPAIRED IMMUNE RESPONSE TO INFECTION

Disease	Associated infection	Basis for suscepti-bility	Mode of inheritance	Evidence for genetic determination of disease	Refer-ences
A. Defective cellular response					
Familial neutropenia	Bacterial infections	Deficient production of granulo-cytes	Domi-nant (?)	Suggestive	79, 124
Chediak-Higashi anomaly	Bacterial infections	Leucocyte abnormality (?)	Autosomal recessive	Suggestive	127, 183, 347
B. Deficiency of gamma globulin					
Hypogamma-globulinemia	Bacterial infections; viral hepatitis	Deficiency of plasma cells	Sex-linked recessive	Convincing	277, 281
C. Unknown mechanism(s)					
Dysgammaglobu-linemia and other syndromes	Bacterial infections	Unknown	Unknown	Suggestive in some pedigrees	2, 391, 424

childhood. Associated genetic stigmata include partial albinism, deficient lacrima-tion, excessive sweating, and skin lesions. Progressive splenomegaly and moderate lymphadenopathy are present in all the patients. The disease is probably trans-mitted as an autosomal recessive. The basis of susceptibility to infection is pre-sumed to be an hereditary, leucocyte abnormality. The polymorphonuclear and staff cells contain unusual cytoplasmic inclusions (Doehle bodies) which are be-lieved to be associated with diminished phagocytic function. However, a case report [200] of the Chediak leucocyte anomaly in a child with leukemia raises the possibility that the leucocytic abnormality is a degenerative phenomenon due to systematic disease. The hereditary predisposition to infection in these children seems probable but the basic mechanism is unknown.

B. *Abnormalities of Antibody Formation*

Hypogammaglobulinemia.—The congenital form of agammaglobuline-mia [277, 281] is an hereditary deficiency of gamma-globulin synthesis which is trans-mitted to males as a sex-linked recessive trait. The primary defect is a deficiency of antibody-producing plasma cells. Bacterial infections develop as the passively transferred, maternal antibodies decline. The microörganisms implicated in these infections include all the common pyogenic bacteria, occasional instances of *Pseudomonas* and *Proteus* infection, and viral hepatitis. The majority of the pa-tients appear to possess normal resistance to most viral infection and resist recur-rent infection with the same viral species. Occasional patients have developed tuberculosis and fungus infections. The delayed type of hypersensitivity which develops to these microörganisms and the clinical course of the tuberculous and fungal infections are not unusual.

Dysgammaglobulinemia appears to be an heterogeneous group of disorders characterized by a pronounced susceptibility to recurrent bacterial infections. Although increased concentrations of gamma globulin are present in many of the patients, there is no evidence that a gamma-globulin abnormality accounts for the disease. Most studies [424] have clearly demonstrated that the gamma globulins are qualitatively normal, so it is reasonable to suggest that the elevation in gamma globulin may be the result of repeated infections and is not causally related to them.

One form of this disease probably has a genetic basis. The familial forms have been accompanied in one series [424] by splenomegaly and in another well-studied pedigree [2] by draining ears, eczematoid dermatitis, bloody diarrhea, and severe bacterial infections. The latter pedigree of six generations indicates a sex-linked recessive inheritance.

The pediatric literature is replete with reports [391] of patients with apparent congenital susceptibility to bacterial infections. In several case studies, siblings were similarly affected. No conclusions are warranted concerning the genetic basis of this group of diseases or the mechanism of increased susceptibility to infection. However, it seems justifiable to conclude that qualitative alterations in resistance to infection have occurred on a familial basis.

INBORN ERRORS OF METABOLISM WHICH PREDISPOSE TO INFECTIONS (Table 30)

Diabetes Mellitus.—Clinically, diabetes may be divided into three disease patterns: [483] (1) lipoplethoric or obese insulin-insensitive diabetics are characterized by normal or even elevated plasma-insulin levels and they rarely experience ketosis unless a complication of their disease is present; (2) insulin-sensitive or brittle diabetics have absent or very low plasma-insulin levels and rapidly develop ketosis; and (3) a rare and controversial group of lipoatrophic diabetics [483] have normal or elevated blood-insulin levels, hyperlipemia, xanthomatosis, hepato-splenomegaly, and marked deficiency of subcutaneous tissue. The recent use of a provocative test employing cortisone [213] with the glucose tolerance test to discover carriers of the diabetic trait has contributed to family studies, although limited by false-negative results. Several genetic studies of insulin-insensitive, stable diabetes and the insulin-sensitive, brittle diabetes conclude that the inherited predisposition suggests a Mendelian recessive,[829, 922] but the complexities of the disease itself make studies on its inheritance exceedingly difficult. Whether there are several kinds of diabetes mellitus with different modes of inheritance is not clear. Neel and Schull,[611] after enumerating pitfalls of genetic analysis in dia-

Table 30

INBORN ERRORS OF METABOLISM WHICH PREDISPOSE TO INFECTIONS

Disease	Associated infection	Basis for susceptibility	Mode of inheritance	Evidence for genetic determination of disease	References
Diabetes mellitus	Skin infections; pyelonephritis; tuberculosis	Unknown	Autosomal recessive (?); secondary factors important in gene expression	Suggestive	188, 213, 216, 437, 483, 611, 656, 829, 922
Gout	Pyelonephritis	Urate deposition in kidney	Autosomal dominant (hyperuricemia)	Convincing	42, 47, 299, 820, 826
Cystinuria	Pyelonephritis (especially in children)	Aminoaciduria with calculi	Recessive in some pedigrees; in others heterozygote shows trait	Convincing	331, 707
Renal tubular acidosis	Pyelonephritis	Nephrocalcinosis and calculi	Dominant (?)	Suggestive	250, 393, 661
Acatalasemia	Oral gangrene and sepsis	Low catalase activity (?)	Autosomal recessive; heterozygote shows trait	Convincing	603, 864

betes, conclude that, although an hereditary predisposition appears certain, the type or types of Mendelian inheritance are not fully established.

Increased susceptibility of diabetics to infection has been recorded since the disease was first recognized. At the Hospital of the University of Pennsylvania,[216] patients with diabetes had a higher incidence of hospital-acquired staphylococcal infection (98 per 1,000) than any other group. The rate of staphylococcal infection in all other patients was less than 14 per 1,000. The reasons for decreased resistance to infections in diabetes is unknown. Experimental diabetes in animals has failed clearly to substantiate the clinical observations of diminished resistance to infection.[656] However, experimental alloxan diabetes in animals may not be an ideal model for investigating susceptibility to infection. The immune mechanisms in human diabetics appear to be normally operative.[656] However, the clinical evidence indicates that there is increased susceptibility particularly to tuberculosis, staphylococcal skin infections,[188] and to recurrent pyelonephritis.[437] Infection is a serious and, unfortunately too often, a fatal complication of diabetes mellitus.

Gout.—Primary gout is a familial defect in uric-acid metabolism. The occurrence of clinical symptoms, however, is not correlated with the elevated uric acid, and undefined "tissue factors" are thought to be of importance in the clinical expression of the disease. The hyperuricemia is inherited as a non–sex-linked, dominant trait,[820, 826] but only a few persons afflicted with hyperuricemia develop clinical gout. Although the inheritance of hyperuricemia is not sex-linked, the clinical manifestations of gouty arthritis are seen predominently in males. The renal complications of gout are not as dramatic clinically as the arthritis; however, many gouty patients die in uremia. The renal lesions are present in about 80 per cent of gouty patients at post-mortem examination,[299] although clinically recognizable, renal damage is less frequent. The pathologic changes are complex and include renal calculi (12 per cent), urate deposition in the walls of collecting tubes, parenchymal urate infiltrates, and vascular sclerosis.[42] As a sequel of renal-tubular urate deposition and obstruction, pyelonephritis is frequently present. The focal medullary tubular obstruction with stasis, dilatation, and a "segmental" pyelonephritis, resembles the induced medullary lesions that Beeson [47] has used in studying experimental pyelonephritis. In primary gout the predisposition to infection depends on a renal lesion which is associated with urate-produced, renal-tubular obstruction and an initial "segmental" pyelonephritis.

Cystinuria.—Cystinuria is a metabolic disease in which excessive amounts of cystine, lysine, ornithine, and arginine are excreted in the urine. There appears to be a specific tubular defect in the reabsorption of these four amino acids. Renal-calculus formation occurs, and pyelonephritis is a common sequel, particularly in children.[707] The inheritance of the disease follows two patterns.[351] Typical recessive inheritance is seen in families in which the intermediate phenotype does not occur. These families contain normal persons as well as grossly abnormal excretors of the four amino acids. The second type of inheritance is an "incomplete" recessive in which relatives (heterozygotes) of the grossly abnormal, homozygous excretors may show an intermediate phenotype with abnormal excretion of the four amino acids, but no clinical symptoms.

The incidence of pyelonephritis in cystinuric patients depends on the presence of stones and secondary urinary obstruction. Infants and children are particularly prone to kidney infections, which makes cystinuria a troublesome disease.[707] Surprisingly, adults are not usually subject to chronic renal infection.[331] The difference in incidence of pyelonephritis in adults and children with cystinuria may be a reflection of the severity of the aminoaciduria, the difficulties in children of maintaining adequate hydration and urinary output, as well as poorer hygiene in children.

Renal Tubular Acidosis.—Renal tubular acidosis [250, 393] is a rare clinical syndrome characterized by the failure of the kidneys, without evidence of advanced glomerular insufficiency, to excrete sufficient acid to maintain normal acid-base equilibrium. Two clinical patterns occur. In the infantile type, patients are usually less than one year of age and have a severe tubular defect. Calcification of the renal parenchyma, calculi, and pyelonephritis are commonly found in them.

Surprisingly, if the child survives the first two or three years of life, the tubular defect in acid secretion gradually lessens. Although this infantile type has been reported in twins and in siblings, it is not certain that a genetic defect is involved. Some authors [661] have proposed that the infantile syndrome might be acquired following pyelonephritis. The *late* or *adult form* of renal tubular acidosis is usually recogized in late childhood or early adult life and presents with a tubular defect in hydrogen-ion secretion. This defect leads to a loss of cations and is responsible for the hyperchloremic acidosis, osteomalacia, and recurrent renal calculi. Several pedigrees show a familial tendency. Three pedigrees involving three generations suggest a dominant inheritance.[393] Pyelonephritis [240, 393] is a frequent complication of the adult form of the disease and is difficult to treat.

Acatalasemia.—Acatalasemia is a rare genetic disease characterized by the absence of catalase from the erythrocytes or its presence in reduced amounts. The disease has been observed only in Japan and is inherited as an autosomal recessive. Heterozygotes have approximately half-normal levels of catalase.[603] The original report [864] described recurrent oral sepsis with ulceration and gangrene in five of six siblings. Subsequent studies have shown a predisposition to oral sepsis, but this is inconstant. It has been suggested that absent or reduced catalase activity may predispose to gangrene because of the failure to inactivate peroxides produced by bacteria.[864] The evidence for this concept is inconclusive.

INHERITED DISEASES ASSOCIATED WITH AN INCREASED SUSCEPTIBILITY TO INFECTION IN A SPECIFIC ANATOMIC SITE

Many genetic diseases primarily affect a particular organ or anatomic region and predispose to infection in that site. As previously described, gout and various tubular defects may damage the kidney, causing stasis and consequent pyelonephritis. A list of genetic diseases which favor the development of infection in specific anatomic sites is given in tables 31 through 39. The discussion will be limited to broad categories, using examples to illustrate the mechanisms of lowered resistance without attempting a detailed and complete description of each entity.

SKIN.—*Ichthyosis congenita* [666] (table 31) is an abnormality in cornification of the skin which is inherited as an autosomal recessive in most pedigrees,[270] although some heterozygous individuals are mildly affected. The tendency for the hyperkeratotic skin to fissure seems to be responsible for frequent acute and subacute bacterial infections of the skin.

Several other familial cutaneous diseases (table 31) present a "locus minoris resistentiae" which predisposes to recurrent superficial infections. *Atopic infantile eczema* [666] is a chronic dermatitis of infancy which is usually engrafted on a familial hypersensitivity. The most serious dermatologic infectious complication is prone to occur in these patients who may develop disseminated infection with herpes-simplex virus or vaccinia virus.

Table 31

HEREDITARY DEFECTS ASSOCIATED WITH INFECTION IN A SPECIFIC ANATOMIC SITE
SKIN AND ITS APPENDAGES

Disease	Associated infection	Basis for susceptibility	Mode of inheritance	Evidences for genetic determination of disease	References
Ichthyosis congenita	Bacterial infection of skin	Fissuring of skin	Autosomal recessive; heterozygote may show trait	Convincing	270, 666
Atopic infantile eczema	Bacterial infection of skin; disseminated herpes simplex and vaccinia virus infections	Chronic dermatitis	Autosomal recessive (?)	Suggestive in some pedigrees	666
Keratosis follicularis	Bacterial infection of skin	Skin fissuring; ulcerating skin nodules	Dominant	Convincing in some pedigrees	666, 799
Anhydrotic ectodermal dysplasia	Chronic rhinitis with ozaena	Absence of sweat and mucous glands; chronic rhinitis	In some families dominant; others sex-linked recessive	Suggestive in some pedigrees	270, 666
Acne and seborreah	Superficial bacterial infection	Sebaceous gland dysfunction	Polygenic susceptibility	Doubtful	844

Table 31 enumerates certain other heritable cutaneous diseases that may be associated with increased incidence of skin infections. The degree of increased susceptibility to infection in these genetically transmitted skin diseases is not fully documented, but is based on reported clinical impressions. Certain dermatologists [453] have been impressed by the infrequency of severe infection in the denuded, macerated, or fissured skin of patients with the diseases described above.

RESPIRATORY SYSTEM (Table 32).—*Cystic fibrosis of the pancreas* [180] is due to an abnormality in the secretions of mucous cells of the respiratory and gastrointestinal systems, as well as abnormal electrolyte composition of the secretions of the sweat, lacrimal, and salivary glands. The clinical disease is probably inherited as an autosomal recessive; however, heterozygous individuals may be detected by the abnormal electrolyte composition of the sweat and salivary-gland secretions. Bronchial and pulmonary infection is virtually a constant feature of the disease. The abnormally viscous mucus cannot be cleared adequately, and stasis with obstruction favors pulmonary infections. Bronchiectasis, bronchopneumonia, and lung abscesses commonly develop. The advent of antimicrobial treatment has prolonged the life of affected children and favored the selection of

Table 32

HEREDITARY DEFECTS ASSOCIATED WITH INFECTION IN A SPECIFIC ANATOMIC SITE
RESPIRATORY SYSTEM

Disease	Associated infection	Basis for increased susceptibility	Mode of inheritance	Evidence for genetic determination of disease	References
Fibrocystic disease	Suppurative bronchitis; broncho-pneumonia	Defect of mucus glands (with failure to clear respiratory passages)	Autosomal recessive; heterozygote shows trait	Convincing	180
Kartagener's syndrome	Bronchiectasis	Situs inversus; ectatic bronchi; sinusitis	Recessive (?)	Suggestive in some pedigrees	146, 683, 959
Lung cysts	Bacterial infection of cyst	Distortion of tracheobronchial tree	Unknown	Doubtful	737, 765
Congenital bronchiectasis	Bronchial infection	Ectatic bronchi	Unknown	Doubtful	150

Pseudomonas and staphylococci as the common respiratory pathogens in these patients. Prognosis depends on the severity of the pulmonary involvement.

Kartagener's syndrome is an association of situs inversus, bronchiectasis, and chronic sinusitis. The number of carefully studied cases is small [959] and the genetic basis of the disease is not established. In support of the heritable nature of the defect is the occurrence of the disease in twins [633] and Cockayne's assertion that transposition of viscera is inherited as a rare recessive trait.[146] The association of the triad with pulmonary infection is presumed to be secondary to ectatic bronchi and impairment of clearing mechanisms.

The familial incidence of *cystic disease of the lung* and *congenital bronchiectasis* is rare. Many earlier reports [765] may have included cystic fibrosis of the pancreas with predominant pulmonary involvement. Other instances of presumed-congenital cystic disease of the lung are probably cysts secondary to neonatal amniotic aspiration and pulmonary infection.[737] Congenital bronchiectasis without Kartagener's triad and developmental pulmonary cysts cannot be convincingly differentiated.[150] A genetic origin for these entities is doubtful.

CARDIOVASCULAR SYSTEM (Table 33).—*Congenital heart disease* does not have a single etiology. It may occur following rubella-virus infection of the mother during the first trimester of pregnancy. The familial or genetic form probably constitutes only a small fraction of the total. Familial transmission [114, 906] in several pedigrees of adequate size displays an autosomal recessive inheritance. One must be cautious in attributing congenital heart disease to

Table 33

HEREDITARY DEFECTS ASSOCIATED WITH INFECTION IN A SPECIFIC ANATOMIC SITE
CARDIOVASCULAR SYSTEM

Disease	Associated infection	Basis for increased susceptibility	Mode of inheritance	Evidence for genetic determination of disease	References
Congenital heart disease	Bacterial endocarditis; tuberculosis; recurrent pneumonia; brain abcess	Shunt; valvular damage; vascular coarctation	In some pedigrees autosomal recessive	Suggestive	114, 906, 264, 622, 731, 823
Marfan's syndrome	Bacterial endocarditis	Valvular abnormalities; shunts; coarctation of aorta	Dominant with variable expression	Convincing	589
Rheumatic fever	Streptococcal infection	Unknown	Autosomal recessive (?)	Questionable	144, 585, 704, 708, 789, 931

heritable factors unless several generations are affected. The importance of the intrauterine environment in the genesis of congenital cardiac lesions makes it hazardous to draw conclusions on the basis of cardiac abnormalities occurring in a single generation of a given family.[731]

Subacute bacterial endocarditis is associated with a wide variety of congenital cardiovascular lesions [264] such as shunts (for example, patent ductus arteriosus and interventricular septal defects), congenital valvular defects (bicuspid aortic valves, et cetera), and inherited, vascular deformities (coarctation of the aorta). A transient bacteremia due to dental or genitourinary manipulation or other cause possibly may permit localization of microörganisms in areas of forceful hemodynamic impact such as the endocardium opposite a shunt or a constrictive vascular lesion.

The incidence of pulmonary tuberculosis is reported to be about 20 times as frequent in all varieties of congenital heart disease [823] as in the general population. Pulmonic stenosis with concomitant, reduced, pulmonary flow is three to four times as likely to be associated with the development of pulmonary tuberculosis as congenital heart disease with normal pulmonary flow.

Recurrent episodes of pneumonia are common in congenital heart disease, particularly those varieties in which pulmonary blood flow is increased. Whether this increased susceptibility to bacterial and viral pneumonia is associated with congestive transudate or an inability to clear the respiratory passages is not clear.

Cyanotic, congenital heart disease is associated with brain abscesses.[622]   The abscesses are generally solitary and are not usually the result of concomitant bacterial endocarditis, suppurative pulmonary disease, or spread from a contiguous source such as otitis media and cranial osteomyelitis.   The mechanism is not known, although it is assumed that a transient bacteremia from an inapparent source may, in the presence of cyanotic heart disease and bypassing the pulmonary filter, produce a hematogenous brain abscess.

Marfan's syndrome [589] is an heritable disorder of mesenchymal tissue characterized by skeletal deformities, dislocation of the lens, and cardiovascular abnormalities.   Coarctation of the aorta, septal defects, and valvular deformities associated with this syndrome are occasionally complicated by bacterial endocarditis. The pattern of inheritance suggests a simple autosomal dominant with varying degrees of gene expression.

Rheumatic fever is a chronic inflammatory disease involving primarily collagen and the ground substance of connective tissue.   The predominant clinical manifestations are observed in the heart, the joints, and the central nervous system.   The importance of streptococcal infections in the etiology of rheumatic fever is established.   The etiologic importance of streptococcal infection is based on clinical,[144] epidemiologic,[788] immunologic,[585] and bacteriologic observations.[704] The manner in which streptococcal infections favor the development of rheumatic fever is unknown.   The importance of heredity in determining susceptibility to rheumatic fever is supported by family studies.[708, 931]   Wilson [931] suggests that the heritable mechanism may be a single autosomal recessive gene.   Because of the role which Group A streptococci play in the etiology of rheumatic fever, it is difficult to assess heritable factors.   Recurrent streptococcal infections in a family environment would be expected to favor the development of rheumatic fever and a genetic susceptibility might be assumed to be responsible when actually environmental factors are at play.   Hereditary susceptibility may provide the "soil" on which rheumatic fever develops, but at present the evidence is inconclusive.

GENITOURINARY SYSTEM (Table 34).—*Hereditary interstitial pyelonephritis* [655] is a rare familial disease, presenting clinically with recurring fever, persistent pyuria, and intermittent bacteruria.   Abnormalities of the renal pelvis, ureters, or evidence of urinary tract obstruction are not demonstrable by x-ray examination.   Pathologic study of the kidneys from two members of an involved kindred showed severe interstitial pyelonephritis.   Males are more severely affected than females and have a more progressive form of the disease.   In one family study,[655] 44 of 134 persons were affected.   Twenty-nine were females and 15 were males.   The inheritance appeared dominant, and the transmission of the disease through fathers to daughters who were mildly affected suggests a possible, incomplete, sex linkage.   Nerve deafness was associated with the renal lesions in some of the affected males.   Clinically and pathologically this disease appears to

Table 34

HEREDITARY DEFECTS ASSOCIATED WITH INFECTION IN A SPECIFIC ANATOMIC SITE
GENITOURINARY SYSTEM

Disease	Associated infection	Basis for susceptibility	Mode of inheritance	Evidence for genetic determination of disease	References
Hereditary, interstitial pyelonephritis	Pyelonephritis	Unknown	Sex-linked dominant with variable expression	Suggestive	655
Hereditary hematuria, nephropathy, and deafness	Chronic otitis media	Unknown	Sex-linked dominant with variable expression	Suggestive	318, 715, 849
Polycystic disease of the kidney (adult form)	Pyelonephritis, especially in females	Collecting tubule obstruction	Dominant	Convincing	168
Familial duplication of upper urinary tract	Pyelonephritis	Urinary obstruction	Unknown	Doubtful	111, 276, 722
Ear malformations and genitourinary anomalies	Pyelonephritis	Lower urinary tract obstruction	Unknown	Suggestive	65, 365, 673

be distinct from the syndrome of hereditary hematuria, nephropathy, and deafness,[849] although the associated deafness and similarity of inheritance suggests they may be variants of the same disease.

The syndrome of *hereditary hematuria, nephropathy, and deafness* [318, 849] is a disease characterized in the male by the presence of hematuria, albuminuria, cylindruria, usually nerve deafness, and progressive, severe, renal disease. The afflicted females have a relatively benign form of the disease with hematuria, cylindruria, and occasional nerve deafness. The disease is transmitted as an incomplete sex-linked dominant.[318, 849] Pathologically, the kidneys show the lesions of chronic glomerulonephritis and interstitial nephritis.[715] Although pyelonephritis is not frequently associated with this disease, chronic suppurative otitis media is often observed. The relationship of such infections to the nerve deafness is not clear.

Bilateral polycystic disease of the kidney [168] presents as two separate genetic and clinical entities. The infantile form is apparently inherited recessively, although in some infants hypoplastic, cystically converted kidneys may be non-genetically determined. The infantile disease is usually fatal within the first year of life.

The adult variety of the disease usually becomes symptomatic in the fourth or fifth decade when the multiple cysts become so large as to encroach and replace the normal renal parenchyma. Death most frequently results from uremia. The inheritance of the adult type of polycystic kidney is probably as an autosomal dominant,[168] although other modes of inheritance have been proposed. The expressiveness of the gene and the clinical pattern are remarkably constant among members of the same family. Pyelonephritis is a frequent complication, occurring in 46 per cent of cases,[168] and may be an important factor in accelerating renal insufficiency. Females with the disease have a significantly higher incidence of pyelonephritis (64 per cent) than males (26 per cent).

Familial urinary-tract malformations, although rare, seem to follow two patterns. The first is confined to the upper urinary tract and includes duplications of the kidney, renal pelvis, and ureters. The information concerning more than one generation is scanty,[276, 722] and no conclusions regarding mode of inheritance are warranted. The frequent occurrence of congenital anomalies of the upper urinary tract [111] suggest that the familial association of urinary-tract malformations may be fortuitous. However, thorough investigation of asymptomatic members of the family in which affected individuals occur may reveal the importance of heritable factors. Recurrent pyelonephritis is the presenting complaint and the most frequent complication of upper urinary-tract duplications.

A second familial variety of urinary-tract anomalies has recently been described. In this group, there is an association between genitourinary abnormalities and gross distortion of the external ears. The original report [673] recorded the association between renal agenesis, a characteristic facies, and soft, flat, abnormally low ears. It is now recognized that familial, asymmetrical malformed ears are associated with a wide variety of congenital renal abnormalities, including polycystic kidney, ureteropelvic anomalies, and a variety of urethral obstructions.[356] The familial occurrence [65, 356] of abnormally shaped ears, combined with obstructive lesions of the urinary tract, is of importance since prompt surgical treatment is imperative if severely affected children are to survive. Pyelonephritis and pyonephrosis are frequently seen with these abnormalities and may be an important factor in causing death. The mode of inheritance is not established, but it is not sex-linked.[356]

ENDOCRINE SYSTEM (Table 35).—*Adenomatosis of the endocrine glands* is a disease associated with multiple-functioning adenomas of the pituitary, the parathyroid glands, and of the islet cells of the pancreas. A familial form of the disease seems probable,[120, 543, 919] although the type of inheritance is uncertain. Pyelonephritis is secondary to the renal calculi or nephrocalcinosis of hyperparathyroidism.

A familial incidence of *congenital adrenal hyperplasia* has been reported.[198, 641] Different biosynthetic defects in adrenal-hormone production are accompanied by clinically characteristic manifestations of this syndrome. The

Table 35

HEREDITARY DEFECTS ASSOCIATED WITH INFECTIONS IN A SPECIFIC ANATOMIC SITE
ENDOCRINE GLANDS

Disease	Associated infection	Basis of susceptibility	Mode of inheritance	Evidence for genetic determination of disease	References
Adenomatosis of the endocrine glands	Pyelonephritis	Nephrocalcinosis; renal calculi	Unknown	Suggestive	120, 543, 919
Congenital adrenal hyperplasia	Increased intercurrent infections (?)	Unknown (decreased glucocorticoids?)	3 distinct genetic abnormalities; recessive (?)	Suggestive	198, 641
Hypoparathyroidism	Moniliasis (?)	Unknown	Recessive (?)	Doubtful	279, 850
Hyperthyroidism (Graves' disease)	Intercurrent infection (?)	Uncontrolled hyperthyroidism	Recessive (?) expressed predominately in females	Suggestive	69, 505, 529

compensated and uncompensated varieties of congenital adrenal hyperplasia are both due to impaired *C*–21 hydroxylation.[198]   A hypertensive form of the disease due to defective 11–β hydroxylation is also recognized.[198]   The three forms of the disease are due to multiple, different, recessive genes. Affected children appear to display increased susceptibility to minor infections, which may precipitate an adrenal crisis. Whether susceptibility is actually increased or the physician's attention is focused on minor infections because of the severe consequences is not clear.

An hereditary tendency for *hypoparathyroidism* has been suggested by two reports [279, 850] of hypoparathyroidism in siblings. In one family,[850] the parents were first cousins and three of ten children were affected with idiopathic hypoparathyroidism. In the second pedigree [279] there was no evidence of kinship between the parents, but all the children (three brothers) were hypoparathyroid. Hypoparathyroidism has been associated with chronic moniliasis.[850] Whether this association is coincidental or due to a predisposition of patients with hypoparathyroidism specifically to develop monilial infection is not clear.

Many reports [69, 529, 565] of cases of *hyperthyroidism* in a single family are to be found in the literature. A large series [529] consisting of patients with both toxic diffuse and nodular goiter were subjected to genetic analysis; toxic, diffuse, exophthalmic goiter appeared to follow a recessive inheritance with expression predominantly in women. No convincing evidence was obtained that toxic nodular goiter was heritable. The possible role of environment producing iodine-deficient goiter, with the development of hyperthyroidism, must be considered. Diffuse exophthalmic goiter, in contradistinction to simple and nodular goiter, is not associated with dietary deficiency of iodine; therefore, familial incidence of this disease can be established with more assurance.

Patients with uncontrolled thyrotoxicosis tolerate systemic infection poorly, and such infections may precipitate a thyroid storm. Whether increased susceptibility to bacterial infections is present in hyperthyroidism is controversial. There is an old clinical observation that active tuberculosis is extremely rare in thyrotoxic individuals.

Table 36

HEREDITARY DEFECTS ASSOCIATED WITH INFECTIONS IN A SPECIFIC ANATOMIC SITE
HEMATOPOIETIC AND RETICULOENDOTHELIAL SYSTEM

Disease	Associated infection	Basis of susceptibility	Modes of inheritance	Evidence for genetic determination of disease	References
Hereditary spherocytosis	Cholecystitis; secondary infection of leg ulcers	Cholelithiasis; supramalleolar ulcers	Autosomal dominant with variable expression	Convincing	698, 868, 955
Gaucher's disease	Tuberculosis (?)	Unknown	Dominant (?)	Convincing in some pedigrees	295, 871
Niemann-Pick disease	Broncho-pneumonia (?)	Infiltration of lungs; impaired reticuloendothelial system (?)	Recessive (?) in some pedigrees; others dominant with variable expression	Convincing in some pedigrees	871

HEMATOPOIETIC AND RETICULOENDOTHELIAL SYSTEM (Table 36).—*Hereditary spherocytosis* is a chronic, familial, hemolytic anemia possibly due to abnormal erythrocytic glycolysis.[955] The disease is inherited as an autosomal dominant with wide variation in the expression of the gene.[698] The chronic hemolysis leads to a marked increase in bilirubin metabolism and a striking tendency for cholelithiasis with occasional cholecystitis. Leg ulcers [868] are a rare complication of hereditary spherocytosis. Secondary infection of the ulcer may occur with progressive suppuration.[154]

Because of the specific susceptibility to Salmonella infection, *sickle-cell anemia* has been discussed previously. However, the frequency of chronic leg ulceration in this disease, with occasional severe, secondary infection, particularly with staphylococci,[154] and the relatively rare [934] occurrence of cholelithiasis and cholecystitis provide two additional cases of susceptibility to infection.

Gaucher's disease or *lipoidosis* [871] is a rare, familial, constitutional disorder of the lymphohemopoietic organs. It is characterized by an accumulation and retention of cerebrosides in the reticular cells and histiocytes. An error of intracellular lipid metabolism appears to be responsible for the disease. An autosomal dominant gene has been proposed as the means of inheritance [295] in several pedigrees. Tuberculosis [871] occurs frequently and often accelerates the progress of the disease and may lead to an early death.

Niemann-Pick's disease [871] is a constitutional, familial, systemic disorder with accumulation of sphingomyelin within cells of the lymphohematopoietic system. An error of intracellular lipid metabolism has been suggested as the cause. The mode of inheritance is not clear. Certain pedigrees have been interpreted as consistent with a recessive type of inheritance; others suggest a dom-

Table 37

HEREDITARY DEFECTS ASSOCIATED WITH INFECTION IN A SPECIFIC ANATOMIC SITE
SKELETAL ABNORMALITIES

Disease	Associated infection	Basis of susceptibility	Mode of inheritance	Evidence for genetic determination of disease	References
Cleft lip and palate	Sinusitis	Maxillary defects	Recessive (?) in some pedigrees; others dominant with variable expression	Convincing in some pedigrees	43, 611
Spina bifida	Pyelonephritis; meningitis	Neurologic deficits; meningocoele	Dominant with variable expression	Convincing in some pedigrees	43

inant with variable expression. The lungs are often infiltrated with Niemann-Pick cells and associated bacterial pneumonia is frequent. Pulmonary complications herald the terminal phase, and death usually occurs in the second year. The accumulation of abnormal lipids in the reticuloendothelial system may impair resistance to infection in these patients, but evidence is not available to support this concept.

SKELETAL ABNORMALITIES (Table 37).—*Cleft lip and palate* [43] is a congenital anomaly in fusion of the two lateral, maxillary processes. A familial incidence is present in about 20 to 30 per cent of the cases. The type of inheritance is obscure and may depend on the particular family. Both a recessive [43] and a dominant gene with variable expression [611] have been postulated, based on different family pedigrees. Distortion of the maxillae leads to a high incidence of purulent sinusitis in affected children.

Spina bifida [43] is a congenital failure of the neural arches to close. There may be an associated meningocele. In some families, the tendency appears to be based upon an autosomal dominant gene showing variable penetrance. Spina bifida occulta is the mildly affected, asymptomatic case. In spina bifida, especially with meningocele, neurologic defects involving the nerves of the cauda equina may lead to neurogenic bladder dysfunction and secondary pyelonephritis. The meningocele, if not covered by skin, readily becomes infected and fulminating meningitis may occur.

GASTROINTESTINAL SYSTEM (Table 38).—The broad concept has been proposed that the *idiopathic malabsorption syndromes* are genetically transmitted, metabolic disorders.[78] Multiple defects of intestinal absorption are present, but the absorption of fat is most severely affected. Presumably because of

Table 38

HEREDITARY DEFECTS ASSOCIATED WITH INFECTION IN A SPECIFIC ANATOMIC SITE
GASTROINTESTINAL SYSTEM

Disease	Associated infection	Basis of susceptibility	Mode of inheritance	Evidence for genetic determination of disease	References
Idiopathic sprue	Tuberculosis and other bacterial infections	Nutritional deficiency	Unknown	Questionable	78, 149, 357
Chronic idiopathic jaundice "mavrohepatic icterus"	Viral and bacterial infections (?)	Unknown	Unknown	Questionable	186
Familial cholelithiasis	Cholecystitis	Cholelithiasis	Unknown	Doubtful	443, 512, 721

emaciation and multiple nutritional deficiencies, infections are common.[357] Pulmonary tuberculosis is a frequent complication and occasionally leads to death.[149] The familial tendency [149] of the disease appears established, although the mode of inheritance remains in doubt.

Chronic, idiopathic jaundice (mavrohepatic icterus) is a variety of chronic or intermittent jaundice in young people, manifested clinically by abdominal pain, fatigue, dark urine, hepatomegaly, and an aggravation of symptoms by intercurrent infection. The hepatic, parenchymal cells contain a lipochrome-like pigment. In about one-half of the patients the onset of jaundice follows in the wake of other diseases, of which viral and bacterial infections are the most numerous. Among 39 patients [186] in which a family history was available, there were 13 with family histories of jaundice. Two of these were siblings. That infection plays a significant role in the fluctuations of this disease is established; whether intercurrent infections are more frequent or physicians are more aware of minor infections in these patients because of their systemic reaction is not certain.

A familial incidence [443, 512] of gallstones has been suggested. It should be noted, however, that the occurence has been among women who, in any case are naturally predisposed to *cholelithiasis*. In one report,[721] males had a higher incidence than females, and chronic pancreatitis was associated with the disease. The pedigree in this family spanned two generations, and four of seven members in the second generation were affected. Acute cholecystitis developed in several members, and symptoms suggesting a mild cholangeitis were present in two of them. The role of infection in producing cholecystitis is uncertain.[70] In the presence of cholelithiasis, coliform organisms are present in a high proportion of inflamed gallbladders.

Table 39

HEREDITARY DEFECTS ASSOCIATED WITH INFECTION IN A SPECIFIC ANATOMIC SITE
NEUROLOGIC DISEASES

Disease	Associated infection	Basis of susceptibility	Mode of inheritance	Evidence for genetic determination of disease	References
Familial syringomyelia	Foot ulcers; osteomyelitis	Trophic skin changes; anesthesia	Unknown	Suggestive	364
Hereditary, cerebellar ataxia of Friedreich	Pneumonia; bronchiectasis	Kyphoscoliosis; disordered respirations	Uncertain; varies with pedigree	Convincing	248, 528, 611
Hereditary, amyotrophic, lateral sclerosis	Pneumonia	Bulbar palsy	Dominant with variable expression	Suggestive in some pedigrees	217, 471
Muscular dystrophies	Respiratory infections	Impaired cough	Depends on type (?)	Convincing	736

NEUROLOGIC DISEASES (Table 39).—In many hereditary diseases of the nervous system, infection is the final event. Diseases such as Tay-Sach's disease, Huntington's chorea, presenile dementias, and many others, are chronic, familial, neurologic diseases that progress to debility and incapacitation. Acute infection is the chief cause of death. The following discussion is concerned with several diseases in which the neurologic lesion may so affect the function of a specific anatomic site that infection is a frequent complication.

Familial syringomyelia or Morvan's syndrome is a familial disease of the spinal gray matter.[364] The onset of the disease occurs during puberty. Manifestations include perforating ulcers of the feet accompanied by osteomyelitis and gangrene. The basis of the ulceration is believed to be trophic skin changes and anesthesia which favor trauma and secondary infection. Extensive suppuration and destruction are the rule. Although the patients continue to walk well on their anesthetic, deformed feet, amputation is necessary in most patients. The details of inheritance are not known.

Hereditary cerebellar ataxia of Friedreich [248] is an hereditary disease of the spinocerebellar tracts, corticospinal tracts, and posterior column of the spinal cord. Clinically, kyphoscoliosis, pes cavus, and hammer toe are typical skeletal deformities. Neurologically, cerebellar tract and posterior column defects are present with marked incoördination. The disease is familial with both dominant [248] and recessive types [611] of inheritances being proposed. The mode of transmission may vary depending upon the particular family. Kyphoscoliosis may be severe. Medullary lesions with disordered breathing favor frequent respiratory infections with bronchiectasis as a complication. Myocarditis [528] is present in some patients, but whether this is an infectious complication or a degenerative, inflammatory myopathy is unknown. The latter seems more likely.

Hereditary amyotrophic lateral sclerosis [217] is an hereditary disease involving the pyramidal tracts and anterior-horn cells of the spinal cord. Either spasticity or muscular atrophy may initiate and dominate the clinical course. The frequent early impairment of swallowing leads to aspiration and to recurrent pneumonia. Late in the disease, respiratory irregularities and failure occur. Pneumonia is the usual terminal event.

The usual type of amyotrophic lateral sclerosis may not be an heritable disease. However, well-documented familial cases [217] do occur. The Chamorro people of the islands of Guam and Rota have an especially high incidence of the disease. The type of inheritance is not certain although a dominant [471] with incomplete expression seems probable.

Muscular dystrophies are a complex group of familial diseases manifested by muscular atrophy and weakness in a characteristic distribution. The mode of inheritance depends on the particular type of dystrophy.[736] Duchenne type is sex-linked recessive; the facioscapulohumeral variety is due to an autosomal dominant; and the limb-girdle group may be due either to a sex-linked recessive or dominant. With involvement of the accessory muscles of respiration used in coughing, the afflicted persons may become susceptible to recurrent pneumonia. Dystrophic myocardial lesions may develop, and congestive failure is a not infrequent contributor to pulmonary infection.

INHERITED RESPONSES TO INFECTIONS IN A PARTICULAR ANATOMIC SITE WHICH INFLUENCE THE COURSE OF THE INFECTION (Table 40)

Certain individuals may, on a familial basis, inherit an unusual response to an infection particularly in the respiratory system. Although there is no clear evidence that increased susceptibility to infection is present, the atypical response may affect the course of the disease.

Table 40

INHERITED RESPONSES TO INFECTION IN A PARTICULAR ANATOMIC SITE
WHICH INFLUENCE THE COURSE OF THE INFECTION

Disease	Associated infection	Basis of susceptibility	Mode of inheritance	Evidence for genetic determination of disease	References
Asthma	Respiratory infections	Allergic diathesis (?)	Autosomal recessive (?)	Suggestive	348, 611, 796
Croup	Upper respiratory infection	Predisposition to laryngospasm	Unknown	Questionable	37

Infectious asthma is characterized by bronchospasm precipitated by an intercurrent infection. Whether the infecting organisms alone elicit the generalized bronchoconstriction or whether the inflammation lowers the threshhold of a sensitized person to a specific antigen that triggers the asthmatic episodes is debatable.[796] Nevertheless, when bronchospasm is superimposed on infectious bronchitis, the disease may become very serious. Retention of secretions, hypoxia, and pneumonitis initiate a vicious circle that all too frequently leads to death. The allergic diathesis upon which infectious asthma may be superimposed appears to be inherited in some families as an autosomal recessive.[611]

Certain allergists [348] are able to elicit a history of allergy in many individuals who initially deny any manifestations of hypersensitivity. A prospective family study of the allergic diathesis might indicate a dominant inheritance.

Croup resembles infectious asthma in that there is airway obstruction precipitated by an upper respiratory infection in susceptible children. The frequency of allergy [37] in the family and the later development of respiratory allergy in many children who have experienced croup suggests that the allergic diathesis may be of importance in the genesis of the disease. The unusual reaction to tracheopharyngeal infection is a disturbing and occasionally a serious consequence of minor infection.

GENETIC ALTERATIONS OF DRUG METABOLISM WHICH INFLUENCE THE RESPONSE OF INFECTIONS TREATED WITH THE DRUG (Table 41)

The rate of inactivation of isoniazid depends in part on the genetic control of hepatic, conjugating reactions.[378, 425] Acetylation of isoniazid is genetically determined, and rapid-inactivators treated with the usual tuberculostatic dosage may not attain levels of the active drug considered optimal for therapy. Approximately one-third of Caucasians as well as American Negroes [425] are rapid inactivators. The frequency distribution of phenotypes is bimodal with no heterogeneity for this character due to sex or age.[677] For slow inactivators

Table 41

GENETIC ALTERATIONS OF DRUG METABOLISM WHICH MAY INFLUENCE THE
RESPONSE OF INFECTIONS TREATED WITH THE DRUG

Drug	Infection	Basis of susceptibility	Mode of inheritance	Evidence for genetic determination of drug metabolism	References
Isoniazid	Tuberculosis	Theoretically, rapid inactivation of drug might lead to inadequate treatment	Slow inactivation appears to be recessive	Convincing	377, 425, 677
Primaquine Sulfonamides Nitrofuradantin	Malaria; bacterial infections	Defect in red blood cells leading to hemolysis	Sex-linked recessive; heterozygote shows trait in variable degree (predominently Negro)	Convincing	131, 578, 677

the family data are consistent with a recessive trait. Heterozygous individuals have an intermediate inactivating capacity. Speculation on an increased susceptibility to tuberculosis in fast inactivators has been entertained, particularly since natural tuberculostatic compounds might be rapidly metabolized. Present evidence is insufficient to conclude that either pattern of isoniazid inactivation is correlated with susceptibility to tuberculosis.

Primaquine sensitive individuals have an heritable defect in erythrocyte metabolism [131] which may lead to an hemolytic episode upon ingesting a variety of chemical agents including sulfonamides, nitrofuradantin, mesantoin, quinines, primaquine, and others. The erythrocyte defect is associated with a relative deficiency of glucose-6-phosphate dehydrogenase and reduced glutathione. The deficiency is most severe in older cells, and it is these that are destroyed when the drug is administered. The exact events leading to hemolysis are not known. The disease is found in Negroes, and about 10 per cent of them are reactors.[677] The primaquine sensitivity occurs primarily in males. It is believed that the gene is located on the X chromosome because the hemizygous male is a reactor and reactor females are homozygous. Heterozygous females show intermediate or normal enzymatic activity.

Hemolysis is rarely severe, and, despite continuing administration of the drug to reactors, younger cells replace the most susceptible, old cells as they hemolyze and the anemia disappears. Reactors, if treated for bacterial infections with sulfonamides or nitrofuradantin, or if treated for malaria with primaquine or quinine, would be at a disadvantage with a hemolytic reaction superimposed on the underlying infection. The wide distribution of this error in erythrocyte metabolism and its geographic distribution have led to the suggestion that resistance to malaria may be afforded by this genetic anomaly.[578]

METHODOLOGY IN STUDYING GENETIC SUSCEPTIBILITY OF HOST TO INFECTIOUS DISEASE

The study of heredity in the susceptibility of man to infectious disease is complex. Not all familial or racial aggregations of infectious disease can be accepted as evidence of a genetic tendency. The presence of a common environment, nutrition, and hygiene may influence the incidence of infection in a manner suggesting a genetic effect, when actually environmental factors are primary. To design methods capable of recognizing and minimizing these limitations is difficult. A sound epidemiologic study of the disease is imperative, and, when the pathogenesis of an infection is unknown, no firm conclusions should be drawn. Leprosy and rheumatic fever are examples of diseases that require better epidemiologic evidence before conclusions regarding their inheritance can be accepted without reservation.

The use of isolates in exploring hereditary influences on human susceptibility to infectious disease is fraught with difficulties because the role of genetic factors is so difficult to distinguish from the effects of environment. The number of consanguineous matings may be difficult to document, and shifts in culture and migration are largely not measurable. The environment within different family groups may have significant factors which predispose to infection but which are not readily apparent. Moreover, comparison of the incidence of a particular infection in two isolates with presumably similar environments is subject to the same objections.

Twin studies, as mentioned in the sections on tuberculosis and poliomyelitis, provide a profitable approach to the problem of human susceptibility to infectious disease. Monozygotic twins, reared separately, although rare, furnish evidence of the relative effects of heredity and environment. Comparisons between identical and fraternal twins within the same family unit can be used to determine a genetic effect within the same environment. However, the tendency of identical twins to be closer and to seek the same environmental patterns has been noted.

Determination of association between susceptibility to specific infections and a recognizable marker is an encouraging approach. The association of sickle-cell anemia with susceptibility to *Salmonella* infections and of the dark skinned races to disseminated coccidioidomycosis are two examples. Another possible example is the occurrence of a high percentage of nonsecretors of blood-group substances *A, B,* and *H* in rheumatic-fever patients.[935] This observation has led to speculations on the role of locally-secreted mucopolysaccharides in influencing the course of streptococcal infection.

Correlation of a number of genetic traits with susceptibility or resistance to infection should provide evidence which may then be used as a basis for predicting to which infectious disease or diseases a person may be most susceptible. In certain instances preventive measures then could be direct and specific. An example is the realization that dark-skin pigment and susceptibility to disseminated coccidioidomycosis are associated. When the Army withdrew soldiers with dark skins from Southern California and other endemic areas of the Southwest during World War II, the incidence of disseminated coccidioidomycosis fell promptly.

Easily recognizable genetic markers of an anthropologic, sociologic, and psychologic nature are not available. However, with increasing precision of measurement in these areas we can expect contributions to genetic associations. It is reasonable to anticipate that some polygenic characters may correlate with susceptibility to a specific disease. Clinical impressions that a particular habitus is associated with susceptibility to certain diseases may, as our knowledge of genetics advances, be placed on a firm scientific basis.

The development of recurrent infections on a familial basis is probably more common than presently recognized. In the cases of agammaglobulinemia and familial neutropenia, the mechanisms of susceptibility seem apparent. However, the basis for the large group of familial syndromes characterized by repeated infections, which have been called dysgammaglobulinemia, is unexplained. Detailed study of the host's defense mechanisms in families with dysgammaglobulinemia may be expected not only to explain the defect in their defenses, but also to illuminate fundamental aspects of host resistance in general.

Measurements of the immune response in families by quantitating the amount of serum antibody produced in response to a defined antigen can be used to assess humoral antibody-producing capacities and to elucidate the genetic basis for observed differences. Studies of this kind have been done in specific diseases such as rheumatic fever,[703] sarcoidosis,[747] and others.

Meyers and Jensen [538] found that the capacity to develop tuberculin skin hypersensitivity and the degree of reactivity in response to a standard *BCG* vaccination is a family characteristic. The children of one family differed significantly in their tuberculin reactivity from children of other families. Similarly, the capacity to become sensitized and the degree of reactivity to dinitrochlorobenzene correlate directly with a family history of rhus allergy.[454] Individuals who develop marked allergy to standard amounts of dinitrochlorobenzene are easily sensitized to other contact allergens and generally have a convincing family history of contact dermatitis. An investigation of capacities for skin sensitization in families predisposed to chronic infection, such as tuberculosis, in which hypersensitivity of the delayed type is important, might supply insight into mechanisms of susceptibility. Experimental techniques are now available for familial studies of immediate and delayed sensitization.

The behavior of human leucocytes from diseases such as Chediak-Higashi anomaly toward various microbes might elucidate the mechanism of increased susceptibility to infection in this disease. Macrophage tissue cultures from patients with a genetic susceptibility to infection, likewise could be of fundamental importance in explaining intimate details of defective resistance.

Errors of metabolism such as diabetes offer a profitable approach to the study of the biochemical details of parasitism and the events that shift the equilibrium to favor the pathogen. At present it is not known in the case of diabetes whether the factors which depress resistance are humoral or cellular in nature, but careful study of these patients should cast much light on resistance to infection in general and the genetic factors involved.

In conclusion, it may be said that, although certain of the approaches noted above have shed some light on genetic factors and susceptibility to infection, the complex nature of the interaction between host and parasite makes a definitive analysis extremely difficult with our present knowledge.

SUMMARY

The natural history of certain infectious diseases indicates cyclic variations in prevalence, morbidity, and mortality. Analysis of the epidemic events is most difficult. Changes in microbial virulence and communicability in experimental infections of animals may produce epidemic fluctuations. On the other hand, evidence that microbial variations are responsible for the cyclic changes in human disease is, with rare exceptions such as influenza, conspicuously lacking. It is probable that the emergence of an infectious strain of high virulence or of new antigenic composition is a rare event rather than a normal process in the fluctuations that occur during the course of infectious disease.

Although heredity is recognized as being important in the individual's susceptibility to infection, the separation of genetic from environmental factors is difficult and in many instances uncertain. The familial tendency of some infectious diseases may be due, in part, to intimate or prolonged contact with the microörganism, a common unfavorable environment, or to other factors such as similar nutrition. Therefore, familial aggregation of infectious disease is not necessarily evidence for genetic susceptibility.

We have elected to classify the heritable tendencies for developing infectious diseases into possible pathogenetic mechanisms. This organization may be premature, since the basis of susceptibility is unknown or uncertain in many of the diseases. However, this classification has the merit of emphasizing the multiple ways in which genetic abnormalities may predispose to an increased susceptibility to human infection. We have included, as examples, diseases where genetic determination is questionable and doubtful, because a final decision concerning their heritability cannot be made at this time. We are impressed with the probability of many unrecognized genetic factors that are associated with an increased incidence or greater severity of infection.

DISCUSSION

DR. BURDETTE: Doctor Motulsky will open the discussion.

DR. MOTULSKY: There are many rare genes in the human population that produce diseases maintained in the population by mutation pressure. Sometimes such genes may act by predisposing the whole organism or a given organ to an infection. This can be a general process, as in hypogammaglobulinemia, with which the patient suffers many infections due to lack of antibodies, or a specific organ can be involved so that the patient suffers an infection in that organ. This approach, of course, has been used for many years by careful clinicians who knew that an infection may not necessarily be the primary event, but who tried to look for something more fundamental such as hereditary predisposition. This approach is embodied in the old concept of "locus minoris resistentiae."

With some of the more common diseases, the situation is more complex. Probably here, as was indicated, multifactorial or polygenic systems may make an individual susceptible or resistant. It is noteworthy that genetic resistance factors have been found in practically all infectious animal disease where they have been sought.[288, 289] Animals resistant to a given microörganism, such as *Salmonella,* can be obtained within relatively few generations [289] by artificial selection. Unfortunately, in most instances of genetic resistance, the nature and the mechanism of action of the genes conferring resistance are unknown.

In man, the evidence for the existence of resistant populations is suggestive,[577] but in no case has it been brought forward in a completely convincing manner. Some of the diseases that might qualify would be tuberculosis, plague, cholera, smallpox, and malaria. In order to act most efficiently in natural selection and thus bring about genetically conditioned resistance, a disease should have a high mortality, should be endemic, and should kill children. A disease that does not kill individuals before or during the period of reproduction will not be a very good selective agent and genetic resistance is unlikely to develop.

More decisive evidence for hereditary resistance to malaria in humans has been brought forward in recent years. Purely from the point of view of theoretical genetics, malaria is an ideal disease to act as a selecting agent. It is a common endemic disease that has killed children for many generations. During the last few hundred years, the most fatal type probably has been falciparum malaria. One might expect, *a priori,* that the occurrence of genetic differences in tissues most necessary for growth of malarial parasites could lead to resistance in persons who inherit such mutation. Since the red blood cell is necessary for proliferation of the malarial organism, any genetic abnormality of the red cell which might curtail proliferation of the malarial organism could protect the individual against death from malaria.

The malarial organism is a typical parasite since it requires hemoglobin and some other metabolites of its host, the red cell, for proper growth. There is suggestive evidence that three genetic abnormalities of the red cell do indeed protect

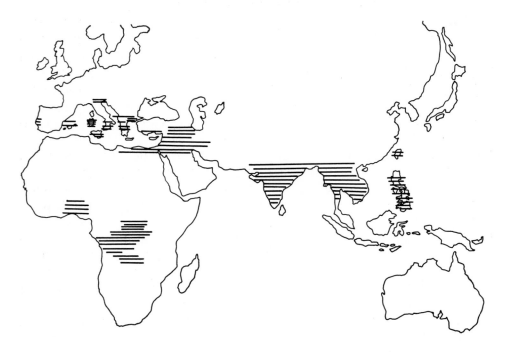

Fig. 4. GLOBAL DISTRIBUTION OF GLUCOSE-6-PHOSPHATE DEHYDROGENASE DEFICIENCY.

from death due to malaria. The situation with regard to the sickle-cell traits is well known, and there is now general agreement that carriers of the sickle trait are less likely to die from falciparum malaria.[6]

It had been suggested earlier, even before the facts were known about the sickling trait, that thalassemia, another abnormality of the red cell, might protect against death from malaria; [307] but studies, although quite suggestive on this point, have not been as extensive as those on sickling. More recently still, another common red-cell trait, deficiency of glucose-6-phosphate dehydrogenase of the red cell,[58] was also found to be present in those populations that had either thalassemia or malaria.[577, 579] This common sex-linked trait renders its carriers susceptible to hemolysis from many drugs, ingestion of fava beans, and infection.[58]

Figure 4 shows the global distribution of the trait, deficiency in glucose-6-phosphate dehydrogenase. These data have been accumulated quite recently.[579] Like thalassemia, the trait is prevalent in the Mediterranean area. It is also found in the Near East, the Phillippines, Formosa, and certainly in Africa. In the Belgian Congo, we were able to obtain data on the combined incidence of

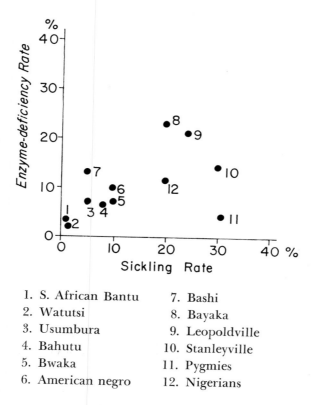

1. S. African Bantu   7. Bashi
2. Watutsi   8. Bayaka
3. Usumbura   9. Leopoldville
4. Bahutu   10. Stanleyville
5. Bwaka   11. Pygmies
6. American negro   12. Nigerians

Fig. 5.   SICKLING FREQUENCY (HETEROZYGOTES) AND ENZYMATIC DEFICIENCY (MALES)
IN AFRICA WITH GOOD CORRELATION BETWEEN THE TWO TRAITS.

this trait and sickling.[577, 579]   Figure 5 shows good correlation between the incidence of the sickling trait and the incidence of the enzymatic deficiency in males in a given population, suggesting as one explanation that a common agent had selected both these genetic traits. Although populations with a high rate of sickling also showed a high rate of enzymatic deficiency, different individuals were usually affected.

In Sardinia, which was one of the most malarial areas in Europe until 1946, we correlated the incidence of thalassemia and the enzymatic deficiency.[55, 579]

Again the correlation between the traits is striking (figure 6). Both thalassemia and the enzymatic deficiency were common in the coastal regions, where malaria was highly endemic and uncommon in the mountainous areas where the malaria did not exist. Thus, three harmful traits of the red cell exist in high frequencies

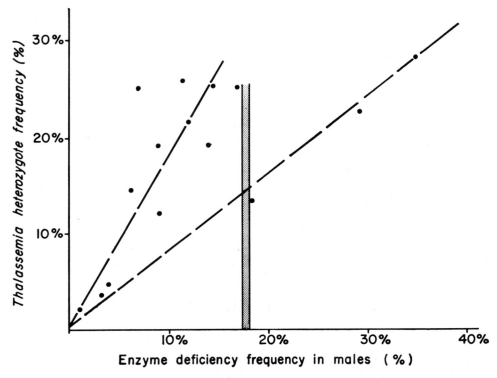

Fig. 6. THALASSEMIA (HETEROZYGOTES) AND ENZYMATIC DEFICIENCY (MALES) IN SARDINIA WITH GOOD CORRELATION BETWEEN THE TWO TRAITS.

Above approximately 188 per cent enzymatic deficiency, a different regression line can be fitted to the data.[579]

in many human populations. The sickle and thalassemia genes are inherited as autosomal traits and are probably maintained by balanced polymorphism with selective advantage of the heterozygote carrier. Sex-linked glucose-6-phosphate-dehydrogenase deficiency presents a more complicated problem since the affected male hemizygotes are favored by malaria and susceptible to hemolytic agents at the same time. It is not unlikely that, in this situation, female heterozygotes are also at a selective advantage.

How can one determine the facts in a more direct manner? The methods to be outlined are discussed in terms of malaria, but they probably could be used to seek the selective effects of other infectious diseases.

(1.) A rough geographic correlation of the trait with the disease is desirable initially.

(2.) A more detailed geographic correlation in a given area with different incidence of the disease under study, as indicated by our studies on sickling and enzymatic deficiency in the Belgian Congo and on the enzymatic deficiency and thalassemia in Sardinia, is of value next.

(3.) Correlation of traits with each other, as well as with the disease, may also give information if more than one trait seems to be implicated as with the various red-cell traits and malaria.

(4.) A gene that is selected by protection against malarial mortality would lose its advantage in a nonmalarial environment. If deleterious in other ways (such as the sickling gene), the genic frequency would decrease in the absence of malaria. The frequencies of genes for sickling and enzymatic deficiency of American Negroes as compared with West African Negroes point in this direction, although admixture with Caucasian genes and lack of ethnic homogeneity are complicating factors. The lowered incidence of sickling in Negroes in nonmalarial Curaçao as compared with malarial Surinam is striking since both groups had a similar origin in Africa. Differences in frequency of the enzymatic deficiency gene between high-frequency Oriental Jews and low-frequency European Jews are also suggestive although difficult to evaluate because of admixture of non-Jewish genes in both population groups.

(5.) The density of malarial parasites in red cells may be correlated with the different traits. Carriers of the sickle-cell trait below the age of five years have lower parasite rates for falciparum malaria than the normal population. Attention to methodology is very important. Children up to six months of age have antimalarial antibodies from their mothers, or passive immunity. After five years of age, active immunity develops and obscures differences caused by genetic factors. Any valid study of this type must be done with children between six months and five years of age. In a study done by Dr. Vandepitte and myself, too many of the children were between five and ten years of age, and our results were not statistically significant.[579] Dr. Allison and Dr. Clyde initiated a study correlating enzymatic deficiency and parasite density in children between six months and four years in Tanganyika; their results indicate that most enzyme-deficient children indeed have a lower parasite density.[8]

(6.) The best and most direct method is to correlate the suspected trait with mortality by the infectious disease under study. The opportunity for this approach is disappearing rapidly with the control of infectious diseases. In Africa, such studies are extremely difficult. No mortality records are kept, and anyone who has been there knows how difficult it would be to accumulate such data. Controlled studies on mortality in villages with and without malarial prophylaxis, however, may be possible by local observers.

(7.) In a population where selection against children still occurs now, or did occur in the recent past, the percentage frequency of trait-carriers correlated with age may provide important information. If trait-carriers are genetically protected from mortality, the population incidence of the trait will be higher in adults than in young children, since the adult population will have fewer normal individuals.

(8.) An alternative to direct observation of mortality might be family studies in which one determined the genotypes of the parents, the total number of children born, and the genotypes of the surviving children, and then compared the expected number of children with and without the trait with the actual number. Significant mortality should disturb the segregation ratios. Peculiarities of family structure in the primitive tribes where malaria is still common raise many difficulties with such investigation; however, this method has not been tested sufficiently yet.

(9.) Inoculation of falciparum malaria *in vivo* into volunteers with and without the sickle-cell trait has been practiced. This approach raises problems of human experimentation, but such studies have been done twice with adults. In one instance, the protection of carriers of the sickle-cell trait was striking; [7] in the other, results were equivocal, although in the right direction.[59]

(10.) One promising approach, not used at all yet, is to grow malarial red-cell organisms *in vitro* to find out whether cells with the different abnormalities support growth of the malarial organisms as well as normal cells.[885]

A general approach, possibly useful in discovering genetic resistance and susceptibility factors to other infectious diseases, is to look for correlation of infectious diseases with some of the polymorphisms that have been detected in recent years. There are newer polymorphisms, such as the haptoglobins and transferrins, that may be related to resistance to infectious disease.[577] Unfortunately, we cannot sample tissues other than the blood easily. If an infectious disease needs another tissue for proliferation, it is unlikely that a polymorphism affecting the blood will be implicated. Fortunately, the formed elements of the blood may contain vestigial models of enzymatic systems in other tissues and could thus be used to detect such polymorphisms.

Genetically-controlled variations in metabolism of drugs probably offer one of the most promising areas in which polymorphisms can be detected. Within recent years, primaquine and other drug sensitivities have been found to be caused by inherited deficiency of glucose-6-phosphate dehydrogenase; sensitivity to succinyldicholine was found to be due to heritable pseudocholinesterase deficiency; [435] and the rate of inactivation of isoniazid, a drug used in treatment of tuberculosis, was found to be due to inherited differences in acetylation of the drug.[207, 329] Figure 7 shows differences between various populations in the isoniazid level in the blood six hours after ingestion. These tests were done by Dr. Mitchell in Denver. Studies in several laboratories, including our own, indicate that the slow inactivators are homozygotes, presumably for the absence of acetylating enzymes.[207, 329] In a mixed white American population about 42 per cent are slow inactivators. The frequency of this gene varies considerably in different populations. For instance, as indicated in the figure, the frequency of

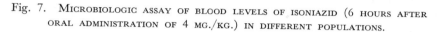

Fig. 7. Microbiologic assay of blood levels of isoniazid (6 hours after oral administration of 4 mg./kg.) in different populations.

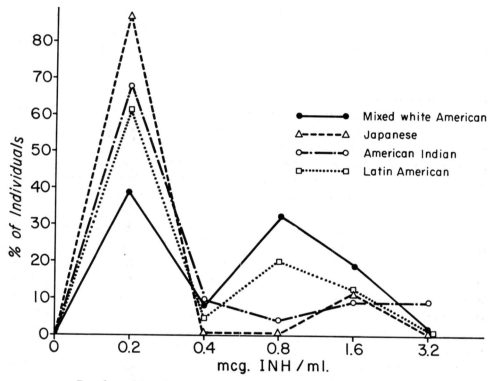

Based on data of Drs. Mitchell and H. W. Harris, *et al.* [455, 540]

acetylating gene is much higher in the Japanese where most people are rapid inactivators. Environmental factors that by natural selection have led to these pronounced differences between the two populations need further elucidation. Since drugs are catabolized by enzymes and since enzymes are under genetic control, it is quite likely that further polymorphisms will be discovered by future inquiry into the biotransformation of drugs in individuals and populations.

Dr. Gowen: The basic concepts included in the paper of Drs. Cox and MacLeod are interesting because they represent a fusion of principles developed from different disciplines focusing on factors that make for health or disease in man. Although the factors are many, they may be grouped under three major elements: the characteristics with which the host is endowed, the innate peculiarities of the pathogen, and the ever-present environment in which these factors must act and interact.

It may seem peculiar for a geneticist to try to extend the discussion of the environmental side of this question, but that side has impressed me as tremendously important to this whole question of the interrelation of genetics to disease and particularly to infectious disease. Dr. Cox and Dr. MacLeod brought that matter into focus indirectly, but I would like to turn the discussion around and make the environment the emphasis and the gene and genetics the background. Indulgence is requested for allowing presentation of personal work, much of it unpublished, because I feel that I know it better than some of the equally significant studies that have been reviewed.

LONGEVITY IN MICE

Everyone is interested in longevity, if for no other reason than extending his own. The span of life represents an integration of all those forces that we have been considering and is perhaps the best total measure of them. Experiments are designed ever so carefully with controls from which valid errors may be calculated and conclusions may be learnedly couched in terms of probability. Suppose, however, that an invisible force is present in this environment, such as irradiation, which is dangerous because it cannot be perceived directly. In a study of longevity, what happens when radiant energy is present in the environment? The results with the animal material probably apply to humans as well.

Eleven strains of inbred mice that are homogeneous within strains, but noticeably different between strains, have been developed in our laboratory. The differences are expressed by survival time in relation to irradiation received, particularly if females are irradiated. The differences in longevity appearing in an environment of irradiation may not be the same as those which occur without irradiation.

The conditions under which the mice are raised may be compared to living conditions of the human population. A new subdivision is established composed of a thousand homes each occupied by a pair of 40-day-old mice. Here their offspring are born, raised, and distributed to other locations. Pregnancies attributable to each pair may be over 20 and young over 100. Early adult years are busy, especially for the mothers even though the food, sanitation, lighting, heating, and laundry services are supplied. Irradiation, in dosages of 20, 80, 160, 320, 480, and 640 roentgens, is given to certain pairs, necessitating an annoying trip into the world. Like man, the mice showed no obvious signs that they recognized any difference between the environments with little radiant energy and those with much; nor would the census taker, for the treated mice were scattered at random over the whole subdivision area.

The result may be illustrated by two of the strains, S and Z. The S male mice had average longevity of 676, 638, 604, 683, 506, 331, and 53 days when respectively exposed to 0, 20, 80, 160, 320, 480, and 640 roentgens of X irradiation (250 p.k.v., 30 ma., and filtered by 0.25 mm. Cu + 1 mm. Al from a maxitron 250). The S females under like condition lived 578, 563, 680, 627, 563, 399, and 55 days respectively; the Z males 645, 799, 627, 585, 506, 212, and 50 days respectively; and the Z females 369, 320, 538 561, 480, 397, and 50 days on the average respectively.

Irradiation effects on longevities of males are minor until 320 r exposure is reached. From that point, irradiation materially reduces survival and is dose dependent. The females of the two strains not only differ from the males but also between themselves. Those without exposure to irradiation live shorter lives than the males, the differences being 85 and 57 per cent. The females also differ between strains, Z females having but 64 per cent the longevity of the S females. Also genotypic differences in span exist when strains and sexes are compared.

What is the effect of the unseen variable, irradiation? The males do not show appreciable effects until dosages beyond 320 r are reached, and shortening of life is rapid and dependent on dosage above 320 r. The females display quite different effects. Beyond 20 r, their span lengthens noticeably and maintains this advantage to beyond 160 r. Beyond 320 r they show dose-dependent and diminished survival but now live longer than their male mates. The change is appreciable, 13 per cent for the S and 37 per cent for the Z strain.

Irradiation up to 320 r may be misinterpreted as beneficial to females, but hidden conditions materially affect this interpretation. The females are subject to risks of death through birth of young that males do not face. The two strains S and Z differ in ability to produce young, the Z strain tending to conceive immediately after every litter birth. Opportunities for organ failure and infection are much increased over those of S females and even more over those of the males. On the other hand, irradiation, even at 80 r, noticeably reduces the potential for bearing litters and risks to life for these females, and their life span is lengthened. This result will be considered favorable by some but not by others, since it represents impairment of a major physiologic function. At higher dosages of irradiation, the life spans of the two strains are comparable instead of showing the wide differences observed in the untreated mice. This case illustrates the significance of invisible, often unappreciated, factors in disease studies where the agents act during only a moment of time early in life.

CATARACT

Cataract is another condition that may be influenced by irradiation. It seemed interesting to test whether or not the vagaries with which cataracts appear in man may in some way be related to the variable genotypes for inheritance of cataract known to exist in animal or human populations, or to unknown irradiation that the individuals may have received. The strain of mice established in our laboratory a quarter-century ago provides a means for investigating the question. Homozygous mice have congenital cataract,[290] whereas heterozygotes have normal vision.

Crosses were made between the strain with cataracts and another of our inbred strains, *BALB/Gw*, which at that time was thought to be essentially free of cataracts throughout its early lifetime. The F_1 individuals from reciprocal crosses and the *BALB/Gw* mice were exposed to irradiation when 46 days of age. Contrary to our studies on longevity mentioned earlier in this discussion, nothing was observed for a period of more than 250 days. At approximately 300 days of age, cataracts began to appear. Progeny of the cross between the two strains had spans up to 1,230 days. The mouse surviving the longest lived 200 days longer than the individual, inbred, parental mouse surviving the longest period of time, and may be used as a measure of hybrid vigor of the offspring.

INCIDENCE OF CATARACTS AND AGE AT WHICH THEY APPEARED

The data on incidence of cataracts have been separated by genotype and by dosages of exposure to X-irradiation. These graphs are shown in Figure 8. Irradiation damage became evident at 475 days in the *Ba* mice and 625 days in the hybrids. Cataracts appeared earlier in life as the doses of X irradiation increased. This was particularly evident in the chart for the *Ba* mice. Mice treated with 160 *r* had an appreciable increase in frequency of cataract at 575 days, those with 320 *r* at 525 days, and those with 480 *r* at 475 days. The natural resistance of younger mice was affected detrimentally earlier and earlier in life as the dosage of irradiation was increased. The shortening of the span free of cataract was more pronounced for the *Ba* mice than for the hybrids.

The third point, attributable to heritable differences in the two groups of mice, relates to the levels of the effects of irradiation on frequencies of cataract. For dosages of 480 *r*, the frequency of cataract was 2.7 times greater for the *Ba* mice than for the hybrids; for 320 *r*, the ratio was 2.6; and for 160 *r*, it was 1.9 greater for the *Ba* than hybrid mice. On the other hand, the untreated genotypes had nearly equal, spontaneous rates. In consequence, the observed differences in irradiation effects on the two stocks do not depend on differences in spontaneous rates of cataract but on their innate susceptibilities to radiant energy. Figure 8 calls attention to still another factor significant to the irradiation-genotype interaction exciting the cataracts. For each irradiation dose and genotype, some fraction of the exposed population became susceptible to cataract. The susceptible fraction increased with the larger irradiation doses, but this fraction appeared fixed for each genotype and dose.

The *Ba* mice are particularly uniform from mouse to mouse. It might be thought that the frequency of cataracts in each group should be all or none, but the same phenomenon of variation between animals within a strain occurs in the measurement of other characteristics such as longevity following exposure to irradiation. This variability apparently arises from chance factors such as alterations in the absorption of irradiation. These chance factors are more important in determining phenotype than was once supposed. The curve illustrating the

Fig. 8. CATARACT INCIDENCE AS RELATED TO AGE WHEN THE CATARACTS WERE FIRST OBSERVED.

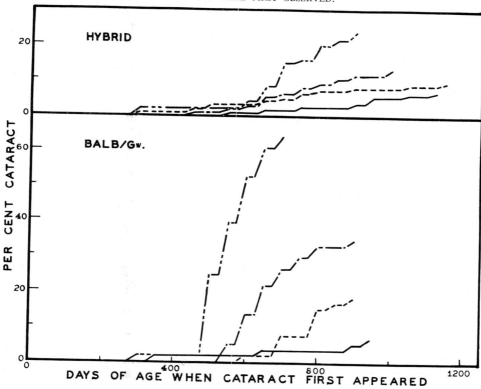

Solid line: 0 *r*, dash line: 160 *r*, dash-dot-dash line: 320 *r*, and dash-dot-dot-dash line: 480 *r*.

relation of total incidence of cataracts to irradiation dosage is, on this account, of particular interest.

CONTRAST OF RADIANT-ENERGY EFFECTS ON VISION AND REPRODUCTION

These voluminous data show how completely the interpretations of irradiation effects depend on the organ system studied, even though the irradiation may be given during the same age interval, 40 ± 3 days. Eyes develop early in life, and a long quiescent period of normal vision follows. Cataracts appear late and are increased exponentially as a power of the roentgen dose. Also they are heavily dependent on the genotypes of the mice exposed to the X rays. Because cataracts are age-dependent, a spurious favorable effect of irradiation on life span appears. Mice having the handicapping cataracts induced by X irradiation live longer than those not having the trait. The frequencies of cataracts are uninfluenced by sex and maternal effects.

The reproductive organs also develop early in life. Irradiation, however, reduced their function immediately following exposure. The reduction in number of progeny was dependent on dosage. In contrast to irradiation effects on the organs of vision, strain differences had less importance than sex differences on the organs of reproduction. A favorable effect on life span came as a consequence of the reduced fertility of the females. Again this effect may be considered spurious, since it is detrimental to the preservation of species while favorable to the individual. Irradiation has an immediate and continuing effect on the testes, ovaries, and reproductive system, whereas the effects on the eyes are held in abeyance for ten months.

A single event, happening in less than $\frac{1}{350,000}$ of the possible life span in the mouse, has increased the expectation of cataract eleven times. This event, brief as it was, happened on the average more than half the subsequent expected life span away from the time when the consequences were expressed as overt cataracts in the exposed mice. These results, when contrasted with those of the first section, show how important environmental experiences may be toward the expression of vital functions as they may occur either early or late in life.

TYPES OF MURINE GYNANDROMORPHS

Gynandromorphs in mice behave similarly in some respects to the cases just discussed, but the significant environmental factors are by no means as clear. Gynandromorphs were observed in one, and only one, of ten strains of mice observed. Mice of the *BALB/Gw, L/Gw, K, Q, S,* and *Z* strains were examined in the largest numbers. Sixty gynandromorphs appeared in either the pure *Ba* strain or their descendants, and the sex-organ pattern was hermaphroditic. A small number of the cases were true gynandromorphs, since the male reproductive tract was on one side and the female organs on the other. A fair proportion of the cases showed mixtures of male and female organs.

The sixty cases were observed as a consequence both of necropsies and examination of thousands of mice. The incidence rate was less than 1 in 200 mice for the *Ba* strain. Within the strain, chance appeared to rule the events leading to a mouse being normal versus hermaphroditic. Family group, treatment, age of mother, size of litter, and so on, were not significant factors. The laterality of the sex organs was fortuitous. The small incidence of the hermaphrodites was possibly increased during the summer months. Yet the condition is obviously dependent on genetic factors. It could be called a case of low penetrance, but strong objection may be registered to this on the basis of insufficient evidence.

The Bagg strain, *Ba,* of which our strain is one branch, possibly has been mated, brother to sister, for the longest time of any strain of mice in existence. Since all individuals should be homozygous for the hereditary factors concerned in hermaphroditism, the very small numbers with this phenotype gives a measure of the accurate fit necessary between the factors of inheritance and those of environment for the condition to appear. In this case and in others, the inheritance may be unique, but the environmental threshold is nonetheless real. Such cases, and they are fairly common in disease studies, emphasize the necessity for looking beyond simple Mendelian ratios and appraising the influence of inheritance and its accompanying interactions with a fluctuating environment if disease processes are to be analyzed.

DISEASES REQUIRING A PATHOGEN

Murine typhoid, initiated by *Salmonella typhimurium,* is a disease with pleomorphic manifestations. Complexity is introduced by the fact that all three disease variables are directly important. The pathogen may exist in several different forms. Difficult to separate, but nevertheless real, environmental differences are ever present. In our work, extensive study has been given to the differences in effect between eleven strains of mice inbred to the point where most of the paired loci should be homozygous for their samples of genes. When these strains were tested for their disease reactions with *S. typhimurium* 11c in a prescribed dose and in the same external environment, almost 100 per cent of one strain survived, another was nearly completely susceptible. In the nine other strains, the percentage of survivals for each fills in the space between 100 per cent to 0 per cent to form steps of roughly equal proportions respectively.[289] The case becomes even more difficult when different pathogens vary in their effects on different species. Both the physician and the geneticist may be equally wrong if they assign the observed symptoms entirely to differences in the host's inheritance patterns. An extreme case is easiest to understand. Dysenteries have common symptoms and the morbidity and mortality from them may erroneously be attributed to the genetic constitution of the host. The pathogens causing the same symptoms may be protozoa or different bacteria. Therefore, it is unlikely that the same genes in the host would be responsible for resistance or susceptibility to the dysenteries of all types. This applies also to guide in the separation of the strain differences between pathogens. Certainly breeding tests with numerous host strains should be combined with tests on known mutants of a pathogen, if conclusions on genotypic effects on the severity of a disease are to be sound.

ENVIRONMENT AND DOMINANCE

In general, progenies of matings between resistant and susceptible strains are intermediate in their resistance. This lack of dominance may add to the difficulties of analysis. One example will suffice to cover the problem. While a student of Dr. Lindstrom, Burdick [108] crossed inbred corn plants characterized by having 4-rowed ears with those having 8-rowed ears, when the plants were grown in the field during the summer. Those inbred by Dr. Lindstrom were known to have but a small variation in their row numbers from plant to plant within the particular line. The hybrid, as normally grown in the field, showed dominance of the larger row number. However, in this case, the parents and hybrids were grown in the greenhouse during the winter with a daily photoperiod of decreasing length provided for the first 20 days. Tassels were removed to induce the ears at the lower nodes to develop. Under these conditions as many as 6 nodes on a plant developed ears. Dominance for low-row number was expressed by the ears on the lowest positions of the plant, but dominance for the high-row number was characteristic of the ears on the top position of the plant. Between these limits there was an arithmetic increase of row number, beginning with the lowest and increasing by steps to the highest ear positions.

These results emphasize the fact often overlooked, especially in studies of disease: dominance is a function of the interaction of gene products and the conditions surrounding them when the products are released and physiologically active. As in the case so well illustrated by the data on corn plants, the degree of dominance may be controlled, not so much by the external environment, as by the environment generated within the organism itself.

In closing, these remarks only increase the basic significance of Dr. Cox and Dr. MacLeod's carefully analyzed material. They may extend the conclusion somewhat in the direction of emphasizing the importance of environmental-genetic interactions and the role they play in genotypic as well as infectious host-pathogen patterns of disease.

DR. NEEL: At the risk of prolonging this discussion unduly, I cannot resist pointing out that the resistance to malaria encountered in the sickling phenomena may involve a most unusual lack of immunologic principle, and it gives cause for pause in some of our thinking. My associates and I have proposed a mechanism for the resistance [539] in falciparum malaria. The parasitized red cell tends to adhere to the capillary wall in the peripheral circulation. Under these circumstances, the oxygen tension may fall to the point at which sickling ensues. The sickle cell is more subject to destruction on mechanical grounds than the non-sickle cell, and, if it contains a malaria parasite, there may be a mechanical interruption in the life cycle of the parasite. This immunity involves not an anatomical predisposition, as was mentioned by Dr. Cox, nor an altered ability to form antibodies, but depends on a mechanical happenstance, and this in turn depends on the fact that sickle-cell hemoglobin is about $\frac{1}{100}$th as soluble as normal hemoglobin in the reduced state.

A somewhat similar situation may result from alterations in glucose-6-phosphate dehydrogenase. It is a fact, I believe, that this enzymatic deficiency becomes apparent only in the older red cell, not in the younger red cell, and one can raise the question of whether the presence of a malaria parasite accelerates the disappearance of glucose-6-phosphate, a disappearance conditioned by a simple genetic mechanism. If this were so, then the life span of the red cell might be shortened thereby, and again the basis of the protection would be a shortened life cycle on the part of parasitized cells.

If this is true, then it suggests the generalization that any condition shortening the life span of the red cell will give an immunity to malaria. There comes to mind immediately one more condition, hereditary spherocytosis. So far, there are no observations on whether or not hereditary spherocytosis confers protection to malaria. If this is so, why a great deal of hereditary spherocytosis is not seen in the Negro is a logical question. The fact is that hereditary spherocytosis due to the heterozygous state for a single abnormal gene carries a considerable hematologic handicap, a much greater handicap certainly than in the case of thalassemia or the sickle-cell trait. There are limits to what selection can overcome.

DR. MOTULSKY: Recently, we placed mice (*Peromyscus* [11]) with hereditary spherocytosis in hot rooms in an attempt to simulate the African environment. Under these conditions, they develop aplastic crises and more severe hemolysis, and many die after a few weeks. At the present time, we are attempting to find out the mechanism of this striking response to heat. It is known that the red blood cells of persons with this disease hemolyze readily when exposed to heat *in vitro*. It is quite apparent that our mice are at a severe disadvantage in a hot environment. If such a mechanism applies to man, as is not unlikely, the gene would be at such a severe disadvantage in a tropical environment that its known rarity in Negroes is not too surprising.

DR. NEEL: Is there a murine malaria?

DR. MOTULSKY: I know of no malaria affecting *Peromyscus*.

DR. COX: We wish to thank Dr. Motulsky for his extensive discussion and also Dr. Gowen and Dr. Neel for their comments. A point worth emphasizing is the difficulty in separating genetic influences from environmental factors that alter human susceptibility to infection. The importance of thorough epidemiologic studies cannot be overemphasized in analyzing the role of genetic factors in man's susceptibility to infectious disease.

We wish also to endorse the comments of Dr. Stern and Dr. Herndon on the importance of collecting empiric data. Our ignorance concerning host resistance to infection is so profound and facts are so scarce that information on families appearing to have increased susceptibility to infection may contribute not only to understanding hereditary disease, but also to interpreting host-parasite interaction.

MUTATIONS

J. V. Neel, M.D., Ph.D.

MUTATIONS *in the* HUMAN POPULATION

The study of the rate at which human genes mutate is certainly one of the most basic aspects of human genetics. That it is today one of the least satisfactory areas of investigation for the human geneticist is equally certain. Barring possible major technical developments, some of which we will be discussing later, it is likely to remain this way for a considerable time to come.

THE STUDY OF SPONTANEOUS MUTATION RATES: THE SPECIFIC–PHENOTYPE APPROACH

Mutations may, of course, occur in both somatic and germinal tissue. We will consider those occurring in the latter tissue first. Rates of germinal mutation are customarily expressed as frequency of mutation per locus per gamete per generation. Thus, an estimated rate of 1×10^{-5} means that, on an average, 1 in each 100,000 genes at the specific locus tested has, as judged by its phenotypic effects, been found to have a different information content than the corresponding genes of the parents from which it was derived. There is an implicit, temporal scale: in the case of *Drosophila melanogaster,* some 20 days during the breeding season; in the case of man, some 30 years. In man, there are essentially two approaches for obtaining germinal mutation rates, namely, the specific-phenotype approach and the population-characteristics approach. The specific-phenotype approach, as the name implies, estimates mutation rates on the basis of the observed or computed frequency with which a specific phenotype is encountered in a population. The population-characteristics approach, on the other hand, involves an attempt to measure a phenomenon that reflects the result of mutation at many loci, followed by assumptions as to the number of loci involved in the phenomenon and the calculation of an average mutation rate per locus.

The Direct Method.—The specific-phenotype approach may utilize either direct or indirect methods. The direct method may only be employed in the case of dominantly inherited traits. One simply determines the frequency of occurrence in the population of isolated cases of a given phenotype which, from previous or parallel investigations, is known to have a high probability of being due to a single dominant gene and computes a rate based on this frequency. A correction factor of 0.5 must be introduced, since each individual scored represents two genes tested. How can so simple an approach possibly go astray? There are at least six well-defined sources of error in this method.

1. *There may be problems in the diagnosis of the trait.* Although one hopes the trait selected for investigation will be clear and unambiguous, this hope is seldom realized. Thus, retrolental fibroplasia can be confused with retinoblastoma; hereditary, multiple, circumscribed lipomatosis with multiple neurofibromatosis; and, until a few years ago, Christmas disease with true hemophilia. In general, misdiagnosis is not a major source of error in estimates of mutation rate.

2. *One must have a reliable estimate of the frequency of individuals who exhibit the trait in the absence of positive family histories.* Both the patience and ingenuity of the investigator are often sorely taxed in attempts to compile a complete roster for a given disease in a given area at any one time or over a period of years. I doubt if any investigator has ever felt absolutely assured his roster of cases for a given area and period was complete. However, with the proper choice of trait, roster inaccuracies should not be a major source of error. The matter of determining which of these roster cases have positive and which negative family histories, that is, which are possibly mutants and which not, is of course routine, and only very minor errors should result on this score.

3. *One needs to know what proportion of isolated cases result from dominant, germinal mutation and what proportion from such other genetic mechanisms as somatic mutation or recessive inheritance or from nongenetic causes.* The acquisition of this knowledge demands not only the usual studies of familial disease but also a meticulous study of the progeny of isolated cases where ascertainment is through the case rather than the progeny. An estimate of the penetrance of the dominant gene in question is essential. Without this knowledge, one cannot evaluate the significance of the ratios encountered among the offspring of isolated cases. Where penetrance is essentially complete, this fact is usually established with relative ease, but where penetrance is low, it may be quite difficult to arrive at a satisfactory estimate, as in the case of retinoblastoma. The possibilities for error here are large.

4. *In making the calculation, one is usually forced to assume that the phenotype in question results from mutation at one locus only.* Testing the validity of

this assumption, which undoubtedly varies from phenotype to phenotype, will generally be laborious in man. One way is through an analysis of certain unusual matings. For example, an analysis of the outcome of marriages between a normal individual and an individual with both the thalassemia and the sickle-cell traits has provided evidence for the genetic heterogeneity of the thalassemia trait.[606, 738] The efforts of a number of investigators have been required to furnish the critical pedigrees, and yet thalassemia and the sickling phenomenon are relatively common traits. A second approach to the validity of the assumption is the demonstration, as in the case of elliptocytosis, that certain genetic linkages exist in some families but not in others.[125, 282, 283, 482, 575] Finally, the increasing recognition of the biochemical basis of many inherited traits will also be helpful. Thus, sex-linked hemophilia has been subdivided into two entities, true hemophilia and Christmas disease. Graham [292] has shown that approximately one-third of what previously would have been designated hemophilia is actually Christmas disease. This necessitates a revision of the mutation-rate estimate for the hemophilia locus. The errors due to the possibility that we may actually be measuring the results of mutation at several loci are essentially indeterminate at this writing.

5. *One must assume that the mutant phenotypes, which it is convenient to use, originate with a frequency representative of mutation rates in general.* It is only necessary here to point out that no one is likely to perform a mutation-rate study with respect to a phenotype which has a mutational origin but once in each 10,000,000 births. Neither is one apt, in man, to come to grips with the problem of the frequency of mutations with small phenotypic effects, which, if mutation rates are in part set by selective factors, might reasonably be postulated to occur with a relatively high frequency.

6. *The final source of error to be mentioned stems from the implicit assumption that one is measuring all the mutation at the locus under study.* However, mutation at this locus resulting in phenotypes other than the one under consideration would go unscored. This source of error is by no means confined to man, but is particularly troublesome in his case.

The Indirect Method.—In the indirect method of utilizing the specific-phenotype approach, one determines the effective fertility of individuals affected with a given dominantly or recessively inherited trait, in terms of per cent of normal and then, assuming equilibrium, calculates how frequently mutation to the gene resulting in this trait must occur each generation in order to maintain a constant proportion of persons with the trait in the population. In the case of a dominantly inherited trait, the appropriate formula is

$$m = (1 - f)q,$$

while for recessive inheritance, the formula is

$$m = (1 - f)[\alpha q + (1 - \alpha)q^2],$$

where

    m = the mutation rate per gene per generation,

    f = the relative fitness of individuals homozygous for the particular gene under consideration,

    q = the frequency of the gene in the population, and

    α = the coefficient of inbreeding.*

An appropriate modification of the latter formula is employed in the event of sex-linked inheritance. The formula is also readily modified to allow for phenotypic effects of the recessive genes when heterozygous.

In addition to the six sources of error listed in connection with the direct method of estimating mutation rates, five additional sources must be recognized for the indirect approach. First, this approach assumes population equilibrium, that is, that the loss of mutant genes through the impairment of fertility associated with the mutant phenotype is exactly balanced, and has been for some centuries, by the introduction of new genes into the population through mutation. The recent dramatic changes in the nature of both the cultural and medical determinants of fertility which render this assumption of dubious validity require little comment. The assumption is patently false in most instances, but it has been employed because it permits an approximation to values that cannot be estimated by this method without the assumption.

Second, in the case of recessives, unless specific allowance is made for heterozygote effects by an appropriate modification of the formula, the assumption is implicit that the gene has no effect whatsoever upon survival or fertility in the heterozygote. The rapidly expanding list of diseases which are nominally recessively inherited and for which heterozygous carrier states can be recognized has been summarized in several reviews.[258, 371, 372, 604, 605, 662] The issue of the extent to which heterozygotes for so-called recessive genes are hindered or helped by the fact of heterozygosity is one of the most controversial in contemporary genetics. Even when heterozygote effects are indubitable, their impact on fertility is often extremely difficult to determine and may indeed vary significantly from place to place. For only a very few genes can allowance be made for heterozygote effects at present.

Third, in the case of recessive inheritance, the assumption is implicit that the distribution of carriers is uniform throughout the area under study. Although there are obvious reasons for suspecting the validity of this assumption, methods for detecting carriers have not been sufficiently accurate or simple to permit anything like an adequate test of this assumption until recently. Some current developments with respect to the recessively inherited absence of erythrocyte catalase known as acatalasemia are of interest in this connection. This trait has thus far been encountered almost exclusively in Japanese. Using the conventional genetic

---

    * In the case of recessive inheritance, the expression $[\alpha q + (1 - \alpha)q^2]$ is equivalent to the trait frequency.

formulae, my associates and I have recently calculated the frequency of the gene responsible for this trait.[865] Carriers of this gene have about half-normal blood-catalase values. On the basis of the above calculated gene frequency, the carrier frequency among Japanese was estimated at approximately 6 per 1,000. The recent development by Hamilton [319] of a simple and rapid method for carrier detection makes it possible to check the validity of this calculation. In the first 7,000 individuals tested in Hiroshima and Nagasaki, the carrier frequency has been only approximately 1 per 1,000. It remains to be seen whether pockets will be encountered in which the gene frequency varies on the high side and what relation the results of a Japan-wide survey will bear to the predicted average values.

Fourth, one must generate a reliable estimate of the relative fitness of affected individuals. While this should be a mere technical detail, in fact, it is often quite difficult. The normal siblings of affected individuals constitute the most readily accessible material for estimates of relative fitness. Unfortunately, as Reed and I [711] have had occasion to discover, the fitness of these normal siblings may not be representative of the socio-economic level from which they are drawn, presumably due to the influence of poorly-understood psychic factors.

Fifth, a source of error for which there was no evidence in man until very recently, occurs when one assumes that, in segregating sibships, the mutant phenotype appears in Mendelian proportions. We believe we now have evidence that there is a significant deficiency of children with aniridia in segregating sibships, a deficiency difficult to attribute to fetal elimination, due to the apparent localization of the phenotypic effects of the gene to the eye.[794]

The magnitude of the bias introduced by the first three of these last five sources of error is essentially indeterminate but could be large. For instance, an heterozygote disadvantage of 1 per cent would more than double the mutation rate necessary to maintain a recessive, lethal trait with a gene frequency of .01.

All the sources of error in the estimation of mutation rates which I have just enumerated have been frequently mentioned in the genetic literature.[45, 305, 595, 598, 600, 651, 824, 899] Every one recognizes them. Nevertheless, tables are continually published containing estimates for both dominant and recessive traits, some of the former being direct and some indirect in their derivation. While this was undoubtedly justifiable during the period when any estimate honestly derived was better than no estimate, perhaps a level of awareness of sources of error now has been reached to advise, for many purposes, that we restrict ourselves to figures derived by the direct method applied to traits dominantly inherited.

Even with this restriction, real difficulties are frequently encountered. The history of efforts to develop estimates of mutation rate for the retinoblastoma phenotype provides an excellent documentary for this effect. Retinoblastoma is a highly malignant, ocular neoplasm that has frequently been shown to be inherited as a dominant trait. From a determination of the frequency of isolated cases of this disease, Philip and Sorsby in 1944 estimated the mutation rate in the population of London at 1.4×10^{-5}. In 1951, Falls and I,[215, 609] on the same basis, estimated the rate at 2.3×10^{-5} in the population of the state of Michigan. This estimate was based on the specifically stated assumption that all isolated cases were due to mutation, an assumption not at that time adequately tested by follow-up studies on the children of isolated cases. That not all isolated cases were due to germinal mutation was suggested by the fact that, when the disease was definitely genetic as proved by the occurrence of multiple affected individuals within a sibship, it was significantly more often bilateral than when it occurred as an isolated event. It seemed possible that the greater frequency of individuals affected unilaterally among isolated cases reflected the occurrence of somatic mutations, and shortly after our first estimate was published, I revised the original estimate downward to 1.8×10^{-5} on that basis.[605]

The next revision of the estimate was that of Vogel.[897, 898] Numerous pedigrees attest to the incomplete penetrance of the gene, although various studies have yielded somewhat conflicting evidence on this point. Utilizing original observations plus a compilation of the literature, and assuming 80 per cent penetrance for the gene, Vogel demonstrated a very significant deficiency of affected children among the offspring of isolated cases of the disease on the assumption that all such isolated cases resulted from germinal mutation. Vogel calculated that only some 17 to 18 per cent of isolated unilateral retinoblastoma was due to dominant mutation, with the corresponding figure for isolated bilateral retinoblastoma being 50 to 100 per cent. On this basis, he reduced the mutation-rate estimate to .6 or $.7 \times 10^{-5}$, a reduction which Smith and Sorsby [818] felt their data also justified.

The most recent study on this subject is that of Macklin,[518] who seems to have exceeded all previous workers in the completeness of her family investigations. Macklin reports that some 10 per cent of cases who, on the basis of preliminary pedigrees, appeared to be isolated actually have affected relatives—first cousins, second cousins, second cousins once removed, and third cousins once removed. The author has interpreted this as indicative of a lower degree of penetrance than assumed by most workers. To me, an alternative interpretation, suggested by the apparently greater penetrance of the gene subsequent to its first phenotypic manifestation, involves the concept of premutation.[24] There may also be significant differences in penetrance from kindred to kindred. If the possibility of premutation is ignored, then, taken at face value, these findings would imply that Vogel [897] has overcorrected previous estimates of mutation rates by making inadequate allowance for nonpenetrance. On the other hand, it is questionable whether Macklin's data provide an unbiased estimate of the true penetrance of the gene, since families in which the gene is poorly penetrant will tend to make a disproportionate contribution to the literature because of the lethal nature of retinoblastoma. Furthermore, Macklin's findings are somewhat at variance with those of other workers.

Stimulated by her paper, I have just reviewed all our own material on this point and am reasonably confident we were not missing possible relationships between apparently isolated cases as distant as second cousins and, in many families, even more remote. In material that now amounts to 90 kindreds in which retinoblastoma occurs, collected in collaboration with Dr. H. F. Falls, we have seen only one of these nonpenetrant kindreds, as illustrated in figure 9. This kindred

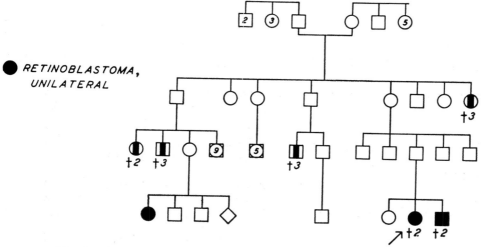

Fig. 9.   University of Michigan Heredity Clinic kindred 4752, retinoblastoma unilateral.

first came to attention in 1956, that is, it is not a modification of a previously published kindred. Macklin herself estimates the mutation rate at between .6 and 1.8×10^{-5}, the lower limit being set by the frequency of isolated bilateral cases, and the upper by the combined frequency of both unilateral and bilateral isolated cases. The propriety of excluding, as she does, from this estimate the examples, described above, of failure of penetrance is doubtful, since, on the theory of the random loss or increase in the population of neutral (nonpenetrant?) genes, nonpenetrance per se does not interfere with estimates of mutation rate.

Although the complications that have arisen in attempting to estimate the rate with which the retinoblastoma gene arises through mutation appear outstanding, it may be that this state of affairs reflects the unusual degree of attention this phenotype has attracted. Already similar problems have arisen in connection with a number of other estimates. Thus, in both Caucasian [554, 841] and Japanese [612] populations, a much higher proportion of isolated than hereditary achondroplastic dwarfs die, leading to questions concerning the genetic nature of a considerable proportion of isolated achondroplastic dwarfs and the appropriateness of basing estimates of mutation rate on such individuals.[805] In the dominantly inherited disease of aniridia, the evidence for a significant departure from a 1:1 ratio which cannot be attributed to nonpenetrance seems conclusive and raises the possibility of an abnormal segregation ratio.[794] Many other examples of such complications could be cited.

Table 42 summarizes the frequency with which isolated cases of various phenotypes judged suitable for studies of mutation rate have been encountered in

Table 42

ESTIMATES OF THE RATE OF APPEARANCE OF CERTAIN MUTANT PHENOTYPES
BASED ON THE DIRECT METHOD

Trait	Locale	Author	Mutant phenotypes *
Epiloia	England	Gunther and Penrose, 1935 [301] Penrose, 1936 [652]	.8– 1.6
Retinoblastoma	England	Philip and Sorsby, 1944 [663]	2.4
	U.S.A., Michigan	Neel and Falls, 1951 [609]	4.6
	U.S.A., Ohio	Macklin, 1959, 1960 [517, 518]	3.6
	Germany	Vogel, 1954 [898]	3.4
	Switzerland	Böhringer, 1956 [71]	4.2
	Japan	Matsunaga and Ogyu, 1959 [533]	4.2
Aniridia	Denmark	Møllenbach, 1947 [544]	1.0
	U.S.A., Michigan	Shaw, Falls, and Neel, in press [794]	1.0
Achondroplasia	Denmark	Mørch, 1941 [554]	8.4
	North Ireland	Stevenson, 1957 [841]	28.6
	Sweden	Böök, 1952 [74]	14.0
	Japan	Neel, Schull, and Takeshima, 1959 [612]	24.5
Partial albinism with deafness	Holland	Waardenburg, 1951 [903]	.7
Pelger's nuclear anomaly	Germany	Nachtsheim, 1954 [594]	5.4
	Japan	Handa, 1959 [321]	3.4
Neurofibromatosis	U.S.A., Michigan	Crowe, Schull, and Neel, 1956 [161]	26.0–50.0
Microphthalmos-anophthalmos without mental defect	Sweden	Sjögren and Larsson, 1949 [803]	1.0
Huntington's chorea	U.S.A., Michigan	Reed and Neel, 1959 [711]	1.0 †

*No family history per 100,000 individuals.
†Maximum estimate.

various populations. Estimates on such traits as Marfan's syndrome and dystrophia myotonica have been omitted purposely because of the marked diagnostic difficulties. The biases just discussed, such as somatic mutation and the occurrence of phenocopies, would tend to result in overestimates when these frequencies are utilized for the derivation of mutation rates, but other biases, also discussed, could result in underestimates. Because of these uncertainties, the appropriate course is not to attempt to apply any corrections. Needless to say, in the light of the foregoing, considerable effort could be devoted to a discussion, largely inconclusive, of just what corrections should be applied to each of these estimates of frequency to convert it into a suitable estimate of mutation rate. If an estimate is obtained of the average frequency of isolated cases of each trait for which several studies have been carried out, and, in turn, an average obtained for all the estimates, the figure is 7.76×10^{-5}. This corresponds to a mutation rate of 3.9×10^{-5}. The figure is greatly elevated, relatively speaking, by the results for achondroplasia and multiple neurofibromatosis. As noted earlier, the possibility has been raised that a considerable proportion of achondroplastics are phenocopies. This does not seem to be the case with respect to multiple neurofibromatosis.

Imperfect though these figures are, they can be utilized for a variety of calculations and comparisons, a full discussion of which falls outside the scope of this presentation. However, it does seem appropriate to direct attention towards the similarity, when several different studies have been performed, of the estimates derived from different populations. This is particularly striking in the case of retinoblastoma. Thus far, there is no evidence for strain differences in frequencies of mutation rate in man such as those which have been encountered in flies and bacteria, although the similarities between relatively large human populations certainly do not exclude the possibility of substrains within the larger population. In view of the widely differing circumstances under which members of the human race live, the careful study of comparative rates can make important contributions to our knowledge of mutagenesis. In this connection, the problem of selection must be borne in mind. There may be a false, apparent concordance among the results of various studies simply because no estimates are published from areas where the rate of mutation is relatively low, since material for a study is not available. Thus, Becker,[46] who has supplied a direct-type estimate of mutation rate of $.5 \times 10^{-5}$ for the gene responsible for the facio-scapulo-humeral type of muscular dystrophy based on German material, comments on the relative rarity of this disease in many other parts of Europe from which no mutation rate estimates have been forthcoming.

THE STUDY OF SPONTANEOUS MUTATION RATES: THE POPULATION–CHARACTERISTICS APPROACH

The principle involved in attempting to estimate mutation rates from certain characteristics of a population is basically simple, but, as in the case of the preceding, fraught with many questions of detail for which answers are not readily available. In essence, one argues from some aspect of a population to the average rate of mutation per locus necessary to account for the phenomenon. Morton, Crow, and Muller [575] have developed an ingenious approach to estimating the average rate of mutation per locus in man based on a study of inbreeding effects. In brief, from the magnitude of the increased rate of death seen in the offspring of various types of consanguineous unions, they compute the number of so-called "lethal equivalents" per gamete and then the mutation rate necessary to support such a number of lethal equivalents in view of their assumed rate of elimination per generation. From this approach, the total rate of mutation to lethals and detrimentals was estimated by these investigators as .06 to .15 per gamete per generation or, assuming 10^4 loci per gamete, .6 to 1.5×10^{-5} per locus.

As the authors point out, there are multiple possibilities for error in this estimate, all basically stemming from inadequacies in the genetic data available, necessitating frequent, difficult choices for the authors. In brief, the more salient possibilities for error are as follows:

1. *The nature of the data on consanguinity effects.* The paucity of data on this point requires no comment. Morton, Crow, and Muller [575] have pointed out some of the deficiencies in the data they were forced to use, which was chiefly that of Sutter and Tabah [853, 852, 851] from France. Recent data from Japan, collected under unusually favorable circumstances, reveal differences between the cities studied (Hiroshima, Nagasaki, Kure) of a magnitude which finds no simple explanation.[785] While the total consanguinity effect in terms of death prior to age 8 for Hiroshima children approaches that recorded by Sutter and Tabah [853] for France, the effect of consanguinity appears to be definitely less in Kure and Nagasaki. An intensive effort to elucidate the reason for these differences is in progress. Recent, less extensive studies by Böök [73] and Slatis, Reis, and Hoene [806] also reveal smaller effects of consanguinity than the data of Sutter and Tabah.[853] Exact comparisons are difficult because socio-economic conditions undoubtedly influence the results.

2. *The assumptions concerning the degree of deleterious effect exerted by the average, so-called recessive gene when heterozygous.* Morton, Crow, and Muller,[575] for want of better data, were forced to base their opinion concerning this point on the behavior of a special class of mutations in an insect, namely, recessive lethals in *Drosophila,* which, on the basis of the data of Stern and collaborators [836, 837] and Muller and Campbell,[575] they found to have between 2 and 4 per cent dominance in the heterozygote, depending on whether one used an harmonic or arithmetic mean. Whether these laboratory-preserved lethals correspond to the type encountered in natural populations is debatable. Furthermore, other types of data from *Drosophila* permit somewhat different estimates of both the proportion of mutations with deleterious effects when heterozygous and the degree of this deleterious effect.[907]

3. *The assumptions regarding the synergism of action of deleterious genes.* The child of a first-cousin marriage, with a coefficient of inbreeding of .06, is, on the assumption of some 10,000 genetic loci in man, at the time of conception rendered homozygous at some 600 loci over and beyond those at which homozygosity would occur in the absence of inbreeding, the actual amount of excess homozygosity depending upon the allelic structure of the population. Morton, *et al.*[575]

assume little or no synergism in the action of these genes. To me, the results of genetic studies utilizing *Drosophila,* especially when the genes concerned bear on the same general characteristics,[596, 597] raise the possibility of considerable synergism.

4. *The lack of data concerning the correlation between the rate of mutation at a locus and the level of expression of the resulting mutations in heterozygous organisms.* There are simply no adequate data on this point at present; for purposes of calculation these authors assume no correlation.

5. *The assumed relationship between the measurable lethal and detrimental rate of mutation (in terms of increased miscarriages, stillbirths, and death prior to the age of reproduction) and the total lethal and detrimental rate of mutation.* The greatest uncertainty here pertains to the amount of essentially undetectable, genetically determined mortality in the very early stages of pregnancy. Thus, the data of Hertig,[345] laboriously acquired but still scanty, suggest that as many as 25 per cent of fertilized ova are lost in the very early stages of pregnancy, usually before this loss would result in a detectable abortion. The role of genetic factors in this loss is not clear.

6. *Assumptions concerning the frequency, in man, of loci participating in balanced polymorphic systems.* Crow [159] interprets the results of inbreeding as indicating that such loci are relatively unimportant. This interpretation depends primarily on a concept of the allelic structure of human populations, wherein it is assumed to be unlikely that the average locus would be represented by as many as a dozen alleles in a population. The need for data on this point needs no emphasis.

7. *The fact that the number of mutable loci in man is unknown.*

In summary, then, the application of this very ingenious approach is seriously handicapped by a lack of critical data concerning a number of basic parameters. The assumptions made by the authors may well be the correct ones. Other assumptions, which would seem of equal validity to some, may, in the extreme case, increase or decrease the estimate by as much as a factor of 5.

Penrose [650] has drawn attention to the fact that, with respect to the parents of isolated cases of a number of traits believed to be of mutational origin, slight increases in both fathers' and mothers' mean ages can be demonstrated, a finding explicable on the basis of the accumulation of mutations with age. On this hypothesis, changes in the sex ratio with advancing maternal and paternal age should be expected and may be utilized in attempts to derive estimates of mutation rate. On the simplest hypothesis, that sex-ratio changes reflect an accumulation of sex-linked mutations, one might expect the sex ratio to decrease with advancing maternal age and increase with paternal age. The paternal-age effect, reflecting only sex-linked dominants, might be expected to be less than the maternal, reflecting both sex-linked dominants and recessives. In point of fact, there is no demonstrable effect of maternal age, but a dependence on both birth order and paternal age with significant interaction between these two factors. This interaction is of such a nature that, whereas sex ratio decreases as birth order increases, it exhibits a negative regression on age of father for low birth orders, but a positive regression for high birth orders.[627, 628] These confusing findings discourage the use of the sex ratio in this context.

A knowledge of the rate of spontaneous mutation is basic to an understanding of the genetics of any organism. The foregoing should indicate how far we are from that objective in man. As one who has contributed to a number of estimates of specific mutation rates in man, I find myself increasingly concerned over the true significance of the values available. Since many of the problems involved in arriving at valid estimates are by no means confined to man, similar questions arise for many experimental organisms. On the other hand, there is no doubt that suitable studies on a laboratory animal can yield results far superior to those obtainable for man. For *Drosophila,* certainly the animal studied best in this respect, an average locus rate per generation of 1×10^{-5} is frequently quoted. The estimates of Russell and Russell [740] of a mean rate of $.81 \times 10^{-5}$, and of Carter, Lyon, and Phillips [119] of a mean rate of $.97 \times 10^{-5}$ for seven selected loci in a particular strain of the house mouse, constitute the animal work most pertinent to human problems. It is apparent that the conditions of these experiments permit the detection of a considerable portion of the mutational spectrum in the mouse.[742] Attempts to extrapolate from such figures as these to man must bear in mind the longer life cycle of the human organism, with the consequent increased exposure of the germ cells to both chemical and ionizing mutagens as well as the increased average number of cellular divisions between fertilization and the functional gamete in man, with the greater possibility for errors in the copying process. To put the matter in another light, if one assumes that the apparent resemblance between rates in *Drosophila* and the mouse is real and if man in turn is assumed to have an average rate of mutation no more than double or triple the rates in the mouse, the implication is that of a truly remarkable genetic control of mutation rates. The further implication would be that mutations only to a limited extent are the result of random errors in the copying process or random responses of genes to noxious agents, but, to a considerable extent, are under cellular regulatory mechanisms. If this is true for the genome as a whole, it must be true for individual loci. However, if this is correct, how can the relatively high rate of occurrence of certain undesirable phenotypes currently attributed to recurrent mutation be explained?

THE POSSIBILITY THAT SOMATIC–CELL GENETICS WILL CONTRIBUTE TO KNOWLEDGE OF GERMINAL MUTATION RATES

In recent years, the prospect has arisen that rates of mutation may be obtained for human somatic cells. This can be done either by means of appropriate studies on the intact organism or by the use of tissue-culture methods. With respect to the former approach, Goudie [286] and Atwood and Scheinberg [23] have demonstrated that about 0.1 per cent of the erythrocytes in young adults of blood type *AB* apparently lack the *A* agglutinogen. Possible origins for such cells, in addition to point mutation, include variability in antigenic sites on cells of unchanged genetic constitution; somatic crossing-over and segregation; and loss of all or a portion of a chromosome. Atwood and Scheinberg have reported a relatively high frequency of such exceptional cells in an individual presumably homozygous for the type *A* gene. From the interpretative standpoint, this is a troublesome observation. If a conclusive demonstration of the mutational origin of these exceptional cells is possible, then, because of the antigenic complexity of the erythrocyte, here is a potent approach to estimating rates of mutation in somatic cells. Unfortunately, the fact that the mature erythrocyte is enucleate precludes the isolation and propagation of these cells by methods of tissue culture with the attendant opportunities to elicit their nature.

With respect to techniques of tissue culture, numerous investigators have reported on the isolation of "mutant" lines of cells.[126] The development by Puck and co-workers of methods of culture which permit handling mammalian cells essentially as suspensions of bacteria and so establishing clones descended from single mutant cells holds enormous possibilities for the precise estimation of mutation rates of somatic cells.[684, 685] Despite great technical advances, some serious problems remain.[637, 862] Theoretically, however, one should ultimately be able to detect, on a quantitative basis, mutant cells varying in serologic and other genetic characteristics, just as in erythrocyte suspensions. Should, for example, the *A* and *B* blood-group antigens prove to be characteristic of somatic cells grown under proper culture conditions and should these somatic cells exhibit the same variations as erythrocytes, a variety of studies become possible. Recently, Lieberman and Ove [505, 506] and Szybalski [863] have demonstrated the feasibility of selecting for mutant cells of a particular type in tissue culture, using as characteristics resistance to the antibiotic, puromycin, and the purine analogue, 8-azaguanine. Using equations developed by Luria and Delbrück in the study of bacterial mutations,[515] Lieberman and Ove have obtained a mutation rate from "no to low" resistance to puromycin of 3.6×10^{-6}, and from "low to high" resistance of $.8 \times 10^{-4}$. Szybalski reports a rate per cell per division cycle of 4.9×10^{-4} for "resistance" to azaguanine, and 1.2×10^{-6} for mutation from the trait, "azaguanine resistance," to the trait, "resistance to a related compound." Since a fibroblastlike cell of unknown chromosomal composition was used in this work, and, since chromosomal studies were not carried out on the mutants, these rates cannot be equated to point mutation rates, but the refinements that would remove these objections obviously can be achieved with the passage of time.

A problem that has thus far received relatively little attention is how to estimate germinal-cell mutation rates on the basis of the findings with somatic cells.

There is no assurance the two rates will be the same, especially when the rates for somatic cells are measured in tissue culture. It would seem necessary to establish a somatic to germinal ratio at selected loci in an experimental animal, such as the mouse, before one can begin to extrapolate to human germinal rates from human somatic-cell rates. This is a large order.

THE STUDY OF INDUCED–MUTATION RATES IN MAN

Studies of induced-mutation rates in man share all the difficulties and uncertainties already listed in connection with the study of spontaneous rates. In addition, there is the added problem of dosimetry, which may be quite complex in the case of human populations. This is true no matter whether one is concerned with radiation or potential chemical mutagens. Thus, the evaluation of gonadal dose has proved extraordinarily difficult in such exposed groups as the survivors of Hiroshima and Nagasaki; [610] French men and women treated for sciatica; [888] Japanese x-ray technicians; [866] and Indians living on the monazite sands of Kerala State.[285]

The same three approaches enumerated for the study of spontaneous rates can be extended to the study of induced rates. The specific-phenotype approach is potentially the "cleanest" approach to the study of induced-mutation rates. However, it can readily be demonstrated that, for most of the exposed groups defined to date, this approach has little likelihood of success. Stevenson [840] and Strobel and Vogel [847] have provided a mathematical demonstration of the validity of this statement even under "favorable" assumptions, that is, the availability of a number of relatively radiation-sensitive, marker loci at each of which mutation gives rise to a specific phenotype that either arises in no other way or that, if not of mutational origin, can readily be identified as such. Given, in addition, the actual present state of knowledge regarding possible marker loci, the specific-locus approach has little to offer.

Of necessity at present, then, one turns to the population-characteristics approach. Here, depending on the level of sophistication of civil and medical facilities, a wide range of possibilities exists. A recent W.H.O. Technical Report [212] defines five levels of successive complexity at which surveys of irradiated populations and suitable controls can be carried out, namely:

(1) accurate demographic statistics,
(2) growth and development data,
(3) patterns of congenital defect,
(4) morbidity and mortality patterns as based on clinical data, and
(5) laboratory studies, including a search for chromosomal aberrations.

To date, no investigations on potential chemical mutagens utilizing human material have been carried out, but there are on record six studies of the characteristics of children born to irradiated parents which achieve one or more of the levels just enumerated.[160, 436, 516, 610, 866, 888] Unfortunately, only the study carried out under the auspices of the Atomic Bomb Casualty Commission meets the requirements of a prospective study.[610] In view of the well-known biases influencing the results of a comparison when one of the groups involved is aware it has been exposed to a potentially noxious agent, retrospective data on such parameters as abortions or malformations seem of limited usefulness. However, sex of child is relatively devoid of emotional overtones and usually reported with accuracy. It is a striking fact that, in three of the four studies in which the sex ratio of the children born to irradiated fathers was studied (and for which controls are available), the ratio appeared to be increased; [516, 610, 866, 888] the three studies yielding this result appear to involve considerably larger radiation exposures than the fourth, that of Macht and Lawrence.[516]

Furthermore, in both studies in which mothers were irradiated and for which controls are available, the sex ratio is decreased. Many factors have been reported to influence the sex ratio; it is certainly not the ideal indicator of induced-mutation rates, especially in view of the unexplained relations to parental age discussed earlier. Schull and Neel [785] estimate the decrease in the proportion of male births following 100 r.e.p. of maternal radiation as 0.0060 and, on the assumption that this effect can be attributed to the induction of sex-linked lethals and the further assumption that the number of lethal-producing genetic loci on the X-chromosome lies between 250 and 2,500, calculate that the probability of a sex-linked lethal mutation per r.e.p. lies in the interval between 2.4×10^{-7} and 2.4×10^{-8}, with the true figure more likely closer to the higher than the lower value. This estimate may be compared with Russell and Russell's estimate [740] of 2.2×10^{-7} per locus per r unit for single-dose (600 r), spermatogonial irradiation, based on the study of 7 specific loci in the mouse. It is noteworthy that, thus far, there is no evidence for an effect of radiation on the sex ratio in the mouse. In view of the apparent paucity of sex-linked traits in the mouse as contrasted to the relative abundance of such traits in man, this difference may reflect a lower genetic content of the X-chromosome in the former.

Thus far, clear-cut effects of parental radiation on stillbirth and malformation rates, birth weights, neonatal and infant mortality, and growth and development during the first nine months of life have not been demonstrated.[610] In view of the relatively small population doses of radiation involved in the study situations to date and the extent to which mechanisms other than newly arisen mutations influence these indicators, the failure to demonstrate radiation effects is not surprising. In the face of our own negative studies in Hiroshima and Nagasaki, I feel obligated to review briefly the recent contribution by Gentry, Parkhurst, and Bulin.[271] Based on birth- and death-certificate information from New York State, these investigators have drawn attention to "a correlation between areas with high malformation rates and geographical locations containing natural materials with relatively high concentrations of radioactive elements" and have concluded that the association "strongly suggests radiation as the primary etiologic agent." The rate of malformation for rural areas classified as probably characterized by relatively high levels of background radiation was 15.8 per 1,000 live births, whereas for areas where "high" background radiation was unlikely, the rate was 12.9.

Dosimetry studies are unfortunately still incomplete, but the difference in gonadal dose between "average" and "high" regions does not appear to exceed a factor of 2 or at most 3, the greatest unknown being the extent to which contamination of the water supply by radium 226 has resulted in the deposition of internal emitters. The demonstration of regional differences in malformation frequency can scarcely be challenged and is, in fact, a well-recognized if poorly understood phenomenon. Thus, Penrose [648] has drawn attention to the striking differences in the frequency of anencephaly in different parts of Europe. Assuming, on the average, five generations of continuous residence in these "high" areas, the mean cumulative excess dose to the gonads should approximate 30 r at most. This, however, is very low-level radiation, which, while undoubtedly mutagenic, may, in the light of Russell's studies on mice [739, 743] yield much less in the way of mutations than the acute high-level dosages customarily used in experimental work. The etiology of congenital malformations is complex. Each generation, the fraction resulting from dominant mutation or simple dominant inheritance with or without complete penetrance, that is, the fraction most responsive to mutation pressure, is unknown but certainly a minority. All in all, to attribute this 25 per cent difference in malformation frequency between the two types of areas to the effects of radiation, genetic or somatic, would appear to imply a somatic and/or genetic radiosensitivity in man substantially in excess of that supported by all the other available evidence, as summarized most recently by the U. N. Scientific Committee on the Effects of Radiation.[890] Other causes must be sought for these interesting correlations, as well as the parallel correlations between background radiation and malformation frequency recently reported on a nationwide basis by Kratchman and Grahn.[462] In the latter instance, the use of a yardstick other than malformation deaths per 1000 deaths is highly desirable, since this statistic is so vulnerable to the differences known to exist in the age composition of the areas studied. Thus, the median age for New Mexico and Utah, the two states with the highest rates, was 24.0 and 25.1 years, respectively, these being lower than any other states except North and South Carolina and Mississippi,[891] whereas the median age for the United States as a whole was 30.2 years in 1950. In these three states the presence of large numbers of Negroes, with a possible lower frequency of congenital malformations [599] plus a lower level of reporting of malformations, complicates the interpretation.

Finally, subject to the problems of interpretation discussed earlier, the study of induced mutation rates can also be extended to somatic cells. Scheinberg and

Reckel [763] have shown an increase in "exceptional" blood cells in radiated pigeons and a man.   Bender [48] and Puck [681] have related frequency of chromosomal breakage to radiation dose in tissue-culture material, and Puck and coworkers [687, 690] have described the appearance, following radiation, of cell lines differing from the parental line in growth and morphologic characteristics.   The crux of the problem here, of course, is the development of methods capable of detecting the spectrum resulting from point mutation at specific loci.

CONCLUSION

Tremendous progress has been made in defining the approximate frequency with which mutation occurs in different plants and animals, including man.   With this progress, however, has come an increasing awareness of the sources of error in estimates.   Human material is especially subject to some of these errors.   In view of the growing number of problems requiring knowledge of spontaneous and induced human mutation rates, the classification of the direction and magnitude of these errors appears as one of the urgent tasks facing human genetics.   Although the recent developments in the field of somatic-cell genetics open exciting vistas, the full exploitation of these developments requires establishing a minimum number of cross-references between somatic and germinal rates.   With respect to the latter, no short cuts are in sight.   Each trait suitable for studies on mutation rate must be subjected to laborious investigations evaluating all the possibilities for error which have been enumerated.   The easily identified traits most suitable for study on a population basis will probably, as a rule, not be found to have a readily identifiable counterpart at the cellular level.   Establishing germinal-mutation rates for serologic and biochemical traits of the type that can be utilized in tissue-culture studies will be extremely laborious, and the acquisition of certain needed types of information regarding the genetics of man will require an expenditure of effort on a greatly expanded scale.

DISCUSSION

DR. BURDETTE: Dr. Neel's paper will be discussed first by Dr. Crow.

DR. CROW: I agree with Dr. Neel about the difficulties of measuring human mutation rates.   He has enumerated a great many of these, and I would like to add one more.   This arises from the fact that the environment has been changing so rapidly in the last few generations that what is lethal now was not lethal a few generations ago.   This makes me want to urge, as Dr. Neel has himself on an earlier occasion,[159] that as much information as possible on mortality and morbidity rates from consanguineous marriages be gathered from primitive populations while they are still primitive.

The only criticism I have of Dr. Neel's paper is that it seems to me one ought to make a larger distinction between the indirect method as applied to autosomal recessive genes and the indirect method as applied to severe dominant or sex-linked recessive traits. The indirect method for the last two kinds of traits has about the same validity as the direct method, whereas a number of rather tenuous assumptions are necessary to make any kind of an estimate of a single-locus rate for autosomal recessives. This means that it may be a long time before one has any reliable estimates for single autosomal loci and that our best information, if we want to make comparisons between dominants and recessives, is to look at autosomal dominants and sex-linked recessives.

I note, as Dr. Neel has implied, that, with greater refinement of methodology, the estimated mutation rate gets lower. Those traits which are most likely to depend on a single locus, such things as Waardenburg's syndrome and facio-scapulo-humeral muscular dystrophy, have quite low mutation rates,[142, 552] something less than 10^{-6}.

Another thing I would like to mention is that, in addition to the direct and the indirect methods, there is a semidirect method yet to be utilized.[569] This has been applied to measuring mutation rates for dominant deafness.[142] In essence, the idea is this: it is sometimes possible to estimate from the residue of what is left after segregation analysis is made how many sporadic cases there are and how many of the sporadic cases may be mutants. This means then that one determines the number, although not the particular individuals, of the mutant cases and from this estimates the mutation rate.

This differs from the direct method in that the individual instances are estimated rather than seen. It differs from the indirect method, however, in that no assumption is made of any equilibrium. For this reason, it has a validity somewhere in between the other two methods and may be used especially for traits when there is some doubt about the proportion of phenocopies as opposed to true mutants. I think the distinction between mutants and phenocopies is what causes the major difficulty and that incomplete penetrance is not a serious problem.

Some of the mutation rates that have been measured are excessively high by standards of what one would expect from studies in other animals. Of course, in maize there is the R locus, which has an especially high mutation rate, but now it is understood as being of a rather special nature;[100] and familiar to all *Drosophila* workers are the *minutes,* which have a high mutation rate simply because of the whole series of genetic changes that can lead to a similar phenotype.

With regard to Dr. Neel's last method, that of estimating a class of mutations, I am as concerned as anybody about the large number of assumptions that are necessary. In the course of future progress there must be a steady refinement of the nature of the assumptions and an increase in the number of ways of testing them. Most of Dr. Neel's discussion of this is based on the methods that were used in 1956,[575] which in some respects have been refined since. It is important to get as many internal checks into the system of measurements of mutation rate as one can. In the attempt to measure mutation rates for mutants causing a certain class of effect, say death or mental deficiency, one does have two kinds of internal checks, both of which are aimed at avoiding the confusion between genes that are maintained by mutation and genes that are maintained by some sort of polymorphic balance between selective forces. One of these mechanisms is the high ratio of the inbreeding effect to the effect of random breeding. I mentioned this earlier and so did Dr. Steinberg. We have restricted our measurements of mutation rate to those that show a high ratio, which are less likely to be loci that are maintained by heterosis.

Another check is given by measuring the segregation load.[159] With low-grade mental deficiency, the fraction of all cases due to simple recessive genes is estimated to be about 15 per cent. Then if one estimates the gene frequency, one can ask how much selection would be required to maintain this many genes for mental deficiency if they were maintained by selective polymorphism, and it turns out to be about 20 per cent. In other words, 20 per cent of the zygotes that are conceived would have to be selectively eliminated one way or another in order to maintain just those cases of mental deficiency that arose in this kind of a study. I suggest that this is excessively high and that this is strong evidence that recessive mental deficiency is maintained in the population not by selective balance but by mutation.

If one is willing to assume that these mutant genes are not held in the population by heterosis, then there does come the question of how they are eliminated. Are they eliminated as homozygotes by inbreeding or are they eliminated through heterozygous effects? Fortunately, from the standpoint of estimating the mutation rate, at least for those genes concerned with mental deficiency, the two estimates lead to similar results: on a per-locus basis something like 2×10^{-5}, assuming they are eliminated by inbreeding and something like 6×10^{-5} if they are eliminated by heterozygosis of the same magnitude that occurs in *Drosophila*. Of course, both of these demand several untested assumptions.

Another criterion that one should use in trying any sort of an indirect estimate of mutation rates for a class of disease is to make sure that the disease is harmful enough to be disadvantageous in all environments. In regard to Dr. Neel's statement that there is an additional assumption involved in trying to estimate single-locus mutation rates after estimating the genomic rate, it is the genomic rate that is of interest for many purposes. I am really more interested in knowing what the total mutation rate is for all genes than in knowing any particular one of them.

Dr. Neel referred to the difficulties of measuring the average amount of dominance in man and made the statement that the data for *Drosophila* leave something to be desired. This certainly was true in 1956, but there have been some more data gathered since that time. An estimate of approximately 2.5 per cent dominance for lethals in *Drosophila* is well established. How true this is for man one can only guess. We have tried seriously in the recent past to see how applicable this is to mutants with deleterious effect less than lethal, and the best evidence we have is that they must have about the same amount of dominance as lethals or perhaps a little more.

DR. STERN: I would like to make two comments. Dr. Neel pointed out that we do not really know how large a part of the spectrum of all the possible mutations at a locus is determined when mutation rates are measured in humans. Most often we measure those having striking effects, usually in heterozygous condition. How many mutations to what one might call normal iso-alleles exist? These are likely not to be fully normal, but to have some selective disadvantage without producing the strong pathologic effects studied in classical cases. The minor effects of such iso-alleles would correspond to the effect of recessives in heterozygotes which were emphasized by Dr. Crow. To take a specific example, is it not likely that many mutations at a locus which gives rise to neurofibromatosis occur to give mutants that do not produce this disease and yet have a selective disadvantage of two or three per cent? If there should be many such mutations, would not the genetic load caused by them be a very considerable one, since they would survive much longer than those with striking effects?

My second comment concerns the question of changes in sex-ratio after induction of mutations, particularly by irradiation. I agree fully with Dr. Neel's skepticism and his reference to the vagaries of the sex-ratio. In addition, I would like to raise a specific point. This has arisen from the discoveries by Jacobs, Ford, and their collaborators [242, 421, 918] and Welshons and L. B. Russell [918] who report that the *Y*-chromosome of mammals is a carrier of male determiners. Let us focus our attention on chromosomal breakage and elimination. After irradiation of a male, it was formerly thought that sperm with a damaged *X*-chromosome subsequently eliminated would give rise to a male zygote instead of a female, and that sperm with a damaged *Y*-chromosome subsequently eliminated also would give rise to males. However, we know now that *XO* individuals, in man and mouse, are not males but females. In other words, elimination of either *X*- or *Y*-chromosome in the zygote will lead to female development so that the sex-ratio would tend to be shifted toward femaleness. Of course, these considerations have to be complemented by evaluation of the effects of other genetic changes of the sex chromosomes in order to derive a full understanding of the effect on the sex ratio. I hope that Dr. Neel will comment on these new aspects of an old problem.

DR. CROW: I do not wish to imply that the earlier studies made by Dr. Stern and others on the average amount of dominance in *Drosophila* are inaccurate in the least. My point is that they are not quite relevant to the question under discussion. What we need to know is the average dominance, not of mutants at the time they occur, but mutants at the time they come to equilibrium in the population. Although one could make an indirect estimate from Dr. Stern's data at the time they were presented, I thought it better to obtain a direct measure of this.[358]

DR. NEEL: Dr. Crow and I are about as close to a meeting of the minds as we are apt to be, and I should probably close now. There are several minor points in the discussion to be examined: I think it is correct to say, as Dr. Crow does, that the indirect method is probably more valid in the case of dominant than in the case of recessive genes. However, even here it is risky, and two examples from our own experience can be used as illustrations.

You will recall the indirect method assumes population equilibrium. Now in the case of retinoblastoma, the selective forces operating against this trait are rapidly diminishing. The child with retinoblastoma today undoubtedly has a much better prospect of survival than a hundred years ago and so will tend to live to propagate and so contribute to an indirect estimate of mutation rate. In the case of multiple neurofibromatosis, this trait is not infrequently accompanied by mental deficiency and these children contribute disproportionately to the inmates of state institutions.[164] In an agrarian society where the pressure for marriage is very high, mental defect, unless very extreme, is not a barrier to marriage. But in our complex society it is becoming more and more of a barrier, and, for this reason, multiple neurofibromatisis is probably more selected against today than it was several hundred years ago in contrast to retinoblastoma, which is less selected against.

Dr. Crow says that what he is really interested in is the genomic rate rather than the locus rate, and I would agree with him that this is a very meaningful rate in which we are all interested. The fact is, however, that very frequently people are interested in rates at a locus, and, when they are, there arises the problem of "segregational" versus "mutational" contributions at the locus in question. Dr. Crow has made the point, in discussing high-grade mental defect, that the analysis of this situation suggests the load of these physical abnormalities is not maintained by balanced polymorphism. This is also true for muscular dystrophies and for deaf-mutism. As far as I am aware, nobody seriously considered that this might be the case. As I see the problem we are discussing, we are not trying to speak in terms of all-or-none mechanisms, that all of something is maintained by mutation or that all of something is maintained by balanced polymorphism. What we have to do is strive to set the proportion of each defect, if possible, which can be attributed to recurrent mutation and the proportion which can be attributed to balanced polymorphism. In demonstrating that mental deficiency is not due to the operation of balanced, polymorphic systems, a straw man has been set up and demolished. Nobody seriously suggested that this was the entire explanation. On the other hand, I would challenge Dr. Crow to demonstrate, for example, that 30 per cent is not maintained by a balanced polymorphic situation.

Finally, I agree with Dr. Stern's comments. Would that we could extend this mutational spectrum to things like the iso-alleles. Here the work must be done on an experimental animal rather than man. In the light of our expanding knowledge of what the *Y*-chromosome does, Dr. Stern's remarks have been instructive as usual.

CYTOGENETICS

C. E. Ford, D.Sc.

METHODS *in* HUMAN CYTOGENETICS

The recent progress in human cytogenetics is the natural outcome of improved technical methods. There had been no lack of pointers to the desirability of examining the chromosomes in various types of human congenital abnormality; the problem was entirely one of devising suitable techniques. Abnormalities of sex development, for instance, were obvious candidates for investigation once the role of chromosomal unbalance in some intersexual types of *Drosophila melanogaster* had been established,[558] as Haldane [314] pointed out. Speculation on the possible occurrence of autosomal abnormalities in man was also current at that time, and it in no way detracts from the discovery that mongolism is a primary trisomic condition [488] to point out that this possibility was put forward, apparently independently, by Waardenberg,[902] and Bleyer.[66] Snell's [821] finding that presumptive chromosomal unbalance in the progeny of mice heterozygous for a reciprocal translocation was frequently associated with anencephaly and embryonic death was another influence, with clear implications for human anencephaly in particular and congenital abnormalities in general. Twenty-one years after Snell's publication, the discovery that the diploid human-chromosome number was 46,[878] and not 48 as had been supposed for so long, may have re-awakened some interest in these possibilities. An independent and important incentive has come within the last decade from the utilization in clinical studies of the nuclear sexual dimorphism found by Barr and Bertram.[41] This led to the discovery that in certain abnormalities of sex development, notably Klinefelter's and Turner's syndromes, the "nuclear sex" is frequently not in accordance with the phenotypic sex, suggesting very strongly that chromosomal anomalies were involved. Finally the finding of highly individual karyotypic changes in primary leukemias of the mouse [244] suggested that a search should be made for similar

227

changes in the chromosomes of human leukemic cells.[240] It is clear that once satisfactory general methods for the accurate determination of human karyotypes had been devised, there was no lack of problems to be tackled.

SOURCES OF MATERIAL

Dividing spermatogonia and spermatocytes from the seminiferous tubules of the testis were the classical objects for the study of mammalian chromosomes, but since they are obtainable from adults of the one sex only, and since biopsy of testicular tissue for other than essential clinical purposes is undesirable, this approach cannot provide a general method in human cytogenetics. Its use in the recent development has therefore been limited.[237] Nevertheless, should testicular tissue become available, it could provide a most valuable adjunct to other material. In the report of the father of the mongol child with 46 chromosomes by Fraccaro *et al.*,[253] for example, it might have been possible to discriminate between the alternative interpretations of trisomy of chromosome 19 or 20 and effective tetrasomy of chromosome 21 by the types of multivalent association found in first spermatocytes, and to examine the possibility of gonosomic mosaicism as well. Furthermore, the characteristic chromomere sequences of individual chromosomes at pachytene [786] provide much more detail for identification purposes than can be seen by present methods in the metaphase chromosomes of somatic cells. Details of a method for making preparations from testicular tissue are given in the section on Method for Testicular Preparations.

For a somatic tissue with high mitotic frequency and reasonable accessibility, the material of choice is undoubtedly bone marrow. In the method developed at Harwell (which will be considered below), the cells seen at metaphase are in the first mitosis after biopsy. The chromosomes observed are therefore those of cells that actually existed in the body.

Tissue cultures provide a third source of dividing cells that can be used for chromosomal preparations.[375] The cells are large and mitoses are plentiful. The one small disadvantage is that an abnormal karyotype identified in a single tissue-culture line can never be referred with certainty to the patient from whom the culture was derived; notwithstanding the constancy of karyotypic features that can be maintained in some instances during continuous sub-culture,[881] the possibility of a chromosomal change occurring during culture, followed by selective proliferation of the descendants of the cell in which the change took place, cannot be excluded in the individual case.[325] It is therefore most desirable that at least two independent culture-lines should be established from separate biopsy specimens whenever possible. The first general method developed was that of Tjio and Puck,[880] using small skin biopsies as the source of living cells. More recently the scabs produced from lightly scarifying a small area of skin have been used successfully.[693] Other methods have been introduced by Ingenito *et al.*,[396]

Chu and Giles,[134] Harnden,[324] Fraccaro *et al.*,[255] and Nilsson *et al.*[623] The interval between setting up the primary cultures and obtaining preparations for study is usually at least fourteen days and often much longer.

The most recent development, and one that shows great promise, is the culture of cells from peripheral blood first utilized for chromosomal studies by Hungerford *et al.*[384] The details have now been modified and the results improved.[553] The great advantages of this method are that 10 ml. of venous blood is sufficient to set up a culture and that dividing cells are obtained in three days. Although one or two cellular divisions may intervene between removal from the body and observation, there is clearly no possibility of a clone with a variant karyotype arising and becoming an important component of the population of cells during the short period of culture. Individual variant cells might conceivably appear, but undoubtedly the karyotype of the culture as a whole would be referable to the patient from whom the blood was taken. The method has a further important advantage: the cells in any sample of blood can be presumed to be derived from all hemopoietic sites. The karyotype observed would therefore be more truly representative of the whole body than if either the tissue-culture methods previously discussed or the bone-marrow method were used.

TECHNICAL REQUIREMENTS AND PRINCIPLES

The development of methods for obtaining high-quality preparations of human chromosomes is a part of the general advance in mammalian cytogenetic techniques that have taken place since the end of the Second World War. The fundamental requirements of a good preparation are, as they always have been, a sufficiency of cell-divisions, good fixation, good staining, and good dispersion of the chromosomes. These requirements are, to a certain extent, mutually compensating; particularly sharp fixation and good staining, for instance, may allow the chromosomes to be resolved and counted accurately even though their dispersion is poor.

The fixation and staining methods that have been employed follow standard procedures fairly closely.[169] They may have been modified in part to suit particular conditions, but no essentially new principles have been introduced and they will not be discussed further.

At the present stage of development of mammalian and human cytogenetics, the positional relationship of one chromosome (or bivalent) to another in natural mitosis (or meiosis) has little significance. It follows that there is no disadvantage in making use of devices to improve the dispersion of the chromosomes within the cell. Indeed, it is the introduction of these devices that is very largely responsible for the present technical advances. There are four of them.

The squash method, originally introduced by Heitz [340] for plant material, was the first to be employed. The principle used is the fixation and staining of tissue in bulk, the cells are distributed into as near a monolayer as possible and then squashed between cover glass and slide by thumb or finger pressure. Any ancillary procedure that promotes the separation of the cells from one another is therefore advantageous. The final aim is a preparation in which the chromosomes in the mitotic cells are spread out evenly and, as far as possible, in a single plane. However, this is rarely achieved without further treatment because the chromosomes of a natural mitotic metaphase are closely congressed at the equator of the spindle; and, since the spindle itself tends to lie with its long axis in the plane of the slide, the chromosomes in most cells are squashed on top of one another.

Independently, and almost simultaneously, Hsu [375] and Hughes [378] discovered that exposure of tissue cultures to hypotonic solutions not only caused the cells to swell through osmotic uptake of water, but also inhibited the formation of the mitotic spindles as well, so that the chromosomes did not congress at metaphase but remained dispersed throughout the cytoplasm. Prefixation treatment with a hypotonic solution has therefore rightly come to be widely used in conjunction with the squash method. It should however be recorded that the hypotonic pretreatment now associated wtih Hsu's name was anticipated by Makino and Nishimura,[525] who deliberately exposed tissues of various animal species in distilled water before making aceto-carmine squash preparations with the aim of getting better dispersion of the chromosomes.

The use of colchicine and other agents capable of arresting mitosis at metaphase has held an important place in the preparative techniques used by plant cytogeneticists for a long time. It is therefore surprising that their use in techniques for mammalian chromosomes came so late.[238, 878] Only two of these agents have been widely used on mammalian material as yet, colchicine and its analogue, desacetylmethylcolchicine (Colcemid, CIBA). Colcemid is much less toxic than colchicine in equivalent concentrations, although apparently equally efficient in arresting mitosis. The action of colchicine, however, seems to persist for a longer period of time. Substances like colchicine and Colcemid are frequently referred to as spindle inhibitors because their action in bringing about the arrest of mitosis seems to be due to failure of spindle formation. The well-known effect of increasing the number of mitotic cells is by no means the only technically important consequence of failure of the spindle to form. The chromosomes are widely dispersed in the cytoplasm as already mentioned; they continue to contract, sometimes far beyond the state observed in a natural metaphase; and sister chromatids come to diverge strongly from one another, remaining attached only at the centromere. When cells are exposed to colchicine or Colcemid before immersion in a hypotonic solution, it is the action of the drug that disperses the chromosomes, but in practice the hypotonic pretreatment is still found to be useful; the chromosomes are better dispersed than if drug pretreatment alone had been used. The cells swell considerably in the hypotonic fluid, and the presumption is that the improved dispersion is simply due to lowered viscosity and increased apparent volume of the cytoplasm.   Probably the most critical part of all techniques that employ hypotonic pretreatment is to maintain the cells in this strongly hydrated state and yet achieve satisfactory fixation and staining.

The other two consequences of pretreatment with drug, increased chromosomal contraction and divergence of sister chromatids, are of value when the chromosomes are less dispersed.   Extra shortening of the chromosomes reduces the amount of overlapping, and the divergence of the chromatids aids interpretation by emphasizing the bilateral symmetry of each chromosome.   In the large cells of tissue cultures, on the other hand, where it should be possible to get the chromosomes dispersed with little or no overlapping, there is no advantage to be gained in this way and possibly some loss.   For instance, overcontraction may be followed by division of the centromeres and partial separation of the sister chromatids, with the consequent risk of mistaking a single chromatid for a whole chromosome.   The better the quality of the preparation, the less the risk of such error; but, even in the best preparations, overexposure reduces the number of mitoses suitable for detailed study.   In my experience, colchicine and Colcemid have no other adverse effect.   For instance, tumor cells and normal hemopoietic cells from mice that have received Colcemid injections can be transplanted successfully.[241]   The objection has been made that the use of colchicine may prejudice interpretation by inducing chromosomal rearrangements.   The evidence for this view is obscure; they have not been observed to occur under the conditions we have used at Harwell, and, in any case, since the agent as normally employed acts for less than a full mitotic cycle, any such effect would be a random one affecting different cells in different ways and making detection straightforward.

As an alternative to squashing, air-drying of cells on slide or cover-glass after alcoholic fixation has recently and independently been introduced by Rothfels and Siminowitch,[734] and Tjio and Puck.[880]   As the fixative evaporates and the preparation dries, each cell spreads into a thin sheet adherent to the glass, the chromosomes of the mitotic cells being dispersed in the process.   This method is now commonly employed for the large and fragile cells in tissue cultures in preference to squashing.

A METHOD FOR BONE MARROW

The procedure we have used at Harwell has already been described in outline [240] and is a direct adaptation of a method developed earlier for use with mice and other experimental animals.[238]   The original decision to convert the marrow into a suspension was prompted by the high-quality preparations obtainable from ascites tumors,[877] the supposition being that the rapid access of fixative and stain to the individual cells were important factors additional to the obvious advantages of large free cells and abundant mitoses.   The animals were injected intraperitoneally with colchicine (more recently Colcemid) at the rate of 0.1 ml. of 0.04 per cent solution per gram body weight. Between one and two hours later the animals were killed, the femurs removed, and the marrow washed from the shafts into 1.1 per cent solution of sodium citrate. The suspensions were incubated at 37° C. for 15 minutes before fixing in acetic alcohol and staining by the Feulgen method, the cells being sedimented by centrifugation to permit fluid changes. This procedure, which has been modified in detail from time to time, has given excellent preparations of bone marrow from several different mammalian species. It could not, however, be used in this form on human material, since Colcemid injection was an essential step. The difficulty was in part overcome by making use of the short-term culture method of Lajtha [473, 474] originally devised for metabolic studies. The cells from a bone-marrow biopsy are first dispersed in Ringer's solution containing heparin and then incubated in a medium consisting of approximately 4 parts human *AB* serum to 1 part of glucose saline. In Lajtha's experience, the proportion of cells in mitosis reaches its maximum at about 7 hours after incubation begins. Therefore it is the present practice to add Colcemid to the cultures after 6 hours of incubation. Thereafter the steps follow essentially as for mammalian material. Although in many instances this method has yielded preparations good enough for an unambiguous definition of the karyotype, even the best of them are disappointing in comparison with the technical quality attainable with mammalian bone marrow, both in terms of the proportion of cells in mitosis and the standard of the individual cells. Full details of the method as used at the present time are given in the section on Method for Preparations of Bone Marrow.

OBSERVATIONS

Although good observations are now much more dependent upon good preparatory technique than good microscopy, it is still important that both should receive care and attention to detail. A little extra time and trouble devoted to the handling of the material and the making of preparations may lead to an enormous saving in observational time later. The maintainance of a microscope in correct adjustment is a relatively simple task with the robust contemporary instruments. Nevertheless a few moments daily spent in routine checks may avoid an insidious deterioration in image quality.

Observations will usually begin with a count of the number of chromosomes in a series of cells. My normal practice is to reject all cells when a quick glance shows that observation will be difficult or uncertain and to count only the remainder. In the cells accepted for counting, the chromosomes will be more widely spread, or at least sufficiently well fixed and stained as to be separately resolvable without ambiguity. Of course the particular technical standard chosen for acceptance will vary to some extent with the general quality of the preparation and the importance of the material. It might be contended that by selecting in this way the final results will not be representative. This is certainly not so in the case of tissue cultures, since the cells are all members of a uniform population and differences in technical quality between individual cells can confidently be attributed to the random accidents of preparation. The complex, mixed, cellular populations from bone marrow present a different situation. The proportion of cells suitable for counting varies considerably from preparation to preparation; in some made from mammalian bone marrow, virtually all may be acceptable. However, whether few cells are rejected or many, the final result, that is, the definition of the karyotype, is unaffected in my experience.

There are several ways in which a count can be made. The method I follow is to divide the cell into a number of sectors by eye, count the chromosomes in each sector separately, check, and then add. This is no less objective, and certainly less laborious, than making a rough outline drawing with a camera lucida. It is subject to errors of two kinds: purely counting errors, in which a clearly resolved, single chromosome may be counted twice in two separate sectors or not counted at all, and errors of interpretation. By accepting, unchecked and uncorrected, all the counts obtained in this way, a high level of objectivity is attained at the cost of an excess of recorded over true variation in chromosome number.

However, it is now probably more valuable to obtain a minimum estimate of true variation. By sacrificing objectivity and setting out to determine what chromosome (or chromosomal type) is missing or extra in cells that depart from the standard number for the individual, errors of counting should be eliminated from the checked group and errors of interpretation very greatly reduced, if not eliminated also. The remaining variation on the low side would then be likely to include at least some cells that had lost chromosomes as a result of damage during the making of the preparations. Only the variation on the high side could be regarded as real. This would be a minimum figure, since true, positive variants erroneously included in the unchecked, standard-number class would be undetected. Artefactual positive errors can arise through the counting of separated chromatids as chromosomes, as already mentioned, and through the incorporation, at the time of squashing, of one or more chromosomes from a broken cell with those of an apparently intact neighbor. However, experience and good preparations should effectively eliminate these two sources of error.

Should preparations of poor quality be obtained from valuable material, there is likely to be a strong temptation to try to obtain counts and carry out an analysis of the karyotype notwithstanding the observational difficulties. Preparations of varying standard are therefore likely to be used in any extended series of observations, and it is clear that there will be continuous variation in the technical quality of the cells examined from the poorest accepted to the best that can be found. It follows that the division of recorded counts into exact and uncertain categories (as I have done myself in the past), must, in the final analysis, be arbitrary. (The practice may nevertheless be of value as an indication of the observer's confidence in his own results.) Since each count will have its own subjective weight (which only the observer himself will know), it seems inappropriate to treat all as of equal value and subject them to statistical analysis as Court Brown *et al.*[153] have done with data obtained from bone-marrow preparations. Significant differences in the frequency of variant cells may well be established between individual specimens, but there will be no measure of how much is real and how much is attributable to error.

Having obtained a sufficient series of counts, the next step is to carry out an analysis of the karyotype. In cells of good quality the chromosomes can be classified into pairs or larger groups by direct microscopic observation. Others, doubtless, will prefer to draw or sketch what they see; I find it simpler and faster to take the sectors used during counting and work around the chromosomes systematically, giving each a tentative identification (to pair or group) and recording it with a tick in a multi-cell table. After analyzing several cells in this way, it is usually clear whether any gross abnormality is present or not. Both counting and analysis are liable to subjective bias, however conscientious the observer, and it is obviously desirable to take whatever steps are possible to minimize it. It cannot be avoided completely because the knowledge of what has been seen in one cell cannot be dispelled during the examination of another; but a great deal can be done, for instance by the use of coded slides. Another method which may be possible in some circumstances is for observers who have confidence in each other to work in pairs and compare results only when they have arrived at a conclusion independently.

Photographs or drawings are necessary for illustration and as a final check on the analysis of the karyotype. I have a very strong preference for photographs on the grounds of speed and objectivity (although even in photographs the appearance of the chromosomes can be distorted by variations of exposure and times of development and uneven illumination). It is significant that virtually all recent publications in the human cytogenetics field have included photographs as illustration. A useful device that has been very widely adopted is to display a karyotype by means of an array of chromosomes in pairs. This has the great merit of bringing any peculiar feature immediately to attention. The arrays are prepared by cutting up suitable prints, trimming around the individual chromosomes, and rearranging them. Ideally, the original print and derived array should be published together. Matchings of the chromosomes can be done very largely by eye alone, although measurement of chromosome arms is sometimes helpful. Again, if two workers carry out the operation independently and their arrangements agree, the confidence to be placed in the result is greatly increased.

The degree to which an individual cell spreads out when squashed or dried is determined in part by the concentration of cells in the preparation. This can easily be adjusted when suspensions are used, and it may often be useful to make a few preparations using a low concentration of cells and heavy squashing especially for photographic purposes, accepting the danger of cell breakage and the tedium of having to scan a greater area for each cell in mitosis.

For photography, I make use of a Leica camera with the Leitz "Micas" attachment and Kodak Microfile Pan film, which gives high contrast and resolution. The principle drawback of this equipment is that dust collecting on the reducing lens in the attachment gives small out-of-focus images on the negative, and cleaning of this lens, unfortunately, is very difficult. In theory, a plate camera should give somewhat sharper enlargements at equal final magnifications, but the 35 mm. eyepiece camera has considerable advantage in economy both of time and material.

In conclusion, no great skill is required to obtain successful results, but patience, persistence, care, and attention to detail *are* necessary. Most of the procedures used at present are highly empiric, and failures not infrequently occur for no apparent reason. Perhaps the single most important factor is that the investigator should be a technician himself, taking pleasure in the purely preparative aspects of his work and not being satisfied with anything less than the highest standards attainable.

TECHNICAL DIRECTIONS *

Method for Testicular Preparations

1. Place fresh biopsy specimen into hypotonic solution. Tease out tubules with blunt needles or seekers and leave for 30 minutes at room temperature. 10 to 20 mm.³ of testicular tissue is ample.

2. Transfer tissue to a 3:1 mixture of absolute, ethyl alcohol and glacial acetic acid as fixative. Leave 1 hour.

3. Hydrate by replacing the fixative first with 30 per cent alcohol and then with distilled water. Leave in each fluid 3 to 5 minutes.

4. Hydrolyze by transferring tissue to normal HCl at 60° C. and leaving for 8 minutes (at 60° C.). Stop hydrolysis by immersion in ice water.

5. Stain in Feulgen reagent 1 hour.

6. Remove Feulgen reagent by rinsing in SO_2 water followed by chilled 45 per cent acetic acid, 3 to 5 minutes each. Transfer to fresh, chilled, 45 per cent acetic acid.

7. Preparations may now be made, or the stained tissue may be stored in 45 per cent acetic acid at −12° to −15° C. and used later.

---

* It is a pleasure to thank Dr. L. C. Lajtha for the help and advice he has given so freely, and to acknowledge the essential part his short-term bone-marrow-culture method has played in the procedure for obtaining preparations from human bone marrow developed at Harwell.

8. Preparations are made by transferring 3 or 4 short lengths of tubule (1–2 mm.) to a clean glass slide, covering with a flat, siliconed coverslip, and then squashing gently between 2 or 3 layers of filter paper. Pressure is applied evenly by both thumbs together and gradually increased, taking care not to impart any sideways motion to the coverglass. The amount of pressure to be applied varies with the specimen and can be judged by making a first, rough preparation from each batch. Best results are obtained if the cells spreading out from each segment of tubule do not come into contact. Squashing should be carried out on an evenly flat surface such as plate glass, or the slides may break. Almost every preparation should contain a few first spermatocytes in diakinesis or metaphase.

9. Make permanent by the dry-ice method.[148]

Notes: The following hypotonic solutions have been used successfully: 0.45 per cent sodium chloride, 1.1 per cent sodium citrate, and a potassium glycerophosphate solution (50 per cent potassium glycerophosphate 93.0 g., 20 per cent glycerophosphoric acid 24.9 g., water added to make up 200 ml., pH adjusted to 7.0). Experience suggests that potassium glycerophosphate is probably the best for human material closely followed by sodium citrate, but there is need for thorough, empiric testing of variations of concentration and time of exposure. Citrate pretreatment gives very good spreading of chromosomes, but renders the cells, particularly the spermatogonia, extremely fragile. Colchicine added to a final concentration of 0.02 per cent increases the chance of finding a good spermatogonial metaphase, but, under these conditions, has little effect on spermatocyte chromosomes. (Concentrations are all expressed as per cent weight/volume).

The treatment times given are not critical except for hydrolysis. Instructions for making up the Feulgen reagent and SO_2 water are given by Darlington and LaCour.[169]

It is frequently advantageous to treat the tubules with 1 per cent papain solution in water for 5 to 10 minutes immediately before squashing. The tubules should be rinsed in water before and after the papain treatment. They shrink considerably in the papain, but swell to greater than the original size after being returned to 45 per cent acetic acid, in which they should remain for about 5 minutes before being covered and squashed. The papain treatment greatly improves spreading of cells, probably by digestion of the basement membrane of the tubule.

Really flat coverslips make a big difference to the final quality of the preparations. Frequently a high proportion of the coverslips received from the manufacturers are far from optically flat. A simple and efficient test is to lay the clean covers on a firm dark surface (such as the back of a cardboard-bound record book) and observe the reflection of an overhead strip light in each one in turn. An undistorted reflection indicates flatness.

Method for Preparations of Bone Marrow

I. CULTURE METHOD (L. G. Lajtha)

1. Equipment

Culture bottles, 25 ml. cylindrical, with screw caps, chemically clean and sterilized. The caps should have cut-away centers to allow injection.

Rubber washers, to fit under screw caps, are used once only, then discarded.

Collodion, for coating washers (rubber is toxic to the cells).

Syringes, all glass.

Needles, $1\frac{1}{2}$ and 3 inches long.

2. Solutions

Heparin-Ringer's: 1 part (weight) heparin in 2×10^4 parts (volume), standard, mammalian, Ringer's solution. Sterilize by Seitz-filtration and store in culture bottles in refrigerator, 16 to 18 ml. per bottle. (1 g. heparin contains approximately 10^5 international units).

Glucose-saline: 0.6 g. glucose plus 0.7 g. sodium chloride in 100 ml. water. Also Seitz-filter and store in refrigerator, 5 ml. per bottle.

Human, AB serum: Monthly supply is obtained from transfusion center. Serum more than 28 days old is unreliable. Patient's own serum is satisfactory.

3. Procedure

Sternal biopsy: Use a large syringe and sudden forceful suction to minimize contamination by peripheral blood. Take about 1 ml. In a normal biopsy there should be about 5×10^6 erythrocytes and 2×10^5 mixed, marrow cells per mm³.

Inject contents of syringe into a culture bottle containing 16 to 18 ml. Heparin-Ringer's, which should have been allowed to assume room temperature beforehand. A long (3 inch) needle and an air vent (a short needle) should be inserted through the rubber washer ready for injection. To insure thorough mixing and to break up clumps of cells, the solution is gently withdrawn into the syringe and expelled again several times, keeping the tip of the needle below the liquid surface to avoid bubbling.

About 15 minutes later, and in any case not later than 30 minutes later, spin down for 10 to 15 minutes at 1500 r.p.m. Replace long needle and air vent. Remove supernatant. Add 2 to 3 ml. glucose-saline at room temperature and mix well by withdrawing into syringe and expelling several times. Avoid bubbling. A count is desirable at this stage and should be in the range 5 to 10×10^4 nucleated cells per mm³. Add more glucose-saline if count is greater than this.

Mix with *AB* serum (or patient's own) at room temperature to give a final concentration of between 70 per cent and 90 per cent serum. Again mix well by withdrawing suspension into syringe and expelling several times. Withdraw into syringe, remove syringe from needle, rock for 1 to 2 minutes in a horizontal position, then dispense into fresh bottles, 2 to 3 ml. each.

The last procedure insures the oxygenation of the erythrocytes. When returned to the bottle, the erythrocytes settle first and form a cushion on which the nucleated cells rest. Slow release of O_2 from the erythrocytes keeps the nucleated cells in good condition for some time.

Suspensions are probably best incubated immediately, but they may be stored cool (*not* refrigerated) and away from strong sunlight for 24 hours and possibly more. At this stage it is possible to transport them over short distances successfully.

II. CYTOLOGIC PROCEDURE

1. Place culture bottles containing bone-marrow suspension in water bath at 37° C.

2. Six hours later inject a volume equal to one-tenth of the culture volume of 0.04 per cent Colcemid solution in glucose-saline.

3. One hour later, remove bottle from water bath and spin down cells gently. Remove supernatant. Add 2 to 4 ml. 1.1 per cent sodium citrate solution. Resuspend cells by gentle shaking. Replace in water bath.

4. Fifteen minutes later, transfer suspension to a centrifuge tube and spin down gently. Remove nearly all the supernatant fluid. Resuspend the sedimented cells in the remaining supernatant fluid by tapping the bottom of the tube. Fix in acetic alcohol (3 parts absolute alcohol to 1 part glacial acetic acid, freshly prepared). Add the fixative slowly down the side of the tube, which should be tapped continuously to insure thorough mixing and to maintain the cells in suspension. (Unless great care is taken at this stage the cells are liable to clump).

5. Thirty minutes later, spin down. Resuspend cells in a small amount of supernatant fluid as in Step 4. Add 50 per cent alcohol. Leave 10 minutes.

6. As Step 5. Change to water. Leave 10 minutes.

7. Spin down. Resuspend cells. Add 1N HCl kept at room temperature. Hydrolyze in water bath at 60° C. After 4 minutes stop hydrolysis by chilling in ice water.

8. Spin down. Resuspend cells. Add Feulgen reagent. Stain 1 hour.

9. Spin down. Resuspend cells. Add chilled, 45 per cent, acetic acid. Repeat this step to wash out Feulgen reagent.

10. To make squash preparations, tap the tube to bring the sedimented cells into even suspension again. Take a drop in a capillary pipette, transfer to a clean glass slide, and place a cover glass on the drop immediately. (The cells otherwise rapidly settle out on the slide in a dense ring). The drop of suspension should be just sufficient to spread to the edge of the cover glass. Gentle warming over a spirit flame at this stage may promote the spreading of the

chromosomes and also help to seal the cover to the slide at the edges. Next, squash between layers of filter paper on a flat surface, for example, a small sheet of plate glass. The whole of one's weight can be used, but it should be applied gradually and released gradually. If the cover moves laterally during the squashing, the preparation will be ruined. The better the suspension and the more hydrated the cells, the more likely it is that the cover will move.

11. Make preparations permanent by the dry-ice method.[148]

Notes: (1) The chances of obtaining successful preparations are greatly reduced if the ratio of erythrocytes to nucleated cells in the culture is 100 or greater and if the concentration of nucleated cells is less than 10^3 per mm^3.

(2) The sodium citrate used is $Na_3C_6H_5O_7 \cdot 2H_2O$.

DISCUSSION

DR. PATAU: Dr. Ford has given a very lucid exposition of a scientific field to which he has contributed so greatly. In particular, the work by him and his co-workers in 1958 [240] on the identification of human chromosomes came remarkably close to what can be said today. I have very little to add to his remarks concerning technique, which I could confirm almost point by point. We were more fortunate than Dr. Ford in that the clinician of our group, Dr. David Smith, has no objection to injecting patients with Colcemid five hours before aspiration of the bone marrow, which then needs only a treatment with a hypotonic medium to be fixed. As Dr. Ford rightly assumed, this method generally gives better results than short-term cultures. However, it does not always succeed; sometimes even an apparently satisfactory specimen of bone marrow will not yield usable mitoses. I should add that the use of Colcemid may involve a slight risk of transient ill-effects. In justification, it should be mentioned that we have applied it only to mentally retarded and otherwise handicapped infants. We feel that the conceivable risk is balanced by the possible benefit to the parents who, if trisomy should be found, can be told that the condition is due to an accident of a kind that may happen to anyone.

The point I would like to enlarge upon is the question of how many chromosomes can be identified in individual cells. Two papers [134, 881] have appeared in which it is claimed that every single chromosome can be identified on the basis of the measured lengths of the arms. The method consists in pairing similar chromosomes that then are presumed to be homologous. I tried this method ten years ago in an attempt to identify plant chromosomes. It turned out that this use of measurements of length is misleading except in those cases in which a pair of homologues is so distinct from all other chromosomes that no measurements are needed to establish homology. It appeared that many of the idiograms with which the cytologic literature abounds are probably worthless.

The fallacy of the pairing-off method is most readily seen when we consider a chromosome that is characterized only by its total length. It must be stressed, however, that essentially the same argument applies also in the case of chromosomes of which the lengths of both arms have been determined. At first glance, the pairing-off method is attractive because it is reproducible. If two competent cytologists measure the same set of chromosomes, they will usually identify the same chromosomes as being the two longest, the two next longest, and so forth. Unfortunately, this kind of reproducibility is wholly irrelevant to the question of whether the pairs thus obtained represent pairs of homologues. To claim that they do implies that the variation in length between homologues is small relative to the differences in true length between nonhomologous chromosomes. All too often the opposite is the case.

Fortunately, there are data which yield estimates of the variation found between homologous chromosomes in favorable material. In less favorable cases, this variation is likely to be still greater. The first set of data was taken from the paper by Rothfels and Siminovitch [735] on the chromosomes of the rhesus monkey, the second set from a recent publication by Levan and Hsu [494] on the chromosomes of man. In either case, the authors have published the measured lengths of ten pairs of the longest chromosomes, which can be identified with certainty. These values yield in each case a sum of squares with ten degrees of freedom. The resulting coefficient of variation happens to be 5.8 per cent in both cases. I suspect that this value approaches the minimum of the variation in the relative length of chromosomes obtainable. This variation goes much beyond the magnitude of errors to be expected in careful measurements on chromosomes that lie flat and are not too small. This variation of 5.8 per cent no doubt represents a composite of variations in length due to different degrees of chromosomal contraction *in vivo* and variation caused by the processes of fixation and flattening during preparation.

The following model will serve to illustrate the absurdities to which the pairing-off method can lead. It is assumed that six chromosomes of identical length (left side of figure 10) are present and that these chromosomes have a coefficient of variation of 5.8 per cent, the value actually found in favorable material. By means of a table of the normal distribution and a table of random numbers, five random samples of six "lengths" each are obtained. These values represent the equivalent of what might have been obtained by actual measurement of six chromosomes in each of five nuclei. If now the pairing-off method is applied to these five sets of "chromosomes," one pair of longest, one pair of medium-sized, and one pair of shortest chromosomes will be found in each nucleus. Finally, the mean lengths of the five pairs of longest chromosomes, those of the five pairs of medium-sized ones, and so on, are obtained. These average lengths are shown in the form of an idiogram on the right side of figure 10. It will be noticed that the differences in length between these three spurious pairs of "homologues" are larger than many of those one finds in published idiograms.   One can hardly avoid the conclusion that quite a few of the results of the pairing-off method incorporated in such idiograms are equally spurious.

That the objections to the pairing-off method remain valid when the lengths of two arms rather than a single length are taken into account has already been stressed.   It has recently become the fashion to transform lengths of the two arms into the length of the whole chromosome and the arm-ratio, namely the length

Fig. 10.   MODEL ILLUSTRATING THE FALLACY OF THE PAIRING-OFF METHOD.

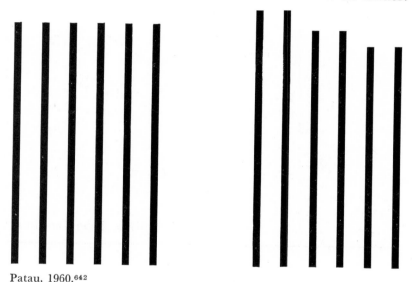

Patau, 1960.[642]

of the long arm divided by that of the short arm.   To this I have no objection, but one should realize that the transformation of one pair of numbers into another pair of numbers does not aid the identification of homologues.   As probably almost everybody does, I find it difficult to handle a set of pairs of numbers.   How does one decide which out of several pairs is most similar to a given pair of numbers?   This difficulty disappears at once when we remember that a pair of numbers is the equivalent of a point in a two-dimentional coördinate system.

If the relative length of the long arm of each chromosome is used as abscissa and that of the short one as ordinate, every measured chromosome may be represented by one point in a "karyogram."   Obviously the identification of two chromosomes as homologous on the basis of their arm lengths alone can be taken seriously only if the two corresponding points in the karyogram are lying closely together in fair isolation from all other points.   We should further require that, in karyograms of other nuclei, an isolated pair is seen consistently at about the same location. Any pairing-off of chromosomes that do not form consistent pairs in karyograms is subject to the same criticism that has been directed against the pairing-off of chromosomes characterized by only one length. Of course, all this would not apply if two chromosomes should possess the same marker, such as a secondary constriction. It should be stressed that a marker manifesting itself only at times, as do the satellites in relation to certain acrocentric chromosomes of man,

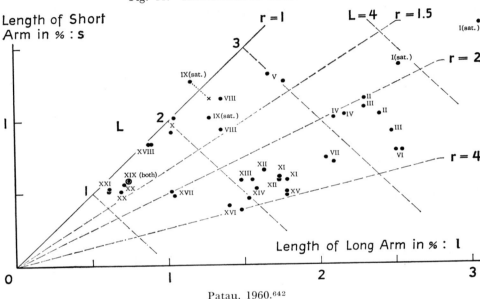

Fig. 11.   KARYOGRAM OF RHESUS MONKEY.

Patau, 1960.[642]

may not be a very sound basis for the identification of homologous chromosomes. This is particularly true when there is a possibility that other chromosomes possess a similar marker. However, only the case of chromosomes for which only the measured lengths of their arms are available will be considered for the moment.

Measurements of the chromosomes of one nucleus of the rhesus monkey as published by Rothfels and Siminovitch [735] are recorded in figure 11. These authors did make reservations in the case of some chromosomes, but by and large they considered the identification of pairs of homologues given by the imprinted numbers as unequivocal. For some of the pairs, such as the two pairs of satellite chromosomes and the chromosome pairs no. V and no. VIII, this is undoubtedly true. However, it becomes equally clear that the authors have carried the pairing-off beyond the limits of biologic meaning if the variation within these pairs is taken into account.

Fig. 12.   KARYOGRAMS OF HUMAN MALES.

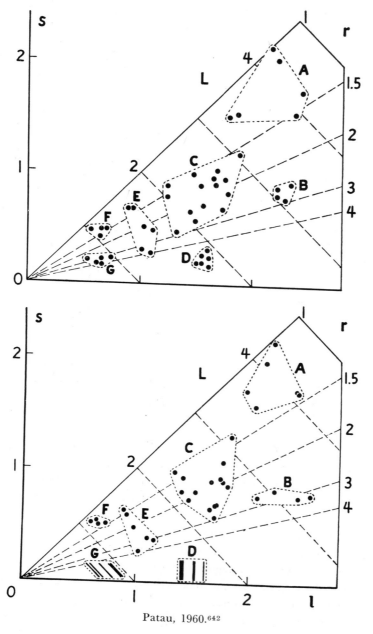

Patau, 1960.[642]

The cytologic situation in man is by no means better than that in the rhesus monkey. The next illustration (figure 12) shows the karyograms of two normal complements of the human male. In both cases, and indeed generally, seven groups can readily be distinguished. They are in fact the same groups upon which the participants of the Denver meeting have agreed (Appendix II). For the sake of simplicity, we use the symbols *A* through *G* for these groups.

Each of the three pairs of *A* chromosomes can easily be identified under the microscope. The corresponding pairs of points in karyograms are usually well separated. This is not true for the points representing the two pairs of *B* chromosomes. Superimposed karyograms of these chromosomes from ten nuclei display such a degree of overlapping of the distributions of the two pairs that a reliable distinction of the latter by means of length measurements is impossible. However, it is probably possible to distinguish these two pairs in favorable plates directly under the microscope. If all *B* chromosomes happen to be about the same length and two of them show somewhat thicker chromatids than the other two, it may be concluded that the former are somewhat more contracted and therefore are somewhat longer than the latter. This illustrates the point that the human eye often can draw valid distinctions between chromosomes that do not differ enough in the true length of their arms to make possible their reliable distinction by length measurements alone. This applies especially to smaller chromosomes.

The *C* group contains seven pairs of autosomes and one or two *X* chromosomes, depending on the sex. In my opinion, none of these chromosomes can be identified in individual cells with any assurance and definitely not by length measurements. However, it ought to be possible to compute the characteristics of the *X* chromosome from measurements of the *C* chromosomes in a sufficient number of male and female nuclei. I have described elsewhere [642] a statistical method appropriate for this purpose and have applied this method to data from six male and four female nuclei in order to obtain a first estimate of these characteristics. The result, a total length equal to 4.8 per cent of the haploid female complement and an arm-ratio of 1.5, may be quite inaccurate because of large sampling errors. Yet I believe that it is the only unbiased estimate available to date.

If it were possible to identify all the eight different chromosomes of the *C* group by measurement or otherwise and if the two relative arm-lengths of each of these chromosomes were determined in a number, *n,* of nuclei, the two mean arm-lengths of any given chromosome should be fairly accurate estimates of the true lengths as each mean would be based on 2*n* values. If this were done by several authors independently and the mean lengths then were plotted as karyograms (which would contain only a haploid number of points), the points of the *C* group, no less than the points of, say, the *A* group, should form similar patterns in the karyograms. Differences in technique between various authors might cause shifts in the karyograms, but it should not noticeably alter the pattern of a group.

Figure 13 shows three such karyograms based on the mean characteristics as published by three groups of authors [134, 494, 881] who attempted, with confidence or, in the case of Levan and Hsu, with grave doubts, to identify every chromosome. (The Roman numerals are the symbols of chromosomal groups as proposed by the

Fig. 13. KARYOGRAMS OF HAPLOID HUMAN COMPLEMENT.

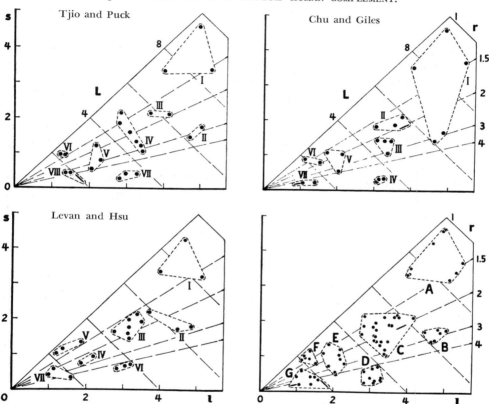

Based on mean characteristics of chromosomes as given by Tjio and Puck,[881] Chu and Giles,[134] and Levan and Hsu.[494] Bottom, right: same, superimposed.

respective authors.) It will be seen that the three point patterns A agree well indeed, but that this is far from true in the case of the C group. The explanation, I believe, is obvious. What these points represent are not the average characteristics of samples containing only homologous chromosomes, but rather the result of a biologically meaningless pairing-off in a two-dimensional set of points. As this kind of pairing-off can be done in any number of different ways, it is hardly surprising that different authors arrived at different patterns of points. These three karyograms may be superimposed in the hope that the various kinds of bias the three groups of authors introduced by their individual methods of pairing-off will more or less cancel each other. Indeed, this seems to be the case, as the diagram in the right lower corner of the figure shows the *C* group as a pattern similar to that obtained by superimposing *C* groups from a number of karyograms of individual nuclei. The seven groups, *A* through *G*, stand out clearly.

I am glad to note that the "standard system of nomenclature of human mitotic chromosomes" as proposed by the Denver Study Group agrees with the one Dr. Therman and I have been using, not only in the recognition of seven groups, but also in the numbering system of chromosomes at least to the extent that we believe chromosomes can be identified in individual nuclei. Considering the many who are not cytologists now taking an interest in human chromosomes it would be well if published karyotypes gave a more realistic view of the limitations of chromosomal analysis than do the karyotypes in which every chromosome is given a number even though the author himself knows, and states in the text, that he cannot identify it with assurance. I suggest that this practice be discontinued and that chromosomes of doubtful identity be labeled with a group symbol rather than with a specific number.

Figure 14 is an example of male karyotype labeled realistically. Chromosomes number 1 to 3 and, with some reservation, 4 and 5 can be distinguished. As stated previously, I do not believe that this can be done with those of the *C* group. I also do not believe that the members of the three pairs of *D* chromosomes can be identified, although it is usually easy to select two *D* chromosomes that are definitely not homologous. There is some doubt in my mind whether the last word about the satellites of *D* chromosomes has already been spoken; at any rate, it is an exceptional cell in which more than two satellites can be seen. Each of the chromosomes number 16 through 18 can be identified and possibly so can number 19 and 20, the former being slightly larger.

In the present karyotype, group *G* is not particularly well represented, although I am reasonably certain that the last chromosome is the *Y*. I agree with Dr. Ford that the latter is the largest *G* chromosome. However, I disagree with him and others about the wisdom of making the satellites part of the definition of autosome number 21. In favorable plates, two of the four autosomes of the *G* group appear definitely larger than the other two. Several authors have stated, as did Dr. Ford today, that the larger *G* chromosomes carry a satellite. Among these authors are Chu and Giles,[134] although in their own figure 8 the two *G* chromosomes with satellites appear to be, if anything, smaller than the one without. (The fourth *G* chromosome is partially hidden.) On the other hand, Dr. Therman and I have seen distinct satellites repeatedly at the smaller *G* chromosomes. So have Levan and Hsu [494] whose drawings of the *G* group agree well with our own findings. Perhaps we should not yet rule out the possibility that both pairs of *G* chromosomes may have satellites. At any rate, in view of the present disagreements, it would seem best to define chromosome number 21 by only one criterion, namely the least ambiguous one which is its slightly larger size as compared with number 22.

I have nothing but admiration for the cytologic technique of the authors I

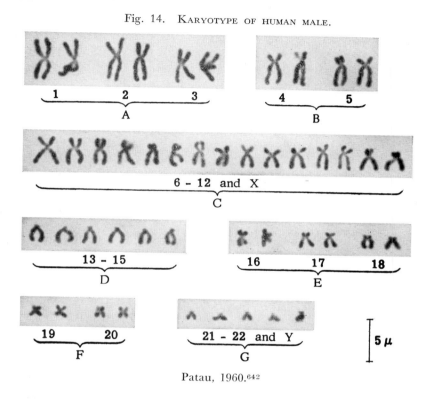

Fig. 14. KARYOTYPE OF HUMAN MALE.

Patau, 1960.[642]

have mentioned, and I take for granted the accuracy of their measurements. My critical remarks concern exclusively their interpretation of the observed facts. We should not forget that a normal chromosome number of 48 was accepted for many years as well established because a consensus had put to rest the original uneasiness about the conclusiveness of the available evidence.

DR. TJIO: In analyzing metaphase plates, we consider not only length but also take into account the position of the centromere and total length of the chromosome, and the arm-ratio is used to locate the position of the centromere. In this way the majority of the chromosomes may be distinguished. In very ideal plates the X chromosome can be distinguished from group 6–12. As for the data we have published, measurements are not claimed to be absolute but are approximations. It is not surprising to find variations in appearance of chromosomes from cell to cell depending on the physiologic condition of the cell and the cycle of mitosis at the time of fixation. The fact that our results are reproducible suggests that our analyses are not so erroneous as Dr. Patau implies.

We have found that not only the longest chromosome of group 5 has a sat-

Fig. 15.   PHOTOMICROGRAPH OF A HYMAN MALE CELL IN COLCHICINE METAPHASE.

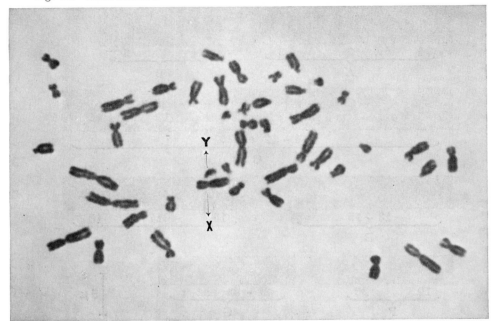

Grown *in vitro;* $2n = 46$/XY. The biopsy was taken from the skin of an adult. The cells were grown as a monolayer on coverslips or slides, pretreated with colchicine and hypotonic saline, fixed in acetic-alcohol-formalin fixative, air-dried, stained with acetic orcein, and mounted in permount.

ellite, but also the smallest has one as well. The six satellites have been seen in a patient with the Marfan syndrome, in another with galactosemia, and also in one of the testicular feminization syndromes.

The accompanying photomicrographs shown in figures 15 through 18 demonstrate analyses of the human karyotype carried out on preparations that have been made with our method. It is necessary to exercise the utmost control over the procedure in order to obtain results that will permit meaningful analysis.

DR. PATAU: I have no question about the quality of the slides of these authors.

DR. PUCK: Dr. Patau's discussion demonstrates many of the difficulties attending chromosomal analysis. While various types of irreproducibility can make such analysis questionable, I believe the situation is not as hopeless as his discussion implies. Dr. Patau's analysis has compared measurements he has made from

Fig. 16. THE KARYOTYPE OF A HUMAN MALE.

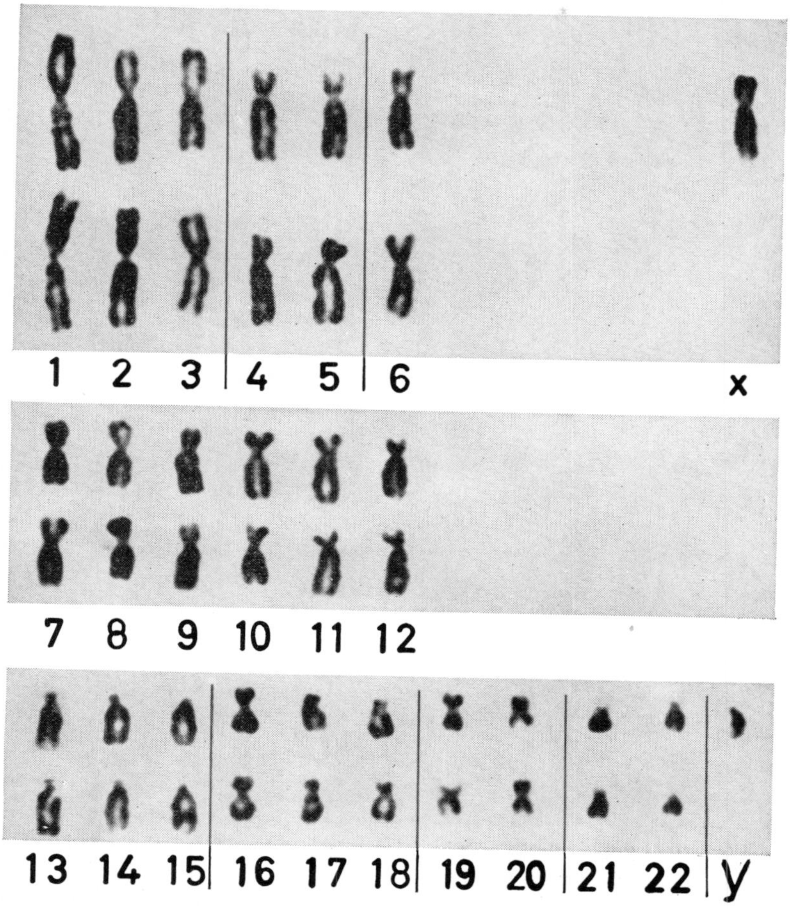

The chromosomes are from an enlarged photomicrographic pint (original enlargement 4,000×) of the cell shown in Figure 15, cut singly, paired, and arranged according to the standard system of nomenclature as proposed at the Denver meeting, April, 1960 (see Appendix II).

published photographs of several investigators as these have appeared in the journals. Such printed pictures are almost always reduced far beyond the size of the original photomicrograph and have been reproduced by a printing process that increases markedly the penumbra around each object photographed. For these as well as for the reasons listed below, measurements obtained in this way cannot be used as a reliable index for comparison of the results by different authors.

Fig. 17.   PHOTOMICROGRAPH OF A HUMAN FEMALE CELL IN METAPHASE COLCHICINE.

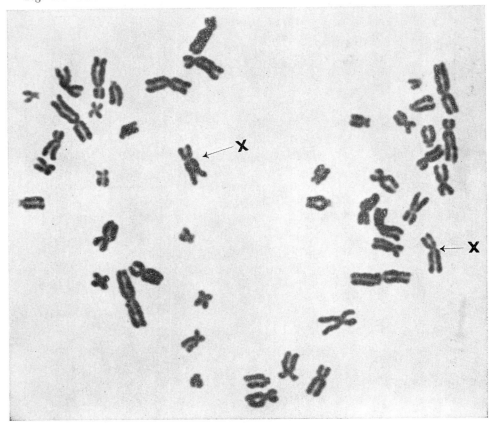

Grown *in vitro;* $2n = 46/XX$. The biopsy was taken from the cervix of an adult. The cells were grown as a monolayer on coverslips or slides, pretreated with colchicine and hypotonic saline, fixed in acetic-alcohol-formalin fixative, air-dried, stained with acetic orcein, and mounted in permount.

In analyzing chromosomal characteristics, the following factors are of greatest importance:

1. Treatment with colchicine is very useful, but results vary with the technique employed. Colchicine causes contraction of the chromosomes to an extent which depends upon the time during which the drug has been present and its concentration. Apparently, the degree of contraction of all the chromosomes is

not constant. Therefore, different investigators who utilize different treatments with colchicine (or no treatment with colchicine) may obtain different chromosomal indices. In this connection, it should be emphasized that the figures ob-

Fig. 18.   THE KARYOTYPE OF A HUMAN FEMALE.

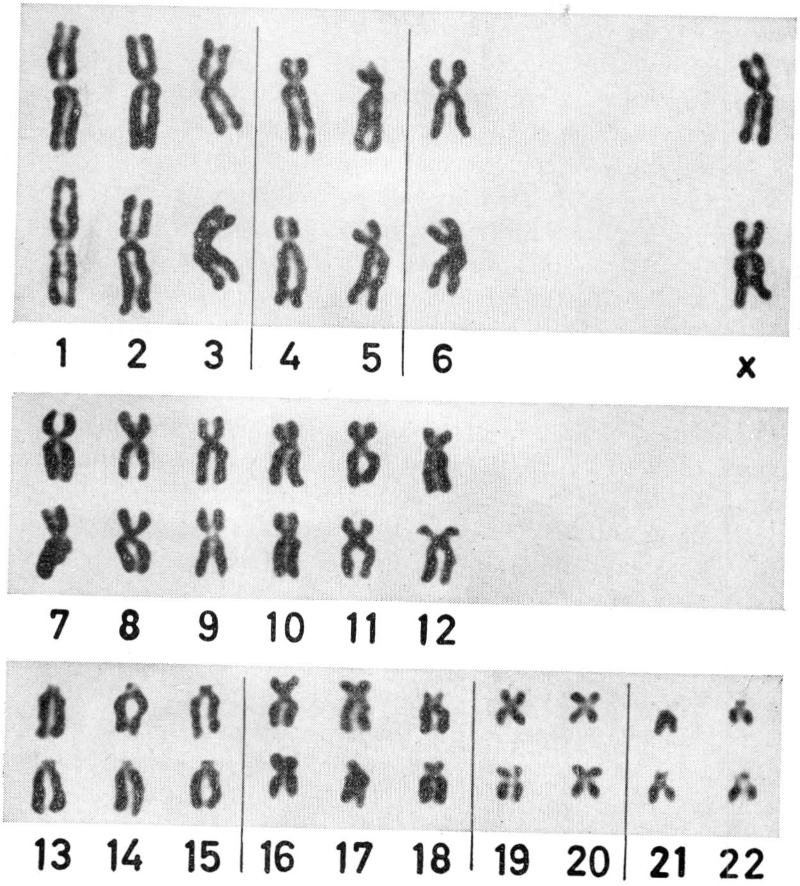

The chromosomes are from an enlarged photomicrographic print (original enlargement 4,000×) of the cell shown in Figure 17, cut singly, paired, and arranged according to the standard system of nomenclature as proposed at the Denver Meeting, April, 1960 (see Appendix II).

tained by a given experimenter are not necessarily more or less correct than those obtained in another laboratory where a slightly different technique is employed. It is necessary, however, to compare only measurements obtained from chromosomal preparations in which exactly the same procedure has been followed.

Sasaki [759a] has recently documented the effect of different colchicine treatments in changing the chromosomal indices. In our laboratory, we have adopted the practice of using a higher concentration of colchicine for chromosomal measurements and a lower one for examination of structures like the chromosomal satellites which are clearer in late prophase or early metaphase. Unless these conditions are carefully controlled, the degree of dispersion obtained in the chromosomal measurements will be very large indeed.

2. It is also necessary to control the growth medium of the cells. Often in the presence of toxic media, chromosomes may tend to become "sticky," and some chromosomes may become partially despiralized and elongated. Such mitotic irregularities are easily recognized, and cells displaying them must be rejected. The cells of a patient with Treacher-Collins syndrome, which we have studied, appear to be much more sensitive to such disturbances of the culture medium and readily display abnormal degrees of contraction of the longest chromosomes. It is obvious that the kind of analysis here described can only be carried out when a sufficiently large number of mitotic figures is available so that, after eliminating all the faulty ones, there is still a good number on which to carry out measurements.

3. The method utilized in obtaining a planar configuration of undistorted chromosomes can also be important. We have the impression that squash techniques are less reliable than methods that utilize air drying of cells grown as a monolayer on glass and fixed *in situ*. This requires rigorous control of the conditions of cellular expansion, fixation, drying, and staining. The chromosomal pictures shown by Dr. Tjio demonstrate the remarkably high quality that can be obtained if sufficient care is employed. The degree to which the different chromosomes can be discriminated depends directly on the quality of these preparations.

4. The photography from which the measurements are taken is equally demanding. Unless great care is taken, some chromosomes may become more heavily exposed than others on the photographic emulsion, with a resulting spread in the area of darkening. Parts of the chromosome may also be twisted out of the plane of focus, and result in a photograph for which the apparent measurements are in error. No chromosomal photograph should ever be analyzed unless the original preparation is examined simultaneously in the microscope in order to monitor the measurements taken from the photograph.

5. The problems of actual measurement of the arm lengths of the chromosomes are not without their complications. The chromosomal arms are rarely straight, and special techniques are necessary for measurement of these curvilinear lengths. It is not sufficient to divide a chromosome arm into segments that appear to be reasonably straight and to add the sum of all such lengths, since this usually will introduce a total error to which all the individual measurements

contribute. Special methods have been developed, and described by us elsewhere, for carrying out such lengths with reasonable accuracy.

Even when all the precautions outlined here are taken, one still obtains a distribution of chromosomal-arm lengths. We have indicated in our publications the standard deviations of such measurements. While increasing the care with which the preparations are made narrows the distribution, it is still necessary to examine large numbers of very good mitoses in order to obtain enough individual measurements to be reasonably sure of the chromosomal characterization. It would be impossible for two laboratories working independently to obtain virtually identical chromosomal indices, as has been the case with the work of Drs. Chu and Giles, on the one hand, and our own laboratory, on the other, if there did not exist a reproducible set of chromosomal characteristics corresponding to these indices. While these two laboratories carried out their work without knowledge of each other's methods, the procedure adopted in each case actually was very similar, even to the methodology for colchicine treatment. There are, of course, some small differences in these two sets of measurements. One of these involved the length of the chromosomes with satellites of the large acrocentric group, 13–15. Consequently, it was very gratifying for us to discover on reëxamination that not one, but at least two, pairs of the chromosomes of this group have satellites, so that even this small discrepancy has been resolved.

Much remains to be done in characterizing the human chromosomes. Doubtless it will be much better to measure the total *DNA* content of each chromosomal arm, rather than simply its length. Studies along these lines are now in progress. In the meantime, measurement appears to be a useful tool for chromosomal identification when carried out with all the precautions described here.

DR. BURDETTE: Thank you, Dr. Puck. Dr. Ford, would you close, please?

DR. FORD: I think it is an excellent thing to have some healthy criticism from time to time, and I prepared to make some sort of reply to Dr. Patau, but Dr. Puck has now done it for me. I also have considerable faith in the eye as an integrating as well as an analyzing instrument, and our own karyotypes are always prepared by means of visual matching. Dr. Patau himself referred to the small differences in density along the chromosomal arm, which may indicate where it is twisted on a broader gyre with slight shortening of the chromatid. Despite Dr. Patau's criticism, I believe that chromosomes 7 to 12 can be identified. I should be very interested indeed to measure the chromosomes in some of our karyotypes and set out the data on this very simple but ingenious system of Dr. Patau's. Dr. Burdette has suggested that there is considerable interest in the present tally of abnormal chromosomal types in man. This has been prepared and is available in summary form (Appendix III).

DR. PATAU: Dr. Ford mentioned the case (Appendix III) published by Edwards and colleagues,[199] of a syndrome caused by trisomy of chromosome 17.

In the introduction of a paper [644] that appeared in the same issue of *Lancet,* my co-workers and I mentioned two cases of a syndrome caused by the presence in triplicate of a chromosome of which we could at that time only say that it belonged to the *E* group. A detailed description of these two cases is in press. Since then, four additional cases of the same syndrome have been discovered. In each case, it could at least be ascertained that the extra chromosome belongs to the *E* group. In one case, the material was adequate to permit a tentative identification of this as chromosome 18. In still another case, this identification was definite. On clinical grounds it appears almost certain that our six cases and that reported by Edwards and colleagues represent the same syndrome. As to the true identity of the responsible chromosome, it should be said that a distinction between 17 and 18 is difficult even with good material. Furthermore, there is the possibility to be kept in mind that a minute deficiency or a small inversion has impaired the true identity of this chromosome in one of the cases. Clinically the syndrome appears to be so well defined that the exact identification of the extra chromosome does not at present appear to be particularly important. We can leave the final decision to the future as there can be no doubt that more cases of this syndrome will be found.

In view of the possibility of encountering an extra chromosome that has come about by translocation, trisomy for a particular chromosome should be considered as definitely established only after at least two independent cases have been found which show not only microscopically the same extra chromosome, but also clinically a similar pattern of anomalies. By this criterion, only three syndromes of autosomal trisomy can be considered as established to date. One, of course, is mongolism. The findings by our group, consisting, besides me, of Dr. Therman on the cytological side and Dr. David Smith on the clinical, concerning the remaining two trisomic syndromes are summarized in figure 19.

You will notice that of the eight mothers involved, no less than four had an age at conception of 37 or more years, suggesting, as does the similar situation in mongolism, that the incidence of nondisjunction increases sharply with the maternal age. The listed anomalies are by no means all that have been observed. They are merely those that were found in every one of these cases, unless stated otherwise. It should be added that the two new syndromes of trisomy are rarer than mongolism, by far. The explanation might be that nondisjunction in the larger chromosomes is very much rarer than in the small chromosome 21. However, prenatal mortality is also a possibility, the more so as *E* trisomy acts almost as a sublethal. No less than four of the six *E* trisomics have already died in early infancy. The two *D* trisomics seem to be more viable. Both are still alive more than one year after birth.

DR. BURDETTE: A written communication from Dr. Hauschka concerning the identification of *X* and *Y* human chromosomes and containing his views about nomenclature has been received.

Fig. 19.   Autosomal trisomy syndromes.

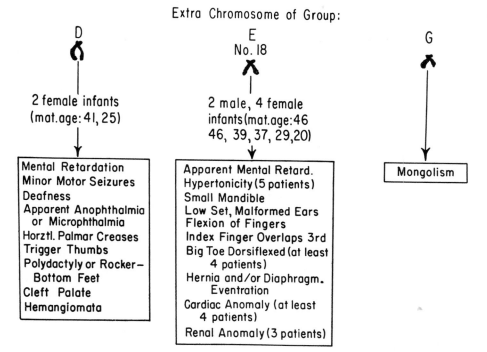

47 chromosomes, one being present in triplicate.

Extra Chromosome of Group:

D

2 female infants
(mat.age: 41, 25)

Mental Retardation
Minor Motor Seizures
Deafness
Apparent Anophthalmia
or Microphthalmia
Horztl. Palmar Creases
Trigger Thumbs
Polydactyly or Rocker−
Bottom Feet
Cleft Palate
Hemangiomata

E
No. 18

2 male, 4 female
infants (mat.age:46
46, 39, 37, 29, 20)

Apparent Mental Retard.
Hypertonicity (5 patients)
Small Mandible
Low Set, Malformed Ears
Flexion of Fingers
Index Finger Overlaps 3rd
Big Toe Dorsiflexed (at least
4 patients)
Hernia and/or Diaphragm.
Eventration
Cardiac Anomaly (at least
4 patients)
Renal Anomaly (3 patients)

G

Mongolism

Dr. Hauschka: Identification of X and Y Chromosomes of Man

Only four years after Tjio and Levan [878] established 46 as the correct diploid chromosomal number of man, detailed knowledge of the human karyotype has progressed far toward identification of individual chromosomes. Most cytologists with experience in this endeavor would agree that, at present, few pairs of homologues can be matched with absolute assurance even in the best-spread metaphases. However, anyone with a discerning eye can arrange the chromosomes in descending order of length and place them in seven distinct morphologic groups according to size and centromeric position. This can be done without the aid of a micrometer.

At the present stage of human karyology, agreement on a uniform chromosomal nomenclature is desirable. It is reassuring that the conspectus,[76] signed by Böök and sixteen other participants (Appendix II), arrived at the same grouping scheme as Patau [642] who, independently, computed the human mitotic karyogram from available measurements. In figure 20, the chromosomes of a normal male have been labeled with letters corresponding to Patau's 7 groups *A* to *G,* and also to the Denver numbers. Under the Denver system, a supernumerary, medium-sized, metacentric unit would be classified as a chromosome of uncertain homology belonging in "group 6–12 and *X*." In the combined Denver-

Fig. 20. CHROMOSOMES OF A NORMAL MALE.

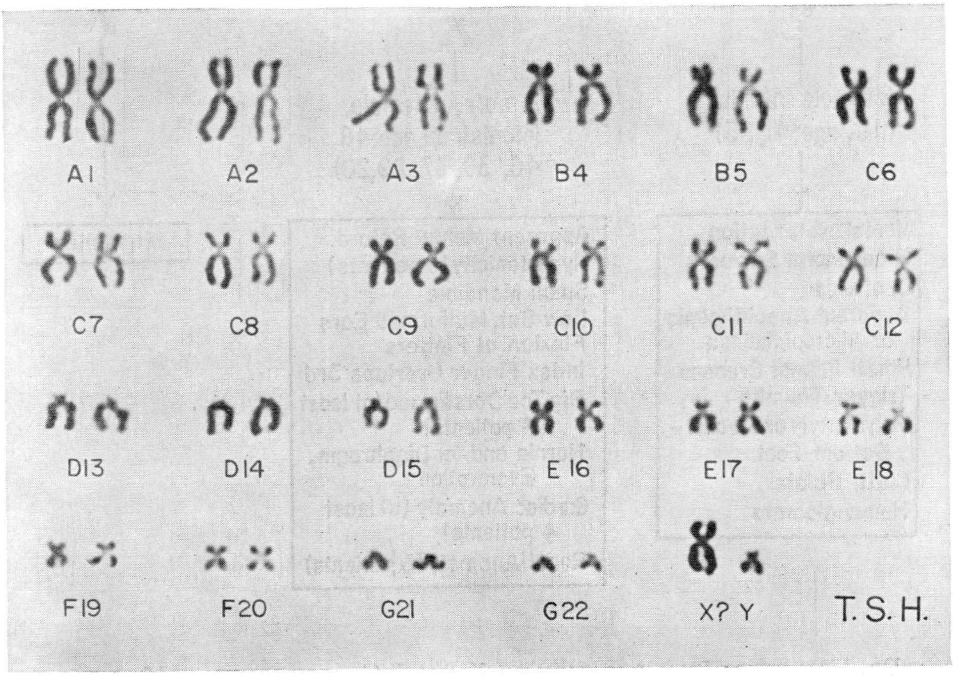

Patau nomenclature this same chromosome can be referred to more simply as a unit resembling *C*8.

Reciprocal translocations that have been established beyond reasonable doubt, as in several recent studies of familial mongolism,[117] can be written as follows: *D*15/*G*21. Abnormal chromosomes, found in many neoplasms and long-term tissue cultures need not be named by "arbitrary symbols, prefixed by a designation of the laboratory of origin."[76] It would usually suffice to designate the abnormality as resembling, but not necessarily identical with, a certain standard unit, *A*3 for example. Where this is not possible, the subversive chromosome can be tagged as an "extra long" or "minute" metacentric, submetacentric, or acrocentric not conforming to any of the seven groups, *A* to *G*. Thus, combination of the Denver numbers and Patau grouping constitutes a flexible nomenclature which could accommodate cytogenetic anomalies of all types.

Table 43

SIZE-RANK (1–24) AND ARM-RATIOS OF X AND Y IN THE HUMAN MITOTIC KARYOTYPE

PARAMETER	Tjio and Puck 1958 [880, 881]	Chu and Giles 1959 [134]	Levan and Hsu 1959 [494]	Lejeune and Turpin [1]	Fraccaro and Lindsten [1]	Buckton et al.[1]	Hauschka and Sandberg [1]	Makino and Sasaki [1]	Patau (computed) 1960 [642]	RANGE
SIZE- } X	6	6	7	6	6, 7	7	7	7	7	6–7
RANK } Y	22	24	22	22	22	23, 24	22	21, 22	22	21–24
ARM- } X	1.9	2.0 [2]	1.6	2.2	1.6	1.7	1.7	2.1	1.5	1.5–2.2
RATIO } Y	∞	∞	4.9	∞	2.9	4.9	⌣3	3.1	—[3]	∞ –4.9

[1] Unpublished data or now in press.

[2] The value of 2.8 which appears in the 1959 paper and reproduced in the Denver conspectus [76] is a misprint (Chu, personal communication).

[3] Computation of arm-ratio for the small Y chromosome would tend to yield a rather subjective value of little quantitative significance.

Identification of the sex chromosomes poses a special problem. There is justified skepticism concerning the recognition of the X among similar metacentrics in group C. Available data on relative size and arm-ratios of X and Y are summarized in table 43. The variability of the somewhat allocyclic X in relation to the autosomes handicaps any effort to pinpoint the X in a karyogram by statistical treatment of relative size or arm-ratios, as evident from Patau's [642] thorough analysis; but one need not conclude that the rank of the X chromosome within group C remains entirely conjectural. Although Levan and Hsu [494] have interpreted the second-largest chromosome of group C (equivalent to the largest of their group III) as the X, they added a note of caution: ". . . in our opinion the identification of the X in somatic metaphases cannot be considered definite as yet, for any of the other chromosomes of group III is just as eligible." In the light of pooled knowledge, it now seems reasonable to eliminate certain group-C chromosomes from the X candidacy. The sizes and/or arm-ratios of several C chromosomes are too unlike the rather consistent values established for the X in different human populations by competent cytologists (table 43).

Referring now to figure 10, $C9$ and $C10$ may be eliminated because their arm-ratios do not fit the X; $C11$ and $C12$ are too short. This narrows the choice to $C6$, $C7$, and $C8$. The chromosome marked X? in the last row ranks seventh in length and closely resembles both $C6$ and $C7$. Hence, one's personal choice for the X in a given karyotype is best labeled with a question mark, as was done by Ford et al.,[242] until we have more precise criteria for identification. The majority of observers has interpreted the seventh largest chromosome with a mean arm-ratio of about 1.8 as the X.

The Y element is more easily recognized than the X, although there is considerable disagreement about arm-ratios. Calculating arm-ratios for the tiny, acrocentric Y, usually a little larger than the two smallest autosomes $G21$ and $G22$, involves rather subjective measurements. A "quantitative" parameter ranging from infinity to 4.9 (table 43, last line) is meaningless for cytologic identification, but may be indicative of Y-chromosomal polymorphism in the human population. Patau [642] states: "The Y does not stand out equally well in the mitoses of all males." He proposes, on genetical grounds, that Y chromosomes with small deficiencies may persist in the human karyotypic pool since they would have little selective disadvantage.

Another possible reason for the apparent polymorphism of the Y is its relatively high content of heterochromatin. This tends to affect its interphase replication rate with regard to the autosomes, which frequently results in heteropycnosis and greater staining intensity. Makino and Sasaki [527] have recognized the allocyclic behavior of the Y; ". . . it is usually well differentiated from others of like shape and size on account of its condensed configuration probably due to its heterochromatic nature." Meanwhile, Ohno et al.[632] had already pictured an intensely heteropycnotic Y in somatic metaphases of the male opossum.

Likewise, the X chromosome of various mammals (mouse, rat, opossum, Chinese hamster) including man exhibits heteropycnosis and intensified staining reactions. This is especially clear in some diploid female cells, where one X may be isopycnotic with the autosomes, while the other is positively heteropycnotic.[631, 682, 745, 870]

The X-allocycly, best observable in prophases and pro-metaphases of material not treated with colchicine, is reflected in the behavior of the interphase sex chromatin. Almost 50 per cent of diploid, female, XX nuclei in certain tissues contain a single sex-chromatin body. Nuclei from XXX females often contain two such bodies. While "the precise relationship between the sex chromatin and chromosomes is an unsolved problem," [40] it is becoming increasingly more certain that Barr's sex chromatin represents a portion of a single X, rather than two somatically paired X's.

The unilateral heteropycnosis of the X in female rodent nuclei has been interpreted as allocycly in the timing of DNA synthesis in the sex chromosomes as compared to the autosomes.[631] That such may indeed be the case also in the human X is suggested by recent studies on the incorporation of tritium-labeled thymidine into the X and Y chromosomes of favorable organisms. Thus, Lima-de-Faria [510] showed that "the heterochromatin of the X synthesizes DNA later than the euchromatin of the autosomes." Similar differential labeling was achieved by Taylor [870] in tissue cultures of Chinese-hamster cells. Autoradiography following brief periods of thymidine-H[3] incorporation showed that, in male cells, the long arm of the X and all of the Y are duplicated toward the end of a 6-hour interphase period of DNA synthesis. Also, in female cells, one X chromosome had the same rate of replication as the autosomes, but the other one was duplicated entirely during the last half of the period of DNA synthesis. These

biochemical determinations coincide completely with our purely morphologic observations on the murine X and the human X.   Thymidine-H^3 labeling of human
tissue cultures therefore offers a very promising key for more critical identification
of the X.   Meanwhile, the rather pronounced heteropycnosis frequently seen in
one of the three largest group-C chromosomes,[745] and evident in the X? unit of
figure 20, may be considered an additional, though not unequivocal, clue to the
identity of the human X.

Theodore T. Puck, Ph.D.

CELLULAR CULTURE APPLIED TO HUMAN GENETICS

Much of the new, detailed understanding of the molecular nature and function of the gene and its control of biosynthetic processes within the cell has come from investigations on independent cells and viral particles. Development of tools for use on somatic mammalian cells, comparable to those that have been so effective in microbiologic studies, appears to offer a new pathway for study of human genetic processes such as gene and chromosomal mutation, gene identification and correlation with specific biochemical processes, genic mapping, and the quantitation of irradiation effects, all of which are so difficult to follow in an animal whose mating is not subject to experimental control.

PLATING OF CELLS

The first of these tools is the procedure that is the basis of quantitative bacteriology, that is, a plating method in which a cellular inoculum can be added to a vessel so that each viable cell in the population forms a discrete colony. This procedure permits accurate counting of the number of cells in any population capable of unlimited growth under given conditions. Thus it prevents obscurity in the measurement of cellular growth with massive inocula, because of situations in which a small minority of a large population of cells is able to grow under test conditions that inhibit the majority of the cells. Such a minority fraction of the population, however small, will eventually overgrow the entire culture vessel in which the experiment is conducted, and so make such tests difficult to interpret. In addition, the single-cell plating technique provides a simple and reliable method of measuring lag periods and generation times of cells growing *in vitro* so that all three parameters, number of cells capable of reproduction, the lag period before exponential multiplication begins, and the generation times attained by the multiplying cells, can be untangled in any experimental situation.

Fig. 21.   CULTURE OF CHINESE-HAMSTER CELLS.

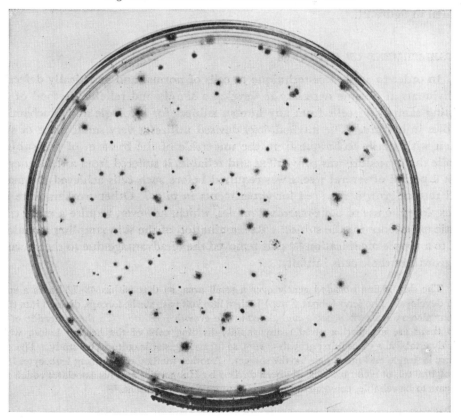

This petri dish has been inoculated with single Chinese-hamster cells and incubated for about 10 days.   About 80 to 90 per cent of such single cells multiply to yield such colonies.

In this way it furnishes a highly convenient and quantitative method for characterization of cellular responses to different physical and chemical environmental conditions.   Finally, it provides means for quantitative isolation even of rare mutants in any population; since, by plating an excess of cells with a stringent selective agent, even the very rare mutant forms grow into colonies from which clonal stocks can be established.[688]

Sanford and her associates [748] had previously shown that single mammalian cells could be grown into clones by a procedure involving sealing the cell into a capillary tube. In order to develop a quantitative procedure, we at first used a feeder layer of irradiated cells to provide additional nutrients and to remove toxic agents from the medium in which the multiplying cells from a single origin were placed.[229] Later it became possible, by improvement of the medium, to obtain single-cell growth without this device, and now large numbers of different cellular types taken from diverse animals can be routinely plated in this way as demonstrated in figure 21.

ESTABLISHMENT OF CULTURES

In order to apply this technique to cells of normal and genetically defective individuals, it became necessary to develop a simple and reliable method of obtaining samples of cells from any human subject for cytologic and biochemical studies in culture. The methodology devised utilizes a very small piece of skin taken with sterile technique from the underside of the forearm of the subject. While the procedure was convenient and reliable, it suffered from a deficiency in that a period of several weeks was required before such cells achieved the maximal rate of growth required for experiments *in vitro*. Other workers were led to explore the use of bone-marrow samples, which, however, require a more traumatic experience for the subject. Reëxamination of the skin-sampling technique led to a simple modification [686] that removed the disadvantage due to the slowness of growth of these cells initially.

The skin is first abraded gently over a small area, so that within 48–72 hours a small scab develops. The scab (about 1 mg.) is then lifted off with sterile forceps, divided into three or four pieces with sterile scissors, and plated in the growth medium. To the underside of the scab there are attached a great many rapidly-dividing cells of the healing lesion, which quickly establish a continuing growth *in vitro*, so furnishing stable cultures for study. The procedure is simple and innocuous for the subject. Another method of culturing leucocytes from about 10.0 ml. of blood has recently been devised by Hungerford and his associates which also appears to be reliable, convenient, and nontraumatic to the patient. [383]

MAINTENANCE OF CHROMOSOMAL INTEGRITY

In the course of these developments, it became obvious from studies in a number of laboratories that most, if not all, of the so-called established strains of mammalian cells in tissue culture possess chromosomal constitutions departing radically from those of the animal from which the cells originated.[548] Such cells were aneuploid and highly hyperploid and often contained abnormal chromosomes. Under these circumstances, many kinds of genetic studies become difficult or impossible. Consequently, it was undertaken to attempt a method of cellular culture which would not develop such chromosomal abnormalities. This has now been accomplished largely by means of careful screening of the media and growth conditions so as to eliminate mitotic-inhibitory actions. Cells have been grown for periods longer than a year and for very large numbers of generations without change in chromosomal constitution.[686] Methods of cellular storage at low temperatures were also adapted from procedures which had been proposed by other workers, and these cells can be stored for periods of many months, at least, and then reintroduced into active growth *in vitro* without change in karyotype.

CHROMOSOMAL STUDIES

In 1956, Tjio and Levan announced the revolutionary discovery that the number of the human chromosomes is only 46 instead of 48. This finding was soon confirmed by Ford and his co-workers, and these workers also characterized many of the human chromosomal morphologies. Studies on cells cultivated with the methods here described were undertaken with Dr. J. H. Tjio to explore further details of the human chromosomal complement. It was shown that the chromosomal constitution of human somatic cells is remarkably constant. In examination of thousands of cells taken from several different tissues of thirteen, normal, human subjects, the only variability found was a relatively small, occasional enlargement in the structure of some of the human chromosomal satellites. The frequency of polyploidy in the cells grown in culture was only about 1 to 3 per cent, which is no more than that normally found for cells *in vivo*.

The method for delineating the chromosomes of cells in monolayers *in vitro* devised by Dr. Tjio produces unusually clear specimens. Moreover, since the growth techniques here described permit long-term cultivation of cells, it becomes possible to reëxamine mitotic figures indefinitely, until any remaining doubts about the chromosomal structure from any particular patient can be resolved. While some laboratories prefer to use techniques yielding smaller numbers of mitoses, so that more cultures can be examined in a given time, it should be emphasized that, at this period when human cytogenetics is in its early stages of development, it is urgent that the data collected be as definitive as possible.

Measurement of the lengths of chromosomal arms were carried out with great care on these specimens by Dr. Tjio, who was able to demonstrate that each pair of chromosomes could be identified uniquely with relatively small uncertainty. For this purpose, however, it is necessary that specimens be of the highest technical excellence and that a sufficiently large number of mitotic figures be available to resolve doubts arising about particular chromosomes whose structures may be obscure in particular mitoses. It is necessary to emphasize that the small differences between some of the chromosomes makes their differentiation not possible in every mitosis, but does permit their resolution with considerable confidence in a large series of chromosomal specimens taken from the same patient and prepared with maximal uniformity and clarity. Another independent quantitative study of the human chromosomal morphologies was published shortly afterwards by Chu and Giles,[134] and their numerical results confirmed those obtained by Dr. Tjio in our laboratory within very narrow limits. Drs. Chu and Giles also were able to differentiate all, or almost all, the chromosomes uniquely. However, one small discrepancy between these first two quantitative analyses of karyotype existed, namely, the identity of the satellited pair of the chromosomes of

Fig. 22. Typical set of human chromosomes showing the six satellites.

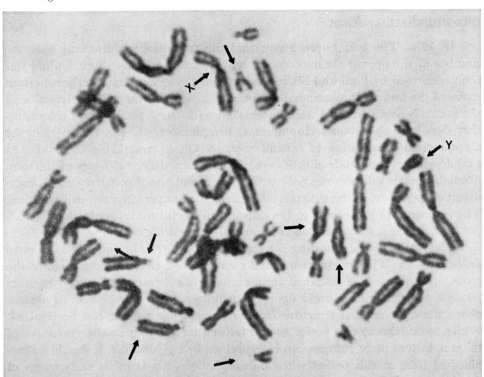

the large acrocentric group. The publication of Tjio and Puck had described the satellited pair as the longest of these acrocentric chromosomes, whereas Chu and Giles reported one of the somewhat smaller acrocentric pairs to be satellited. Since the difference in size between these particular chromosomes is quite small, this divergence was not considered serious and was ascribed to the uncertainties inherent in the identification of the chromosomal-arm lengths. It was with considerable satisfaction therefore that, on reëxamination of this question, we found not one, but two, of the large acrocentric chromosomal pairs to be satellited, so that even this relatively small discrepancy is now resolved (figure 22). Satellite delineation was improved by decreasing the concentration of colchicine to which the cells were exposed and increasing the period of exposure before fixation.

The measurement of chromosomal arm lengths is a difficult and critical procedure because of the many close similarities to be found in the members of the human karyotype. Yet without it, identification of some of the chromosomes, and particularly the X, is not possible. Mitotic figures often present chromosomes with curved, bent, or twisted arms in which partial despiralization may have occurred. These considerations underlie the requirement that many mitotic figures of maximal clarity be available, in order that the final measurements can be carried out on specimens offering minimum uncertainty.

The findings of chromosomal abnormalities in a variety of genetic diseases have been reviewed by Dr. Ford (Appendix III) and will not be discussed here. Our laboratory has analyzed the chromosomes of a patient with Turner's syndrome (XO); a case of testicular feminization (XY); and two cases of Marfan's syndrome which displayed enormously enlarged satellites, in the one case on one of the large acrocentric chromosomes, in the other on one of the small ones.[883] Further investigation is under way to determine whether satellite abnormality is, indeed, regularly associated with this particular disease. No chromosomal abnormalities have been found in cells of patients with a variety of other genetic diseases such as Gaucher's disease and galactosemia.

In April of this year, an International Chromosomal Conference [76] convened in Denver and a study group, composed of persons in various countries who had published a complete analysis of the human karyotype, agreed on a common system of notation for the human chromosomes. This achievement should end the confusion that had arisen in the literature, due to each investigator's using a numbering system different from that of all other workers in this field. A copy of this group's statement is included as Appendix II.

NUTRITIONAL REQUIREMENTS OF SINGLE CELLS

Studies have been undertaken to define the molecular constitution of a medium which will permit single cells to multiply in high efficiency to form discrete colonies. Such a medium permits isolation of nutritional mutants in a manner particularly important to microbial genetics. It soon became evident that molecular nutritional requirements of mammalian cells are vastly different, depending on whether the cells are grown as massive populations, in which case individual cells can support each other and pool the effects of submarginal biosynthetic activities, or whether they grow as isolated individuals.[229] At this point, two media have been developed which contain synthetic micromolecular components and two highly purified proteins, albumin and fetuin.[225, 226, 228, 317, 760] These media, which differ in the micromolecular but not in their macromolecular complements, respectively support the growth of single S3, HeLa cells which are aneuploid and hyperploid and of cells of the Chinese-hamster strain which are much more nearly euploid and diploid. Studies on nutritional requirements of mutants of these strains and of other cells such as the normal diploid, human cell are continuing. It is worthy of note, perhaps, that our first attempt to demonstrate a biochemical-genetic difference in the diploid cells of a patient with galactosemia as compared to those of normal human cells, has not been successful. The addition of even large amounts of galactose has so far produced only marginal, if any, impediment to the growth of the cells from the galactosemic individual.

IRRADIATION STUDIES

Many attempts have been made to study the radiosensitivity of mammalian cells. Observations based on experiments with massive cellular populations always have had to cope with the disadvantage of being unable to unravel the extent to which the given behavior was influenced by the following, separate parameters: irreversible reproductive death of individual cells; temporary mitotic lag in individual cells; regrowth of cells reproductively undamaged by the irradiation experience; and migration of cells not irradiated into the irradiation area where they could initiate new reproduction. As a consequence, the literature dealing with the dosage needed to destroy completely the ability of individual cells to reproduce indefinitely is most confusing, and the various estimates that have been proposed for this figure for human or other mammalian cells vary from several hundred to several hundred thousand roentgens. Studies on the effects of ionizing irradiation on single cells *in vitro* revealed that mammalian cells have an exceedingly high radiosensitivity, much greater than that of microörganisms such as *E. coli* or paramecium which have at one time or another been considered to be reasonably good models for animal cells. Virtually all the mammalian types of cells studied have exhibited survival curves classical in shape with hit numbers greater than 1 and with a mean lethal dose varying from 50 roentgens in the case of the normal human diploid cell to 96 *r* for the aneuploid-hyperploid cell like the S3 HeLa and 160 *r* for the Chinese-hamster cell.* [486, 563, 680, 681, 682, 688, 689, 884]

While the general magnitude of these $D°$ values is well established, the meaningfulness to be attributed to differences of 20–40*r* may be open to some question. This uncertainty arises as a result of new developments in several laboratories which have shown that the $D°$ value for any given cell may be significantly affected by the following factors: the contribution of back-scattered photoelectrons may introduce an appreciable correction factor to the absorbed dose for cells attached to glass, and this factor may well be different for different cells which stretch out to different degrees on the glass surface; the degree of oxygenation of cells in culture can influence their radiosensitivity; and finally, the method em-

---

* Values are expressed as exposure dose in roentgens because of uncertainty in the value of the conversion factor needed to give absorbed dose in rads for glass-attached cells exposed to X rays. A number of laboratories are working on the problem of determining this factor, which may lie somewhere between 1.1 and 1.5.

ployed to estimate the D° value from a given set of data can lead to somewhat different values, depending in part on the manner in which theoretical assumptions are allowed to influence the type of curve to be fitted to the data. These considerations will be dealt with in detail elsewhere. It is clear that this field has reached a stage of development in which a higher degree of experimental accuracy and precision are now needed, both for theoretical purposes and for practical reasons, as in the determination of relative radiosensitivities of various normal and malignant cells.

A variety of experiments, including demonstration of a high proportion of mutants among the survivors of relatively low doses of irradiation (5 to 7 lethal hits), indicated that the mechanism of the action, lethal to reproduction, involves damage to the cell genome. Experiments undertaken to quantitate chromosomal damage showed that such changes are induced by irradiation at doses so low as to make the chromosomes easily fit the requirements for the site of primary damage in mammalian cells. Conversely, all experiments testing for irradiation damage in systems unrelated to the cell genome, such as protein synthesis, energy-source utilization, ability to carry out active transport, and even synthesis of complete viral particles, have revealed these functions to be unaffected even by doses of thousands of roentgens.

The estimate of the amount of irradiation required to initiate any change in a chromosome is not yet susceptible to exact measurement. This uncertainty is due to action of a variety of factors difficult as yet to control, but which appear to operate so as to decrease the number of chromosomal abberrations which can be scored. Thus, chromosomal damage may be of submicroscopic dimensions; visible breaks may heal before the cell is fixed and stained; more highly damaged cells may be delayed in coming to mitosis so they may not be adequately represented in the harvest of mitotic figures collected; lethal chromosomal damage may end the cell's reproductive career before the time of fixation and staining; and, finally, undamaged cells reproducing normally may outgrow the others and so dilute them disproportionately in the mitotic harvest collected.[486, 563, 680, 681, 682, 688, 689, 884] Yet with all of these factors operating to diminish the yield of visible abberrations, the number of these obtained is still considerable, even at low irradiation doses.

Thus, if normal human cells are fixed and stained within 3 hours after irradiation, they display 1 chromosomal break per 18 to 30 r of irradiation on the average. These cells have come into mitosis very soon after irradiation, so the breaks have had minimal opportunity for repair. With increasing time of incubation after irradiation, the number of visible chromosomal aberrations drops steadily, as expected from consideration of the factors listed earlier. When the number of aberrations was scored for cells incubated for varying times up to 3 days following irradiation, the dose required to make visible an average of one chromosomal aberration per cell was still only 40 r. Since the cell in question displays a mean lethal dose (D_{37}) of approximately 50 r, it is clear that chromosomal damage, even though much of it may remain invisible, is well able to account for the loss of reproductive ability.

Some question has been raised as to how reliable these counts of chromosomal aberrations may be regarded in view of all the various aforementioned factors that can seriously affect the result. The fact that the disturbing influences tend to decrease the number of scored aberrations is sufficient to make these numerical results valid as a lower limit for the number of aberrations occurring at various doses. Moreover, consideration of the frequency with which complex translocations occur supports this conclusion. Only occasional translocations, such as those pictured in figure 23, will remain recognizable even after the resealing process is completed because of the bizarre configuration they produce. These can only occur as a result of simultaneous breaks in the same cell, in which fragments from different chromosomes have re-united in a fashion which clearly identifies their complex history. From the frequency of such multi-hit aberrations, it can be calculated that, on the average, about 110 r will produce a break capable of participating in a recognizable multi-hit complex such as a dicentric or a ring chromosome. This would mean that about 1 in 3 of the single visible breaks will recombine with a similar break in a fashion recognizable microscopically, a ratio that appears quite reasonable in view of the dynamics of the process. These figures are to be regarded as rough first-approximations.

There remains to be determined the extent to which different kinds of chromosomal damage (single gene mutations, multiple gene mutations, position effects, damage to centromeres, chromosomal deletions, altered balance phenomena due to alterations in genic ratios, aneuploidy, and the production of dicentric and other complex chromosomal figures) are required to be lethal reproductively for different types of cells. Considerations bearing on this subject have been discussed elsewhere.[486, 563, 680, 681, 682, 688, 689]

These methods [486, 563, 680, 681, 682, 688, 689, 884] have also made possible accurate measurement of the irradiation-induced lag in cellular reproduction which is repairable. For the S3 HeLa cell irradiated with 230 k.v.p., it has been demonstrated that the surviving cells, which retain their ability to multiply so as to form a colony, undergo a temporary inhibition of growth which is a linear function of the dose and equal to 0.12 hours per roentgen.[205]

It is sometimes assumed that, because it is reversible, this particular inhibition of reproduction represents an action of the radiation on the nongenetic structures of the cell. This need not necessarily be true. On the contrary, there is some reason to consider that this effect also may involve nuclear structures like the chromosomes:

1. The very small dose needed to effect a significant mitotic delay indicates that, unless complicating phenomena like chain reactions are involved, the volume of the sites at which this phenomena can occur must be of the same order of magnitude as that involved in chromosomal changes.

Fig. 23.   Complex chromosomal aberration.

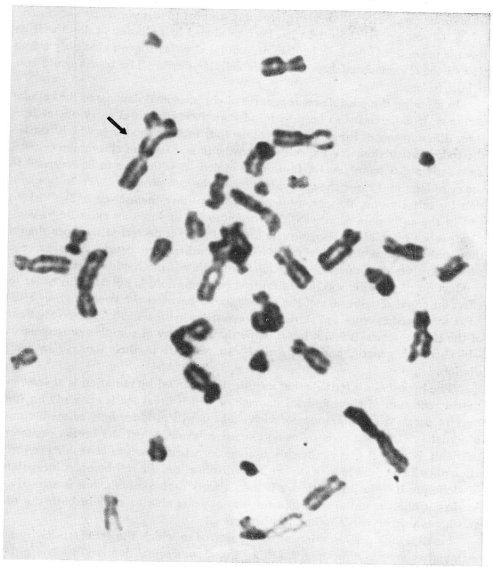

This example of a complex, chromosomal aberration produced by reunion of a chromosomal break into the gap produced in a chromatid break of another chromosome. As expected, such complex aberrations appear to exhibit a multiple-hit dependence on dose in contrast to simple chromosome breaks, and are visible only for irradiations greater than 50 to 75 r.

2. Failure to observe radiation inhibition of all the specific nongenetic structures such as those involved in active transport, and the various biochemical transformations previously alluded to, gives some support to this view.

3. Chromosomal aberration has been produced by radiation at doses well below the mean lethal dose. The data accumulated so far suggest that only a fraction of the chromosomal breaks produce cellular death. The others could contribute to lag.

In assessing the possible consequences of chromosomal damage to the cellular economy, it is necessary to keep in mind that these structures carry out at least three different sets of functions. (1) Since they include the genes which contain the coded instructions for the cell's biosynthetic activity, the chromosomes must constantly act as templates for fashioning molecules which are to be returned to the cytoplasm and cause the synthesis there of the cell proteins. (2) In preparation for cell division, the chromosomes must replicate themselves. (3) Prior to and in the early part of mitosis, the chromosomes undergo an exceedingly complex process of condensation, in which their length is decreased by more than a thousand-fold, and their thickness is correspondingly increased. This process, which involves development of chemically-mediated, mechanical forces and the production of a highly specific, spirallized configuration, is necessary in order to effect an equal distribution of the chromosomes between the two daughter cells. After mitosis, this process must then be reversed. In view of these diverse actions of the chromosomes, it is not improbable that a variety of complex consequences —biochemical, genetic, and mechanical—can result from their damage by irradiation.

The hypothesis is tenable that mitotic delay caused by radiation is at least in part a reflection of chromosomal damage of which breakage is certainly an important part. A small amount of such damage will produce only reversible mitotic delay. In those cells suffering extensive breaks, when the breaks occur in particularly critical sites of the chromosomes or when processes of repair produce anomalies incompatible with indefinite multiplication, the cell becomes incapable of clone production. This theory is only a working hypothesis, but is attractive in that it offers a unitary mechanism that appears able to explain both the reversible and irreversible radiation phenomena.

Finally, the question arises as to the degree to which the great sensitivity of mammalian cells to reproductive killing and chromosomal damage *in vitro* faithfully reflect processes which occur *in vivo*. By this time, a variety of studies have been completed which indicate a high degree of correspondence in these situations. In these are included survival-curve data *in vivo* for murine, leukemic cells; [346] murine, fertilized ova; [692] and normal, murine, bone-marrow cells.[587] Moreover, measurements by Brauer [97] and his associates have demonstrated that chomosomal damage is demonstrable in the cells of regenerating livers of irradiated mice to an extent which, while numerically uncertain, is at least qualitatively in agreement with the experiments *in vitro* just described.

Tissue culture began as a purely qualitative discipline, designed to keep tissues alive and, in some cases, multiplying outside the mammalian body. Following a pattern of evolution similar to that exhibited by bacteriology, uses of mammalian cells have, in the last decade, evolved to the extent that highly quantitative studies can be carried out and many applications of the technique to genetic and genetic-biochemical processes now appear possible.

DISCUSSION

DR. BURDETTE: Thank you, Dr. Puck. Dr. Sanford from the National Cancer Institute will discuss Dr. Puck's paper.

DR SANFORD: As usual, Dr. Puck has presented a very stimulating paper. Since this is a symposium on methodology, I wish to comment on some of the methods that have been presented and suggest some others that can be used.

First, with respect to the methodology of cloning, there are two general types of procedures, the plating procedure that was first applied to tissue culture by Moen [542] and has been further developed for cells of the fixed tissues by Puck, Marcus, and Ciecura [688] and the single-cell isolation technique first successfully used by Sanford, Earle, and Likely [748] for the growth of somatic cells of fixed-tissue origin. Several objections to the plating technique for the critical isolation of clones have been raised,[280, 764, 947] particularly for genetic studies, in which it is absolutely essential to know that the clone is derived from one cell. In handling the petri dishes containing fluid medium, cells can come loose from the glass and settle on other areas of the flask. Also cells can migrate during the week or more required for development of macroscopic colonies. Migration rates as high as 10 mm. in 4 to 5 days for strain *L,* murine cells have been measured in this laboratory from microcinematographic analyses, and rates for other types of cells such as monocytes are certainly much higher. To avoid possible contamination of the clone with other cells in the culture during handling, observing, and subculturing, Goldstein [280] described a modified plating technique, and other alterations in the plating technique have been suggested by Puck.[685]

One advantage of the capillary, single-cell technique is that one cell is completely separated from all other cells in a separate flask. At first this technique seemed arduous and inefficient, but, with modifications [751] and with improved culture media, the procedure has proved reliable for many cellular types of human, simian, murine, and avian origin. Other single-cell isolation techniques have been developed by Wildy and Stoker,[927] Wu-Min,[947] Schenck and Moskowitz,[764] and Aronson and Kessel.[20] In these procedures, the cells have been isolated in microdrops or on glass squares or beads.

For many years, we have been interested in following some of the changes that occur in populations of cells descended in culture from single cells of somatic-tissue origin. At the same time, we have been working in our laboratory particularly toward a more precise control of environmental factors in an effort to understand and control the changes observed. Parallel lines of murine cells originating from the same cell in culture, and carried under similar culture conditions and in the same lots of culture fluid, have been found to undergo diverse neoplastic changes as determined by their capacity to grow as malignant tumors when returned to mice of the inbred strain of origin.[752, 753] A clonal analysis of two of these variant lines of cells differing in neoplastic properties [754] has revealed hereditary differences in morphology, metabolism,[936] number and type of chromosomes,[135] and enzymatic patterns.[789] In another study, within a tumor-cell clone, Dr. Westfall has found differences as great as 400-fold in the arginase activity of the cells.[920] A clonal analysis of these lines of cells has shown the differences in enzymatic activity to be heritable, persisting not only during culture *in vitro* but also *in vivo* when the cells are grown as tumors in mice.[750, 755]

These studies on populations of cells derived *in vitro* from one cell have shown clearly that mammalian cells can and do undergo genetic change in continued, serial propagation. The mechanism of these alterations, whether resulting from genic mutation, chromosomal aberrations, or from other unknown factors, must still be explored. With better stabilization of numbers and types of chromosomes, further studies, especially on enzymatic activities following deliberate changes in culture conditions, should be of value in elucidating adaptive mechanisms of alteration. Clones grown in chemically defined media and under environmental conditions better controlled provide ideal systems for investigating these fundamental problems. Without question, however, we consider that a major problem now confronting the tissue-culture worker is better control of culture conditions to stabilize cells of clonal strains.

A second aspect of methodology discussed by Dr. Puck was the use of certain serum fractions to induce attachment, stretching, and growth of cells on glass surfaces. A number of years ago, with Dr. Westfall,[756] we attempted to isolate and identify those components of the macromolecular portion of serum that appear to be essential for the rapid proliferation of strain *L,* murine cells on glass substrate, in order to have an identified, defined component that could be added to a chemically defined medium for increasing the rate of cellular proliferation. With the introduction of quantitative methods for the handling of cellular suspensions, preparation of replicate cultures,[210] and measurement of cellular proliferation by enumeration of nuclei,[749] it was possible to define accurately the influence of different fractions of serum on known members of cells planted as suspensions of single cells or small clumps of cells. By means of these procedures, we found certain commercially-prepared bovine plasma and serum fractions entirely unsatisfactory and sometimes toxic, apparently because of partial denaturation of the proteins or use of toxic precipitating or purifying agents.

Dr. Westfall then fractionated horse serum by a special modification of the Cohn low-temperature separation. The difficulties of removing zinc ions from protein to be used on living cells precluded use of the complete Cohn procedure.

The fractions were freed of water-soluble diffusibles by washing with saline and ultrafiltration at constant *pH* and low temperature. With this modified procedure, all protein fractions isolated were found to promote rapid growth of strain *L* cells when tested as supplements in a protein-free basal medium which failed to support growth of the cells when unsupplemented. Attachment and stretching of single cells on the glass occurred with all fractions (except possibly globulin fraction III). Globulin fraction I used as supplement gave as good growth of strain *L* cells as the whole unfractionated serum,[225] and it has been used in our laboratory as the only supplement to chemically-defined medium *NCTC 109* for rapid growth of certain strains of cells which, so far, we have been unable to grow on completely defined medium.

Three years later, Fisher, Puck, and Sato [225] also fractionated both adult and fetal serum to find the active component required for attachment and stretching of HeLa cells on glass. Fractionation was carried out by a salting-out procedure using ammonium sulfate. During fractionation, they found the activity to follow the alpha-globulin fraction (or fetuin of fetal serum). Albumin and gamma-globulin displayed no cell-stretching activity. However, in obtaining negative results with the latter two fractions, no experiments are reported on the recombination of these fractions to establish whether the loss of activity resulted from denaturation of the protein during the fractionation or not. In the absence of such tests, one could not conclude that any one fraction is the factor required, since all fractions, if carefully isolated without denaturation or contamination with salts of heavy metals or other injurious materials, might be effective, as was found in the studies carried out in our laboratory. Further, we observe that, in the recent studies by Puck and his associates [227, 685] albumin is also added to the culture medium along with the fetuin fraction in order to enhance the growth of the cells. The large-scale preparation of protein fractions that are well defined and in no way denatured or toxic is beset by numerous difficulties, as is illustrated by the literature during recent years.[671] Even electrophoretically-homogeneous protein fractions may actually consist of many different components.[647]

DR. EVANS: The group might like to know that there are several different cellular strains derived from different species, mouse, monkey, and human, growing in chemically-defined medium *NCTC 109* [208, 209, 592] (table 44) without the addition of any large-molecular material. No serum-protein fractions, proteoses, or peptones are required. One of these cellular strains, *NCTC* strain 2071,[209] a subline of *NCTC* clone 929 [748] of Earle's strain *L*,[193] has been growing at a rapid rate in this medium without the protein for five years. When the population of replicate cultures [210] are counted,[749] we have often seen approximately 13-fold weekly increase [224] of the inoculum numbers of this particular strain of cells in this medium in static cultures. By analysis of cinemicrographic records [591] of the growth of these cells, the population was found to have increased threefold within the period of 33.9 hours during which the film was run. By conversion of these data to

Table 44

COMPONENTS OF PROTEIN-FREE CHEMICALLY-DEFINED MEDIA OF *NCTC* SERIES

Item		Source of chemicals	Concentration (mg./100 ml. medium)	
			NCTC 109	*NCTC* 117
1	*L*-alanine	S	3.148	3.148
2	*L*-α-amino-η-butyric acid	C	0.551	0.551
3	*L*-arginine *	S	2.576	2.567
4	*L*-asparagine *	S	0.809	0.809
5	*L*-aspartic acid	N	0.991	0.991
6	*L*-cystine	M	1.049	1.049
7	*D*-glucosamine *	C	0.320	0.320
8	*L*-glutamic acid	N	0.826	0.826
9	*L*-glutamine	C	13.573	13.573
10	glycine	C	1.351	1.351
11	*L*-histidine *	S	1.973	1.973
12	hydroxy-*L*-proline	C	0.409	0.409
13	*L*-isoleucine	S	1.804	1.804
14	*L*-leucine	S	2.044	2.044
15	*L*-lysine *	M	3.075	3.075
16	*L*-methionine	S	0.444	0.444
17	L-ornithine *	S	0.738	0.738
18	*L*-phenylalanine	C	1.653	1.653
19	*L*-proline	S	0.613	0.613
20	*L*-serine	C, S	1.075	1.075
21	*L*-taurine	N	0.418	0.418
22	*L*-threonine	S	1.893	1.893
23	*L*-tryptophan	S	1.750	1.750
24	*L*-tyrosine	N, M, S	1.644	1.644
25	*L*-valine	S	2.500	2.500
26	thiamine hydrochloride	M	0.0025	0.0025
27	riboflavin	M	0.0025	0.0025
28	pyridoxine hydrochloride	M	0.00625	0.00625
29	pyridoxal hydrochloride	M	0.00625	0.00625
30	niacin	M	0.00625	0.00625
31	niacinamide	M	0.00625	0.00625
32	pantothenate, calcium salt, dextrorotatory	M	0.0025	0.0025
33	biotin	N	0.0025	0.0025
34	folic acid	G	0.0025	0.0025
35	choline chloride	M	0.125	0.125
36	*ι*-inositol	N	0.0125	0.0125
37	*ρ*-aminobenzoic acid	M	0.0125	0.0125
38	cyanocobalmin	M	1.00	1.00
39	vitamin A (crystalline alcohol)	N	0.025	0.025
40	calciferol (vitamin D)	C	0.025	0.025
41	menadione (vitamin K)	N	0.0025	0.0025
42	α-tocopherol phosphate disodium salt (vitamin E)	D	0.0025	0.0025
43	glutathione-monosodium salt	S	1.00	1.00
44	ascorbic acid (vitamin C)	M	5.00	5.00

* Concentrations of A.A. are given without water of hydration of hydrochloride.

Components of Protein-Free Chemically-Defined Media
of *NCTC* Series (Continued)

Item		Source of chemicals	Concentration (mg./100 ml. medium)	
			NCTC 109	*NCTC* 117
45	cysteine hydrochloride	*M*	26.00	26.00
46	diphosphopyridine nucleotide (cozymase, coenzyme I)	*S*	0.70	—
47	triphosphopyridine nucleotide sodium salt	Si	0.10	—
48	coenzyme A	*P*	0.25	—
49	cocarboxylase	*S*	0.10	—
50	flavin adenine dinucleotide	*Si*	0.10	—
51	uridine triphosphate—sodium salt	*P*	0.10	—
52	deoxyadenosine	*C*	1.00	—
53	deoxycytidine · HCl	*C*	1.00	1.00
54	deoxyguanosine	*C*	1.00	—
55	thymidine	*C*	1.00	1.00
56	5-methylcytosine	*C*	0.01	—
57	glucuronolactone	*C*	0.18	0.18
58	sodium glucuronate	*C*	0.18	0.18
59	sodium acetate	*M*	5.00	—
60	sodium chloride	*M*	680.00	680.00
61	potassium chloride	*M*	40.00	40.00
62	calcium chloride	*M*	20.00	20.00
63	magnesium sulfate	*Ma*	10.00	10.00
64	sodium monobasic phosphatehydrate	*M*	14.00	14.00
65	sodium bicarbonate	*M*	220.00	220.00
66	dextrose	*M*	100.00	100.00
67	tween 80	*H*	1.25	1.25
68	phenol red	*A*	2.00	2.00

SOURCES OF CHEMICALS

A = Allied Chemical and Dye Corporation, National Aniline Division, New York 6, New York
C = California Corporation for Biochemical Research, Los Angeles, California
D = Distillation Products Industries, Rochester, New York
G = General Biochemical Company, Chagrin Falls, Ohio
H = Hilltop Laboratories, Inc., Cincinnati, Ohio
M = Merck and Company, Rahway, New Jersey
Ma = Mallinckrodt Chemical Works, St. Louis, Missouri
N = Nutritional Biochemical Company, Cleveland, Ohio
P = Pabst Laboratories, Pabst Brewing Company, Milwaukee, Wisconsin
S = Schwarz Laboratories, New York, New York.
Si = Sigma Chemical Company, St. Louis, Missouri

a growth period of one week and assuming logarithmic growth continued, the population may be said to have increased 30-fold. Such increases as these indicate that this medium supports growth and does not merely keep cells alive.[645, 800] One strain of epithelium [657] from a 62-year-old man has been adapted [32] to cultivation in the medium and grows rapidly. It, too, is derived from a single cell.[658] This particular subline, designated *NCTC* strain 3075, has been cultured in this medium for approximately two years. A strain of human bone cells, *NCTC* strain

3354, as well as 3 simian, renal, epithelial lines of cells,[211] grow well in this medium. These latter are designated *NCTC* strains 3196, 3206, and *NCTC* clone 3526 respectively. In all these strains the rate of proliferation is consistent and rapid. They have not merely been kept alive; the rates of proliferation are extremely good.

The simian, renal cells are interesting because they never went through any stage or period of adaptation. As soon as they were removed from the medium containing serum and were put on the chemically-defined medium, they began to grow rapidly without any loss of cells. As far as these renal cells are concerned, the point of view that cells grown on chemically-defined medium must be adapted is hardly valid, nor can it be said that they grow only on chemically-defined medium because they are variants from their cells of origin. Cells on chemically-defined medium, in particular, are not more or less variants than are any cells grown *in vitro,* no matter what their diet may be, defined or undefined.

Furthermore, fresh murine tissue and fresh suspensions of simian renal cells have been cultivated in medium *NCTC* 109 for intervals of time up to a year. This seems to controvert the concept [720] that freshly isolated, naked cells lyse in protein-free medium in the course of a few hours and appear to differ in their growth behavior from whole explants and from some strains of cells that grow in chemically-defined medium only because the latter have been "trained" to grow in chemically-defined medium *in vitro.* This past week I learned from Dr. Allen Rabson [695] in the Laboratory of Pathology at the National Cancer Institute that he has been growing *P*388,[172] a murine line of lymphoma cells, in medium *NCTC* 109 for almost a year without the addition of any protein.

Two of these lines that I have mentioned are interesting because they can be grown in mass amounts in suspended, agitated cultures.[102] These two lines of cells are the human skin epithelium [32] and Earle's strain *L* fibroblasts.[102] The procedure for growing these cells in our laboratory follows. The inside of the culture flask in which these cells are grown is coated with Dow *DC*200 silicone at a concentration of 5 per cent in benzene. This silicone is baked onto the flask at 250° C. for three hours. Such treatment prevents the cells from adhering to the glass surface of the flask. Methylcellulose is added to the medium, usually in 0.12 per cent concentration of low viscosity grade; one of its functions is to aid in keeping the cells suspended in solution. The cells in this suspension are shaken at about 8640 r.p.h. In addition to this method briefly described for cultivation of cells in chemically-defined medium in agitated, suspended cultures, there are numerous other methods for growing cells in suspension; [103, 126, 195, 196, 285, 291, 467, 590, 636] some are more complex, some less so.

I do not necessarily suggest that the strains we have growing in chemically-defined medium are ones other investigators would particularly want to use. I do, however, suggest that what has been accomplished by us represents what has been done by other investigators [338, 555, 556, 915] as well and may yet be done by further investigators with types of cells in which they have a particular interest for a particular job, be it one involving radiation, biochemistry, or immunology, as they may relate to genetics.

A word of caution is owed to the beginner. The use of cells in tissue culture should not be entered into without considerable thought and a carefully calculated program. Studies done with cells *in vitro*, provided they are done well, are expensive in physical equipment, time, and manpower. Vigor and a sense of mission in approach to and in completion of a problem, as well as brain power, is not complete substitution for experience and/or some training. Failure in some small detail that experience alone teaches may result in an indictment of the method as a tool for investigation. This becomes particularly true as the complexity of the methodology increases, as it must, for example, where pure lines of cells are grown in chemically-defined, bactericide-free medium in mass amounts. These are stringent requirements. Investigations may have to be repeated when current data are suspect because the information has been obtained from cultures in which inadvertent admixture of lines and types of cells has occurred. Also, pleuropneumonialike organisms introduced into tissue cultures by unknown means continue to be present when cultures are maintained in certain antibiotics used for controlling bacterial contamination. Conditions such as these complicate results. With tissue cultures more precisely maintained, they could be avoided, but increased complexity in procedures is required for these standards.

I have no intention of being discouraging, even to the slightest degree, but suggest only that the greatest acumen be observed in the selection of problems in relation to the wide variety of types of tissue culture that can be used,[102, 194, 218, 274, 537, 557, 584, 636, 640, 645, 668, 688, 720, 766, 859, 886, 923, 930] whether short term or long term, small cultures or large cultures, semi-organ or organ cultures. One must be prepared to make contributions to technical advancements in tissue culture such as others have made in this field in order to make tissue cultures a suitable tool for a particular type of job.

Passing to another point in Dr. Puck's interesting paper, there is one very interesting detail that arose in his presentation showing the effect of fetuin [225] on the appearance of cells. In our experience with the strain *L*, murine cells growing in the chemically-defined medium with no supplementation of protein, the cells adhere more tightly to and spread out on glass better when no horse-serum proteins are present in the medium (figure 24). The same strain of murine fibroblasts in static cultures in horse serum and embryo extract has a tendency to become spheroidal. We have noticed, however, that there are certain strains of cells which, when cultured in chemically-defined medium free of added protein, do round up. Subsequently, these have uniformly been found not able to grow in the protein-free medium. In other words, the medium is inadequate in some respects for the particular type of cell, perhaps physical, perhaps nutritional. We have found rounded, free-floating cells to be associated with nutritional inadequacy because we have seen this phenomenon occur in cultures on chemically-defined medium deficient in an essential vitamin. Success utilizing chemically-defined media with some strains of cells certainly indicated to us that there are

Fig. 24 A. *NCTC* 929 CELLS CULTURED TWENTY YEARS.
  B. *NCTC* STRAIN 2071 CELLS CULTURED IN CHEMICALLY-DEFINED MEDIUM *NCTC* 109.

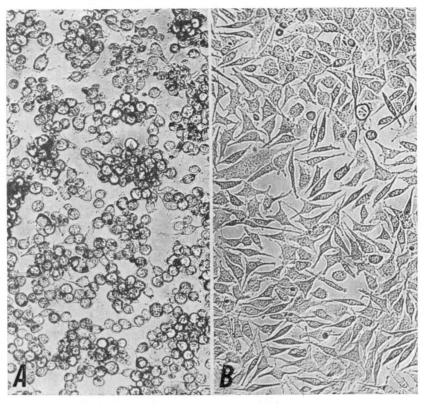

A. These cells were cultured in chick-embryo extract, horse serum, and saline medium. Note spheroidal shape of cells. 180×.

B. This subline has been growing at a high rate of proliferation in a chemically-defined medium free of added protein for five years. Contrast flattened, glass-adherent, multiprocessed, fibroblastlike cells with those in Figure 24A. 180×.

strains of cells that do not require whatever it is that protein or proteinlike material contributes. Nevertheless, we recognize that a factor or factors which chemically-defined media used so far appear to lack is still required for some strains. Obviously, those that grow in the absence of fetuin or fetuinlike materials disprove the universal necessity for this material.

Although the problems of the effect of the presence of protein materials have been studied by us, as noted in Dr. Sanford's comments, no evidence from chemical analysis has been obtained to indicate that any protein or fetuin-like material is produced and discharged into medium *NCTC* 109 by actively-growing prime cultures. Also, strains of cells or tissues newly explanted requiring protein for whatever factor it contributes have not been found to grow on used medium *NCTC* 109 recovered from rapidly growing, prime cultures of *NCTC* strain 2071, even after appropriate fortification. This again suggests that no protein is produced by the growing cell other than that required for its own internal needs. Furthermore, careful examination of hundreds of feet of film on *NCTC* strain 2071, rapidly growing and in prime condition, indicate that death of cells that could contribute protein is practically nonexistent.

Cultures planted with populations sufficiently sparse to be analyzable by cinemicrographic techniques (of the order of 150 cells/0.3 mm.[2]) proliferate rapidly; close analysis of such films shows that practically no death of cells occurs while this rapid proliferation is taking place. Therefore, death of cells is so slight it can hardly be the mechanism that provides the source of protein to supplement the medium. Strains of cells requiring protein, and specifically cultures of single cells requiring anything like 2.0 mg. of fetuin and 2.0 mg. of serum albumen [225] per ml. of medium, could hardly be supplied in this way. Thus, it is inconceivable that these few dying cells could supply this amount. In those instances where culture conditions or specific strains of cells still require protein, it would appear that we must look at the protein as supplying something not yet completely fathomed.[190, 209] Perhaps, since its function is still so ephemeral, we may be allowed to say that it functions as a "detoxifying agent," [209] whatever that loose term may mean or may ultimately prove to mean.

DR. YERGANIAN: There is one aspect of cultures of Chinese-hamster tissues that may be of some service when evaluating diploid lines of human cells. Following a period of classic diploidy, a number of rapidly proliferating sublines may appear in the form of quasidiploids throughout the parental population. Instead of the $2 \times 2 \times 2$, paired, chromosomal elements of the diploid state, quasidiploids have, as an example, a $2 \times 2 \times 1 \times 3$ relationship. This slight alteration in genic balance is sufficient to be reflected in increased proliferative rates in the absence of structural changes, followed by complete displacement of the original diploid population. The shift is even more dramatic during so-called controlled procedures of clonal isolation. Quasidiploids, with excellent properties of proliferation and clonal efficiency, have been utilized successfully in quantitative experimentation.[202, 203] One line of cells employed by these researchers happened to be female quasidiploid of clonal origin, having only one *X* chromosome and an extra chromosome *III* (trisomic) which strongly resembles the *X* chromosome. Thus an *XO*, trisomic *III* state resulted in selective proliferation and increased plating efficiency when compared to the parental stock, while still retaining the diploid feature of 22 chromosomes. D. K. Ford of Vancouver [246] was unaware of the presence of such a quasidiploid component within the parental culture at the time he provided it to Elkind, who initiated the clone in question. It has been our experience in Boston that the most unusual, and unsuspected, karyologic types can be isolated immediately following cloning procedures. In fact, cloning seems to release the storehouse of variant forms that would otherwise fail (negative selection) to become established as sublines within the parental population.

When projecting observations stemming from the karyologically simple, Chinese-hamster cell of normal and malignant origin to situations evolving from lines of human cells having 46 chromosomes continuously propagated, one must always be wary of the possible strength of the quasidiploid state, particularly when selecting for proliferative ability and higher plating efficiencies. A number of quasidiploid and diploid cell lines of the Chinese hamster have generation times as rapid as 10 hours.[203]

In the standard diploid karyotype, distinct point or genic mutations will require lengthy periods of propagation before becoming evident, unless these alterations are dominant. Monosomaty, particularly for the sex elements (XO) is noted very frequently among sublines of Chinese hamsters, and clonal isolates of both sexes and monosomic autosomal types have greater potential for future genetic studies than diploids. Knowledge of cytogenetics of plants and lower animals has been hastened by structural variations about the diploid norm. A similar approach may be profitable among mammalian systems *in vitro*.

During the past few years, there has been a gradual shift toward the utilization of lines of Chinese hamster and human cells for quantitative studies, primarily because of the relative constancy with which chromosome numbers are known to persist in these two species. During the course of selection and trial at various independent laboratories, it appears that the mouse, as another potential source for analyses *in vitro,* fails to retain its diploid or near-diploid component of chromosomes soon after the initiation of primary cultures. After a period of dormant-like adaptation, aneuploidy and polyploidy set in, and the final result is one of karyologic chaos.[493] The extensive formal genetics of the intact mouse appears to be limited the moment cells are grown *in vitro* and depart from the normal and expected limits of aneuploidy.

The actual identification of the sex elements in man and in the mouse remains to be verified unanimously. Also, extensive cytologic recordings of the murine complement of chromosomes during the past ten years have failed to provide methods for routine identification of the X chromosome. Its unique morphology has been confused as an expression of viral action [879] and has remained unnoticed in the most favorable of cells, the ascites tumor cell, for many years.[335, 631]

In my opinion, the most pressing task connected with studies on the human karyotype is identification of the sex chromosome. This must be accomplished as soon as possible in order to minimize reëvaluations of previous reports and to eliminate the need for providing alternate explanations when autosomes are readily mistaken for whole or aberrant sex chromosomes. It is the present task of our laboratory to attempt an unerring identification of the human sex elements by means of physiologic differences in the uptake of tritiated thymidine by autosomes and sex chromosomes. I refer to the recent report by Jacobs *et al.*[419] as an inadequate consideration and approach to the problems relating to the proper identification of the X chromosome in man, although this most helpful aspect of nucleic-acid synthesis has been noted very clearly in examples of aneuploid [870] and diploid cells [954] of both sexes, and the technique has greatly augmented our present knowledge concerning the role of sex chromosomes in oncogenesis, transplantability, and inheritance of sex-limited patterns.[954]

DR. BURDETTE: Dr. Herzenberg has been working with murine cells in culture. Perhaps you wish to comment, Dr. Herzenberg.

DR. HERZENBERG: One of Dr. Yerganian's final statements astonished me when he mentioned that perhaps the mouse is useless for genetic analysis in tissue culture. I have been working with cells of the mouse for about two and a half years now, and I think there are quite a few advantages to working with murine cells in culture. It is certainly true that no one has a diploid or quasidiploid line of murine cells, but many genetic studies utilizing tissue culture do not have this very stringent requirement at this point.

The mouse offers many advantages for genetic work such as the very rich genetic lore that is known in the mouse itself and the particular kinds of markers available in mice which can also be used in culture. Much is known about the inheritance of histocompatibility markers in mice. Some work is now being done on studies of these histocompatibility markers and changes in them in culture. I think that within a reasonable time one may relate histocompatibility antigen markers in culture with the known genetics of these markers in mice.

DR. BURDETTE: Thank you, Dr. Herzenberg. Dr. Ford.

DR. FORD: Perhaps I should apologize to Dr. Puck for referring to his skin method as traumatic. Originally I thought it was *the* answer and that nobody would be concerned about having a little fragment of skin taken from them, but some of the biopsies submitted convinced me of the trauma involved. This new and ingenious method that Dr. Puck mentioned of scarifying a little skin and waiting for a scab will probably be less harmful and even more acceptable to the individual patient than taking 10 to 20 ml. of his blood.

DR. PUCK: Use of the technique of skin sampling, in which a preliminary abrasion produces a small scab which is removed, is a mild procedure and can be carried out with no discomfort and without leaving any scar.

In discussing the role of proteins in cell nutrition, it is important to distinguish between cells growing in isolation and those surrounded by a massive population of other cells, or which are growing under conditions where the total amount of fluid is sharply restricted. In the latter two cases, the cell may not display the requirement for certain molecules needed by single cells in a large amount of medium. Thus, the nutritional requirements of a given type of cell depend strongly on the conditions of growth. In any case, however, it is important to identify these molecules, whether they can be synthesized marginally by the cell, or not at all, but which do play a vital role in metabolism. Fetuin exerts its characteristic action, promoting glass attachment and stretching, on hyperploid human-cancer cells, on the quasidiploid cells of the Chinese hamster, and on euploid human cells. It would seem, therefore, that it is a molecule with an important biological action. It may also be worth noting that possibly all cells contain the potentialities for making this molecule, but that some cells, and particularly the hyperploid ones, such as the L cells, may be able to synthesize much more than diploid cells.

We have carefully investigated all the protein fractions obtained from serum by chromatography, by ammonium-sulfate fractional precipitation, and by Cohn fractionation. We have never found activity of the specific kind I have described, except in preparations containing fetuin or a molecule not yet differentiable from it.

Dr. Sanford: First, I think there is a little misunderstanding about our cells in chemically-defined medium. These are planted as suspensions of single cells. With light inocula, the single cells can continue to grow.

Second, from our limited experience, it seems probable that all the fractions, other than fetuin, could have been adversely affected by the separation procedures mentioned by Dr. Puck.

Dr. Evans: With respect to my suggestion for some caution in the use of tissue culture itself and Dr. Puck's earnest plea that geneticists use tissue culture because of their great need for careful biological experimentation, I can concur with him that we do differ, but only, as he suggests, in a matter of philosophy. We concur that tissue culture can be done reasonably simply; at least for certain jobs, in order to get some answers, some types may be done reasonably simply. One must keep in mind, however, if investigations, after appropriate contemplation and planning, are to be made with cells in tissue culture, that there are many types of tissue culture and many conditions under which these might be done. There is probably one method and one condition that may be superior for a given problem. All I ask is that, when a decision is made to use tissue culture, it be the type that fits the problem. One must keep in mind that my thinking is oriented most often along the lines of massive types of tissue culture: large, static cultures and even agitated, suspended cultures. When and if one thinks in these terms, and one certainly should not exclude this experimental approach, one is also obliged to think in terms of (1) a certain amount of experience and training in tissue-culture technology and (2) specialized instrumentation, precise, intricate, and possibly costly if the job is to be well done. This is, however, my plea, as earnest as is Dr. Puck's: each investigator must recognize that not all tissue-culture procedures can be carried out simply and that the procedure you require to answer your particular problem may be one of the less simple ones. This would be particularly true if one were interested, for example, in certain facts of (1) biochemical genetics [371] or (2) metabolic studies as they might relate to genetics. It would be an advantage in such studies to work with masses of tissues arising from single cells and to grow them in defined protein solution and solution free of nucleic acid. By all means use tissue cultures for your work, on a small scale or a large scale, simple procedures or intricate procedures, but recognize that they all exist and that one of these may be the more nearly perfect and the only way for a solution of your experimental problem.

BIOCHEMICAL GENETICS

H. Eldon Sutton, Ph.D.

METABOLIC DEFECTS *in* RELATION *to the* GENE

The methodology of human genetics, certainly as applied to biochemical traits, can be approached from at least two points of view. First, one may consider the organization of human material to be investigated, that is, individual pedigrees, twins, or populations. This aspect of methodology is basically the same as for nonbiochemical measures. A second consideration has to do with the utility of various types of measures in the detection and elucidation of inherited variation. In this case, the methodology is intimately associated with the measure itself, and discussion will necessarily be concerned with a number of nongenetic aspects.

Geneticists are unique among biologists in that they cannot examine the object of their interests directly. Only by studying the effects of gene variation as expressed some steps removed from the primary action can they infer the existence of genes. In order to understand fully the effects of genetic variation on metabolism, it is necessary to understand the mechanisms by which genes regulate the kind, amount, and location of proteins. The past five years has seen tremendous expansion of knowledge in this area. Any attempt at review must be both incomplete and out of date. Nevertheless, this is the point at which a discussion of metabolic variation must begin.

GENIC ACTION AND ENZYMATIC VARIATION

It is now generally accepted that deoxyribonucleic acid (DNA) is the substance which is primarily responsible for the transmission of genetic information. It is thought that the information is contained in the sequence of base pairs along the polynucleotide chain, although the manner in which this information is coded has not been established.[156] Furthermore, a mutation may involve an alteration in the primary sequence of bases, either an exchange of one base for another or a deletion of one or more base pairs.

The many studies on human hemoglobin have now demonstrated clearly that mutation, at least the class of mutations which produce abnormal protein products, leads to the substitution of one specific amino acid for another amino acid in the primary structure of the polypeptide chain.[400] Thus, in hemoglobin S, valine is substituted for a glutamic acid residue at one of the 280 amino-acid positions of human hemoglobin. A similar specificity has been observed in the other, abnormal hemoglobins that have been studied in detail. As yet, other "mutant" proteins have not been analyzed at the same structural level. It seems probable, however, that all abnormal proteins will be found to have alterations in the primary structure and that changes in the folding of the polypeptide chains will be secondary.

A number of discussions have been presented on the relationship of DNA code to amino-acid sequence. Whereas it has not proved possible to establish the exact nature of the code, there seems to be little doubt that two effects of mutations can occur with respect to protein products. The first is the "non-sense" mutations: those mutations resulting in the complete absence of a protein product. This would be expected if part of the code were deleted and should also occur for some mutations involving base-pair substitutions. The possibility that this type of mutation is involved in thalassemia has recently been discussed by Ingram and Stretton.[405] The second type is the "mis-sense" mutation. This type of mutation causes the wrong protein to be formed, as in the abnormal hemoglobins.[156] The altered protein may or may not be readily distinguishable from "wild type" protein.

The preparation of cell-free extracts capable of incorporating amino acids into proteins has led to the elucidation of some of the steps involved between coded DNA and completed protein product.[361, 956] A diagram of the intermediate steps is given in figure 25. It seems most probable that the DNA code is transferred to an RNA template, which in turn serves in the alignment of amino acids. The amino acids at this state are bound to small-molecular-weight RNA molecules, the latter presumably serving as adaptors.

The quantitative regulation of protein synthesis is a topic that has only recently yielded to experimental efforts. It is an obvious fact that organisms or cells do not synthesize at maximal rates all the proteins for which they have information. One of the first good experimental systems for studying the control of protein synthesis was the induction of enzymes or "adaptive enzyme formation." In this case, it has been possible to demonstrate that inducing substances act to trigger specific protein synthesis, the information and mechanisms for which already exist in cells.[546] Of genetic interest is the observation that occasional mutants arise in these adaptive strains which are constitutive, that is, the mutants produce the "induced" enzymes continually regardless of the presence or absence of inducing substance. Present evidence indicates that the regulation of induced enzymes occurs through the synthesis of repressor molecules by a specific genetic locus.[638] The function of the inducer is to overcome the effects of the repressor, thereby permitting the locus for the enzyme to function.

Fig. 25.   SCHEMATIS DIAGRAM OF AMINO-ACID ACTIVATION
AND INCORPORATION INTO PROTEINS.

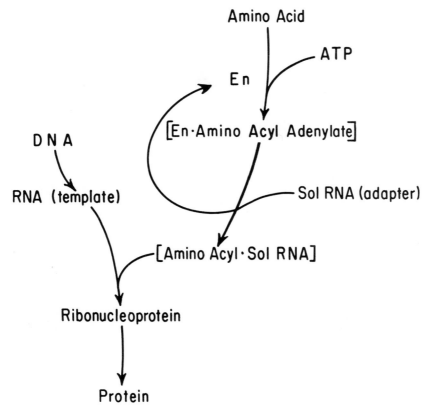

The origin and nature of the repressor molecules influencing induced enzyme synthesis are unknown. In the case of β-galactosidase in *Escherichia coli,* inducibility is regulated by a portion of the chromosome adjacent to the locus for galactosidase. Mutation at this position results in a failure of galactosidase synthesis to be repressed, presumably because of the absence of functional repressor molecules.[638]

The repressors just discussed appear to be intimately associated with the genetic information of the enzyme itself. Another class of repressors also play an important role in regulating enzyme synthesis: these are the end products of biosynthetic pathways. With increasing frequency, examples are being found of feedback repression, in which the product of a pathway represses the synthesis of enzymes specific for the pathway. In an example recently described, the synthesis of four enzymes on the biosynthetic pathway leading to histidine is repressed by histidine.[9]

The importance of feedback mechanisms in the regulation of protein synthesis is only beginning to be recognized. The mechanism by which specific segments of a chromosome regulate the function of other segments is completely unknown. One must be prepared to encounter the type of mutation which influences the quantitative synthesis of protein without influencing the nature of the protein product that is synthesized.

METABOLIC BLOCKS

The usual action of mutant genes in metabolism is the specific decrease in activity of enzymes. According to current thought, a particular mutant locus will affect only one enzyme directly. That is, the primary structure of only a single protein will be altered by a single mutation. Recent work on hemoglobin has demonstrated that the same α-chain appears in more than one hemoglobin,[385, 429] so one might expect occasionally to find one locus affecting several enzymes or the converse, several loci affecting one enzyme. Such examples would not appear to be common among the systems of enzymes which have been investigated in detail.

Inherited blocks in enzymatic activity can be either primary or secondary. By primary is meant a decrease in activity caused by an alteration in the primary structure (amino acid sequence) of the enzyme in question. Secondary blocks are those resulting from the inhibition of enzymes by metabolites that accumulate because of some primary block. In the latter case, the blocked enzyme will be normal and may be present in normal or above-normal amounts, but its activity is decreased.

Several examples of primary blocks are known in man. Perhaps the best understood is phenylketonuria, in which the enzyme, phenylalanine hydroxylase, is missing.[541, 908] This enzyme, responsible for the conversion of phenylalanine to tyrosine, is normally found only in liver. Several workers have demonstrated a complete absence of activity in individuals homozygous for the abnormal gene. It is possible that inactive protein homologous to the enzyme is produced in these patients. Such has not been detected, however. Other examples of primary blocks in man are found in galactosemia (galactose-1-phosphate uridyl transferase),[10] alkaptonuria (homogentisic acid oxidase),[472] and acatalasemia (catalase).[624]

The distinction between a primary and a secondary block is not always easy. In the absence of a demonstrated change in protein structure of an enzyme, it is difficult to be sure that decreased activity is primary. In primaquine sensitivity, glucose-6-phosphate dehydrogenase is relatively unstable and decreases in activity much faster with age than the enzyme from normal individuals.[115] Yet purified enzyme preparations from both affected and normal individuals fail to show this difference.[451] Whether some subtle change exists in the enzyme of patients leading to denaturation, but only within the environment of the erythrocyte, has yet to be established. Quite possibly the primary defect lies in some system influencing the stability of the enzyme, but genetically distinct from the enzyme itself.

Examples of secondary blocks are also found in phenylketonuria. The many features of this disease (mental deficiency, seizures, decreased pigmentation) cannot be explained by a deficiency of the product of the blocked reaction, tyrosine, since this common amino acid is supplied in more than adequate amounts in the diet. Instead, one must invoke an effect of the high levels of phenylalanine which accumulate. That such an interpretation is correct is indicated by an increase in pigmentation when affected individuals are placed on a low-phenylalanine diet, thereby reducing the blood levels of phenylalanine to near normal.[61] A more precise indication of the production of secondary blocks can be demonstrated in 5-hydroxyindoleacetic excretion in phenylketonurics. Uncontrolled phenylketonurics excrete less of this serotonin metabolite than do normal individuals.[639] Removal of phenylalanine from the diet causes the excretion to increase to normal.[33]

The nature and origin of the secondary blocks in phenylketonuria have not been firmly established. One of the interesting suggestions involves the inhibition of amino-acid-decarboxylase reactions.[171,857] Decarboxylases play an important role in the metabolism of nervous tissue. For example, the four amino acids occurring in highest concentration in brain are glutamic acid, gamma-aminobutyric acid, glutamine, and aspartic acid. The gamma-aminobutyric acid is derived from glutamic acid, and the enzyme glutamic acid decarboxylase has been shown to occur at high levels in brain.[723] Among the other compounds formed by decarboxylation reactions are serotonin, dopamine (and indirectly epinephrin and norepinephrin), and histamine. These decarboxylations have been studied in some detail and have been shown to be catalyzed by enzymes requiring pyridoxine as a co-enzyme.[219]

In vitro studies on the inhibition of decarboxylases show that some of the aromatic compounds that accumulate in phenylketonuria are inhibitors of decarboxylases.[323, 867] Phenylalanine itself does not act as an inhibitor. Some results obtained with glutamic acid decarboxylase are shown in figure 26. Interestingly enough, it can be seen that the inhibition is competitive for the glutamic-acid-enzyme site, even though glutamic acid and ortho-hydroxyphenylacetic acid are not closely related structurally.

An interesting corollary emerges from this type of observation. If the phenotypic expression, such as found in phenylketonuria, is caused by inhibition of an array of enzymes, then one might expect that the same enzymes could be inhibited by other compounds, structurally related to the phenylketonuric inhibitors but different in origin. Thus a block at some other point in metabolism might cause the accumulation of aromatic compounds which would also inhibit decarboxylases but which would not be chemically the same as in phenylketonuria. Whether or not one could easily detect these compounds would depend very

Fig. 26.   LINEWEAVER-BURK PLOT SHOWING COMPETITIVE INHIBITION OF RAT-BRAIN GLUTAMIC ACID DECARBOXYLASE BY AROMATIC COMPOUNDS.

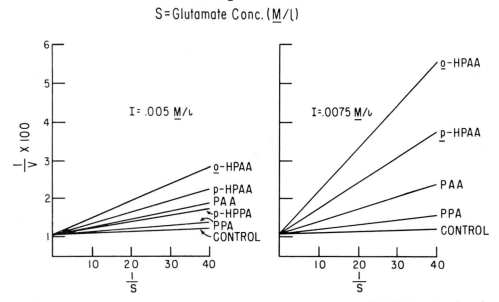

PPA: phenylpyruvic acid; HPPA: *p*-hydroxphenylpyruvic acid; PAA: phenylacetic acid; *p*-HPAA: *p*-hydroxphenylacetic acid; *o*-HPAA: *o*-hydroxphenylacetic acid.

much on the site of their formation, whether nervous tissue or liver, and whether or not the kidney reabsorbs them to be further metabolized or whether they are readily excreted. Thus a variety of inherited enzymatic blocks could lead to very similar, if not identical, phenotypic expression by what is essentially an identical mechanism.

A possible example of this type of disease group has been suggested by Bickel for the diseases grouped together as Fanconi Syndrome or Vitamin D-resistant rickets.[60] In its purest form, the Fanconi renal rickets consists in an increased excretion of amino acids, glucose, and phosphate, with concomitant skeletal changes similar to those of rickets. The rachitic changes can be reversed partially by massive doses of Vitamin D. This complex is seen not only in its pure form, but as part of several other disease entities. The usual pattern of inheritance of these diseases is as a recessive trait, although they would appear to be genetically different. Bickel has suggested that, although the primary biochemical lesion may differ in the various diseases, there may be a secondary block that is identical in all the diseases, thus giving rise to the same phenotype.

METABOLIC ACCELERATION

A third type of enzymatic defect must also be considered as a consequence of mutation. Earlier the role of enzymatic repression was discussed as a means of regulating synthesis of protein. In addition to enzymatic repression there is also feed-back inhibition. In this case, the activity of the enzyme is inhibited by some product of the enzyme itself or by a product of some enzyme in the same pathway. This acts to maintain a stable chemical milieu.

The susceptibility to inhibition is a specific property of the enzyme molecule. One may therefore anticipate mutations that cause a lack of inhibition of enzyme by a normal inhibitor. A model for this is found in the atypical cholinesterase of human serum, which has been investigated by Kalow.[434] In this case, the atypical enzyme appears to have not only an altered catalytic activity, but also to have a changed substrate specificity as well as a changed susceptibility to inhibition.

In man, a possible case of mutation to lack of susceptibility to inhibition has been suggested by Wyngaarden.[948] Wyngaarden and Ashton have demonstrated that the first step in purine synthesis, the reaction between glutamine and phosphoribosylpyrophosphate, catalyzed by PRPP amidotransferase, is inhibited by the end-products of the metabolic pathway, adenosine triphosphate and adenosine diphosphate.[949] On the basis of evidence that hyperuricemia is a defect resulting from overproduction of purines, Wyngaarden has suggested that the basic defect in this disease is a failure of inhibition of PRPP amidotransferase by purine products.[948] This could result from a mutation that effects only susceptibility to inhibition and not catalytic activity. Such a genetic defect might well express itself as a dominant trait as does hyperuricemia.

Whether or not this proposal proves to be true for hyperuricemia, the possibility of this type of mutation must be recognized and must be considered in the genetic and biochemical analysis of inherited disease. It is not a block in the usual sense of the word; it is the opposite of a block or what might be called an accelerated enzyme.

THE PHENOTYPIC EXPRESSION OF METABOLIC ERRORS

The existence of a metabolic error does not result per se in a pathologic condition. Whether or not deleterious effects are produced depend upon a variety of factors, such as the nature of the products of the blocked reaction, the transport mechanisms influencing the concentration of substrate and product, and the excretory mechanisms for the products or substrates which accumulate. For example, the inherited high excretion of β-aminoisobutyric acid (BAIB) is a normal trait.[853] The enzymatic basis for this variation has not been determined, but it would appear to be somewhere along the catabolic pathway of thymine. Since BAIB is not reabsorbed in the renal tubules, high concentrations of it cannot accumulate in the body tissues. Therefore it is not able to inhibit other reactions. Furthermore, the products of this breakdown would appear to be nonessential in human metabolism.

A similar example is found in alkaptonuria. Again the block is in the breakdown of a substance, homogentisic acid, which is a waste product and which is not reabsorbed by the kidney. Hence it is impossible for high levels of the compound to build up in the body, and the absence of products of homogentisic-acid oxidation are not deleterious. To be sure, in this case there is a very slow accumulation of abnormal products of homogentisic acid in connective tissue, leading slowly to an arthritic condition.[456]

By contrast, one may consider phenylketonuria, in which deficiency of phenylalanine hydroxylase prevents the formation of the essential metabolite, tyrosine, giving rise to individuals who, in microbial terms, would be considered tyrosine auxotrophs. This particular fact becomes important only if one is treating such individuals with an artificial diet lacking tyrosine. More important for the etiology of the disease is the fact that phenylalanine is a substance that is efficiently reabsorbed by the renal tubules, resulting in accumulation of very large amounts of phenylalanine in blood and body tissues. These very high levels cause the activation of reactions that are normally of little significance, resulting in accumulation of several other compounds which are more injurious than phenylalanine itself.[857]

The preceding comments have referred primarily to blocks in metabolism which are essentially complete. Several examples have been established in which blocks are incomplete, as in the carrier state of phenylketonuria[373] and galactosemia.[374] One may assume that most abnormal genes are relatively benign in the heterozygous condition. In many cases, one would expect normal regulatory mechanisms, such as feedback repression and inhibition, to compensate entirely for any partial defect in genic function or efficiency. In other cases, although there may be a difference in the amount of enzyme present, only rarely would this become a seriously limiting factor. In phenylketonuria it is possible to show that heterozygous individuals are deficient in the amount of phenylalanine hydroxylase in liver. This is reflected even in the blood levels of phenylalanine of such individuals.[457] However, the distinction between heterozygous carriers and normal individuals is very small, since so many other factors act to maintain a balanced homeostatic system.

METHODOLOGY IN HUMAN BIOCHEMICAL GENETICS

Many of the problems of methodology are primarily problems in biochemical technique, but these will not be considered. Rather, the type of human material which has and will continue to prove useful in detecting new biochemical variation and the types of chemical approaches which should prove useful when applied to such human material will be examined.

LEVELS OF METABOLITES *IN VIVO*

The most prevalent experimental approach to detecting biochemical variation in man has been the time-honored one of collecting random, or haphazard, samples of various biologic fluids, notably urine and blood, and subjecting them to a variety of assay procedures. Popular among the procedures applied is paper chromatography, making use of ninhydrin for revealing amino acids and various other reagents such as diazotized sulfanilic acid, which reveals phenolic compounds. The utility of paper chromatography for screening purposes has long been recognized, since it reveals simultaneously and by simple procedures a variety of recognizable compounds, not only those occurring at normal levels, but also those perhaps seen only in abnormal conditions.[929] As an approximate means of detecting variation in a number of compounds, paper chromatography is unexcelled and will continue to prove very useful.

The subjects from whom haphazard samples have been collected have varied widely depending upon the interest of the individual. Normal variation has been studied both in unselected individuals and in specific groups of individuals such as twins or individuals of different racial backgrounds.

One interesting inherited trait detected by this approach has been the high excretion of β-aminoisobutyric acid. This substance was first detected independently by two groups, one working with normal individuals [165] and the other working with cancer patients.[223] The large range of excretion of this substance among individuals makes variability in this substance easy to establish. A considerable variation in lysine excretion has also been detected by paper chromatography, although the high excretion of this substance appears to be restricted to a few individuals.[855] Beyond this there seems to be relatively little marked variability in excretion which has been revealed by chromatographic examination of normal individuals.

Chromatographic techniques have also been applied to selected subjects, such as twins. Two studies in particular could be mentioned in this respect, one carried out at Columbia University several years ago by Berry, Gartler, Dobzhansky, and others,[57, 269] and one carried out at the University of Michigan.[856] These studies each have revealed several amino acids to be influenced by genetic factors. However, they are discordant with respect to which amino acids show the greatest

Table 45

SUMMARY OF CHROMATOGRAPHIC STUDIES OF URINE FROM A VARIETY OF PATIENTS

Diagnosis	Number	Amino acids	Phenolic acids
Phenylketonuria	10	High phenylalanine	High *oHPAA*, *pHPLA*, etc.
Fanconi's syndrome	7	General aminoaciduria	Normal
Lowe's syndrome	3	General aminoaciduria	Normal
Muscular dystrophy	10	9 Normal	
		1 High valine, leucine	5 Normal
Hereditary ataxia	2	Normal	Normal
Pseudohypoparathyroidism	1	High glycine	Normal
Pseudopseudohypo- parathyroidism	1	Normal	Normal
Congenital glaucoma	7	1 Normal	
		6 Intermittant general aminoaciduria	Normal
Congenital cataracts	7	Normal	Normal
Parkinson's Disease	6	Normal	Normal
Gout	23	20 Normal	
		3 Slightly elevated	Normal
Amyotropic lateral sclerosis	27	Normal	Normal
Sickle-cell anemia	8	Normal	Normal
Diaphysial aclasia	5	Normal	Normal
Mongolian idiocy	53	High *BAIB* (?)	Normal
Hereditary telangiectasia	1	Normal	Normal
Werdnig-Hoffman	1	Normal	Normal
Huntington's chorea	1	Normal	Normal
Achrodermatitis	1	Increased peptides (?)	Unidentified spots
Polyuria	2	Normal	Normal
Neonatal convulsive disorder	2	Normal	Normal
Gaucher's disease	1	Normal	Normal
Von Gierke's disease	1	High valine, leucine	Normal
Hallermann-Streiff	1	Normal	Normal
Tyrosyluria	1	High tyrosine	High *pHPPA*
Non-nutritional anemia	4	3 Normal	
		1 Mild aminoaciduria	Normal
Autism	6	4 Normal	
		2 High valine, leucine	6 Normal
"Failure to thrive"	19	14 Normal	
		5 General aminoaciduria	19 Normal
Congenital nephritis	1	Normal	Normal
Renal glycosuria	4	Normal	Normal
Diabetes mellitus	2	Normal	Normal
Albuminuria	3	2 Normal	
		1 Mild aminoaciduria	3 Normal
Niemann-Pick disease	4	Normal	Normal
Narcolepsy	1	Normal	Normal
Myoclonic seizures	85	61 Normal	61 Normal
		22 Mild aminoaciduria	12 Increased
Eales's disease	1	Normal	Normal
Behavioral disorders	3	Normal	Normal
Chorioretinitis	1	Normal	Normal

genetic variability. In part, this may be explained by the differences in age groups of the twins and in environmental factors. There seems to be one other deficiency of the twin approach which should be pointed out, however. The strong genetic influence on excretion of β-aminoisobutyric acid as a normal trait has already been emphasized. In our own twin sample, this amino acid failed to show significant genetic influence. Upon inspection of the data the reason is obvious. Only about five to ten per cent of Caucasians fall into the clearly high-excretor group. In our sample of some eighty sets of twins, there were only two sets who showed appreciably elevated excretion. In the analysis for heritability, these few individuals contribute very little to the variance of the population. Hence a nonprevalent trait will fail to be revealed by the twin-study method. It seems likely that twin studies will prove most useful in studying variation that is polygenic and for which polymorphism is very common. Such studies will continue to prove useful, but cannot be used to exclude the existence of variation that is less common.

The chromatographic analysis of biologic fluids has also been used extensively in investigations of various disease states. Several entities have been recognized or further characterized on the basis of chromatographic studies. Examples include Hartnup syndrome,[38] cystathioninuria,[330] cystinuria,[328] Lowe-Terrey syndrome,[514] and galactosemia.[365] In all these examples, the excretion is so aberrant that the deficiencies of chromatographic procedures do not interfere with recognition of the abnormality.

It is characteristic of this type of investigation that only positive findings are usually reported in the literature. A number of investigators, including our own group, have searched extensively for variations in excretion of amino acids and phenolic acids in a variety of disease conditions, both classifiable and unknown. A summary of these investigations is given in table 45.

At least two facts emerge from examination of this table. In the first place, most diseases do not show abnormal excretion either of amino acids or phenolic acids. In the second place, we have obtained negative results in some cases for which other individuals have reported positive results. In part, this may be due to heterogeneity of the clinical material. Other reasons may also account for this discrepancy. To enumerate some fairly obvious reasons, one may invoke differences in severity of the disease and differences in the manner of expressing results. If one is to attach importance to elevated excretion of amino acids, it is essential that this elevation be present at a time when the pathologic process has not led to a general, metabolic inadequacy. It will not always be easy to recognize the point at which this occurs, but generalized aminoacidurias associated only with the advanced states of a disease must be treated circumspectly.

The other source of discrepancy, the means of expressing results, should not require comment. However, the persistence in the literature of reports of elevated amino-acid excretion based upon volume of urine, rather than upon a timed sample or upon creatinine excretion, indicates that this fairly basic aspect of measurement is sometimes disregarded. At best, volume is a poor means upon which to express results, and it is valueless in disease conditions.

The foregoing comments refer primarily to gross aberrations in excretion of amino acids. A factor that undoubtedly plays a role in the excretion of amino acids and definitely plays a major role in the excretion of phenolic acids is diet and environment. Of the many chromatograms that we have prepared for phenolic acids in various disease states, very few have indicated anything of persistent interest. I am confident that, in part, this is due to our inability to control the environment of such individuals. Many of the phenolic acids are known to respond dramatically to dietary changes. In spite of this there would appear to be a residuum of variation which is genetic and which, in some cases, may be associated with inherited disease. From the experimental point of view, it is difficult to distinguish the interesting variation of this sort from environmental variation. As a result, we have undoubtedly missed some variation that would be interesting to pursue simply because we cannot detect it in the presence of large environmental variation.

The answer to this problem is obvious and expensive. Increasingly we must rely upon studies carried out in a metabolic laboratory with proper controls in order to eliminate undesirable environmental fluctuations. Such studies will require the development of highly standardized diets, preferably preparations that will be uniform when used in various laboratories. The recent establishment by the National Institutes of Health of several centers for metabolic studies should alleviate this problem somewhat.

ARTIFICIAL INTERNAL ENVIRONMENTS

An experimental approach useful for overcoming environmental fluctuation is the tolerance test or loading test, long in use as a means of recognizing diabetics or prediabetics. This type of test is based upon the fact that an enzymatic deficiency, which may not be clearly expressed at random normal substrate levels in the body, can be fully expressed if its substrate is supplied artificially at greatly increased levels. In such a case, environmental fluctuations in substrate level are effectively overcome. In addition to their use for aberrations in carbohydrate metabolism, tolerance tests have been developed for such conditions as phenylketonuria and have proved effective in revealing variations between homozygous normal and heterozygous individuals who are otherwise phenotypically normal.[373] From the theoretical point of view, tolerance tests should be a very useful means of recognizing minor variation in many substances. Gartler has used what is essentially a tolerance test in attempting to identify the metabolic defect in β-aminoisobutyric-acid excretion.[268] A number of authors have used tryptophan loading to try to reveal metabolic variations among mental patients, particularly schizophrenics [444, 859] and Jepson *et al.,* have used monoamine-oxidase inhibitors with phenylalanine to study amine formation in phenylketonurics.[426]

One obvious deficiency of the tolerance test lies in application to healthy individuals. While it is true that some individuals are coöperative to a gratifying extent, it seems unlikely that a tolerance-test situation, with the demands which it makes upon time and comfort of the individual, will become a very widespread tool for the study of normal variation. Where it is feasible, it is an approach that deserves to be considered.

STUDIES *IN VITRO* OF BIOCHEMICAL POTENTIAL

Most of the deficiencies of the methods previously discussed could be avoided were it possible to carry out studies of the biochemical potential of tissues *in vitro*. In this way one can control completely the environment of cells or of cellular extracts and can thereby avoid those factors that are part of the recent and sometimes random history of the individual. Such techniques have not yet been applied extensively to studies of genetic variation.

The approach proved effective in a few cases is that of direct enzymatic assays. This has been most notably successful in the study of galactosemia, in which the enzyme, galactose-1-phosphate uridyltransferase, is missing in erythrocytes of galactosemic individuals.[10] Furthermore, individuals heterozygous for this gene have only about half the normal complement of enzyme in their erythrocytes.[99, 452] Thus it becomes possible to trace with some accuracy the presence of the abnormal gene in otherwise normal individuals. Another well-known example is primaquine sensitivity, in which erythrocytes are deficient in glucose-6-phosphate dehydrogenase.[115]

In both these examples, the disease is characterized by an enzymatic deficiency that is expressed in readily available tissue, erythrocytes. While there may be other cases in which enzymes occur in erythrocytes, it seems probable that at best the amounts of many of the enzymes will be so small as to prove of little value in genetic studies. This may be particularly true in the case of those enzymes with primary function carried out in other special organs, such as liver. There is little reason to hope, for example, that phenylalanine hydroxylase can be detected in tissues other than liver. Thus we are virtually limited to those enzymes found in blood cells, either erythrocytes or leucocytes.

A possible way of overcoming this difficulty has presented itself with the development of methods for tissue culture. Some problems of working with differentiated tissues may be overcome as knowledge of the regulation of cellular function increases. Earlier in this review, quantitative aspects of protein synthesis were discussed. Whether or not a gene is expressed depends very much on the level of various metabolites in the surrounding medium. Thus it is possible to inhibit protein synthesis by supplying repressors or to induce protein synthesis by overcoming the effects of repressors. An interesting application of this has recently been reported in a case of megaloblastic anemia accompanied by excretion of orotic acid.[382] In this patient there appeared to be a block in the conversion of orotic acid to other intermediates leading to the synthesis of pyrimidines. On the basis of observations made in *E. coli*,[951] the authors administered cytidylic acid and uridylic acid to the patient. As in bacteria, these end-products repressed the synthesis of an enzyme preceding orotic acid, thereby preventing the accumulation of orotic acid. If it is possible to manipulate the expression of genes in a complex system *in vivo* such as that of man, then one should not be too discouraged at the prospects of manipulating cultures of tissues in an even more effective manner.

That the biochemical potential of differentiating tissues may not always be lost is also indicated in a report on phenylalanine hydroxylation in HeLa cells.[191] Normally these cells are unable to hydroxylate phenylalanine to give tyrosine. In a variant culture that arose several years ago, it was observed that in the absence of tyrosine, the culture could hydroxylate phenylalanine readily. In the presence of tyrosine, the enzyme that carried out this function was not produced. Although this must be recognized as a variant culture, it is enough to give hope that chemical manipulation of genic expression in tissue culture will become a feasible and productive approach.

CONCLUSION

If a conclusion can be drawn from the preceding discussion, it might be that there are no absolute methods for approaching the study of biochemical variation. On the purely laboratory side, technical limitations prevent the attainment of certain obviously desirable objectives. At the other extreme, one must overcome the difficulties of working with living human systems, which are buffered against extreme chemical fluctuations. Between these two extremes one must devise methods useful for detecting the small amount of variation permissible. Above all, as in any scientific investigation, one must have adequate controls, a problem that is sometimes unusually acute in working with human traits.

DISCUSSION

DR. CHILDS: Doctor Sutton has given a comprehensive summary of the scope of human biochemical genetics. It has been said that human, biochemical genetics involves the conversion of a "good recessive" into a "bad dominant." I should say, "It involves taking a 'bad recessive' and making of it a 'good dominant,'" because this indicates the first step one takes to follow the characteristic down to the point of study of a protein substance that has the closest affinity to the gene involved.

This is a definition of the complementary employment of genetic techniques and biochemical methods to elucidate human variation. Genetic analysis circumscribes the problem or poses questions answerable only by application of chemical methods. Obviously the converse obtains also, and the chemical techniques can be used to discover variations, the genetics of which can then be worked out. There are many examples of this. Among these are the serum-protein variants, and some of the hemoglobin variants also fall into this class of characteristics which were discovered without any knowledge of hematologic or

Table 46

A LIST OF SOME GENE-DETERMINED QUALITATIVE DIFFERENCES IN PROTEINS

Type	Method	Reference
Transferrins (β-globulins)	Electrophoresis	Smithies [819]
α-2 globulins (haptoglobins)	Electrophoresis	Smithies [819]
Fast albumin	Electrophoresis	Fraser [263]
Hemoglobins	Electrophoresis Amino acid sequence Spectroscopy	Itano [408] Ingram [846] Gerald [272]
Double albumin	Electrophoresis	Earle [192]
Blood groups	Immunologic	
Pseudocholinesterase	Inhibition	Kalow [434]

other abnormality. The glucose-6-phosphate dehydrogenase characteristic is also one of these. That is, the biochemical characteristics of the sensitive erythrocytes were already well worked out before any genetics was done at all.

I wish to define the present status of our knowledge with regard to genic action. In seven or eight conditions there are qualitative differences in proteins for which the inheritance is known (table 46). Not all, but most of these are associated with very common genes. Most were discovered by application of electrophoretic methods without regard for the function of these particular proteins. This suggests to me that one of the most important reasons for the recent expansion in our knowledge of human genetics is the extraordinary development of techniques which one can employ, rather than that modern medicine is no longer concerned with the major problems of infections and nutritional disorders. The latter is an explanation frequently given, but I would guess it is the technical developments that are more responsible.

When we pass to the next level of gene action, we find a dozen or more diseases or characteristics in which some alteration in the activity of a protein has been shown to be present (table 47). This requires study *in vitro* of the activity of enzymes and employing tissues in which one would expect to find the enzyme and to which are added the appropriate substrates and cofactors. The reaction is carried out under optimal conditions of temperature, *p*H, et cetera. These are usually rare conditions, and knowledge of their precise genetics is much less than is that of the variation in protein mentioned. These diseases are likely to be recessive, and employment of special techniques usually detects some characteristic measurable among the heterozygotes. Through application of such techniques, one finds the major problem to be that of discrimination between the two kinds of homozygotes and the heterozygotes.

Table 47

SOME CONDITIONS IN WHICH EVIDENCE *in vitro* OF LOSS OF ENZYMATIC
ACTIVITY HAS BEEN DEMONSTRATED

Condition	Enzyme system	Reference
Galactosemia	*P*-Gal-transferase	Kalckar [432]
Phenylketonuria	Phenylalanine hydroxylase	Mitoma [541]
Adrenogenital syndrome	21-hydroxylase	Bongiovanni [72]
Familial jaundice	Glycuronyl transferase	Arias [19]
Hypophosphatasia	Alkaline phosphatase	Fraser [260]
Primaquine sensitivity	G-6-*P* dehydrogenase	Carson [115]
Glycogen-storage diseases	G-6-phosphatase	Cori [151]
	Brancher	Cori [151]
	Debrancher	Cori [151]
	Phosphorylase	Mommaerts [545]
Alkaptonuria	Homogentisic oxidase	La Du [472]
Goitrous cretinism	Deiodinase	Stanbury [694]
Acatalasemia	Catalase	Takahara [864]
Methemoglobinemia	Diaphorase	Scott [790]

The next step involved is to move all the examples in this second list into the first list, that is, to attempt to show qualitative differences between the enzymes involved in these conditions and in the normal. There are many methods available to do this. I have mentioned electrophoresis; there is also immuno-electrophoresis, combinations of chromatography and electrophoresis, and, if there is not already an immuno-chromato-electrophoresis, perhaps there will be soon.

Not much has been done in the area of quantitative evaluation of these enzymatic disturbances. Most of the efforts so far have involved their characterization. However, examination of quantitative differences between families and between individuals and between racial groups may be most interesting and useful. In connection with the glucose-6-phosphate dehydrogenase characteristic, for example, there is among susceptible red cells of American or African Negroes considerable residual activity of this enzyme, perhaps 10 to 20 per cent of the normal activity. But upon examination of this same characteristic in the red cells of white individuals, these usually being of Mediterranean origin, Oriental Jews, Sardinians, Greeks, and such, one finds that the enzymatic activity in the

susceptible cells is nil or exceedingly low.[131]   A third group also show this characteristic—people whose racial origin is different from either of these others and
who have an hereditary, non-spherocytic, hemolytic anemia.   These people, and
they number very few, are of European origin, and the enzymatic activity of their
erythrocytes has been shown also to be zero or very nearly zero.[957]   Here is a clear
difference that may provide the first step to the detection of alleles or genes at different loci or, conceivably, these differences may be due to the imposition of the
same gene in different genotypes.

One further area that is providing interesting information is the study of the
development or induction of enzymatic systems in fetal and newborn tissues.   Several systems of enzymes have been studied in tissues of fetuses or newborn infants
and have been found to be deficient at birth.   Very little is actually known about
the genetic control of this aspect of tissue differentiation and development, but,
if investigators can be patient enough to await the births of all the children in
families, a good deal may be learned about the genetic control of normal development of enzymatic systems.   Perhaps this can also be studied profitably in
animals.

Dr. Sutton: Dr. Childs has given interesting, additional examples of the
mechanisms that act to maintain metabolic systems in their normal range.   One
of the very important studies for the future is the question of enzymatic induction
and repression in man.   According to the microbiologists,[638] it is more a question
of repression than induction, and I am sure that the study of such systems will
help very much in understanding some of the metabolic defects which are less
obvious but no less important than phenylketonuria.

Robert L. Hill, Ph.D.

METHODS *for the* STRUCTURAL ANALYSIS *of* HUMAN HEMOGLOBINS*

INTRODUCTION

Two kinds of experimental approach have been used to determine inherited differences in the structure of proteins. One is the study of the variation, from species to species, of proteins of similar or identical function. A second, which has been used more recently, is investigation of the variation in a specific protein in an individual or between individuals of the same species. Hemoglobin has been studied in detail from both of these aspects, because it is not only available from a wide variety of animal species, but also it is susceptible to a large number of gene-controlled alterations within a single species. However, regardless of the experimental approach, it should be recognized that the extent of our knowledge of inherited differences has been almost directly proportional to the development of methods for the structural analysis of proteins.

In keeping with the theme of this volume, some of the historically important developments that led to the use of the human hemoglobins as a system for determining hereditary differences in proteins and some of the methods currently used for measuring these differences will be described. Before doing this, a short account of the methods that have been used for judging species differences in hemoglobin will be given. Whereas species differences are not a primary concern

---

* I would like to thank Drs. Emil Smith, Albert Light, and Robert Swenson for their help in preparation of this manuscript. Much of the original work that is referred to here has been performed in collaboration with Drs. Robert Swenson and Herbert Schwartz. We are indebted to the United States Public Health Service for financial support of this research.

304

in this presentation, it should be remembered that this experimental approach gave chemical geneticists their first ideas about genetic control of the structure of proteins. In addition, some of the methods that recently have been useful for evaluating human hemoglobins were first used in this kind of study.

Species Differences in Hemoglobins.—Reichert and Brown,[713] in 1909, were among the first investigators to study differences in the hemoglobins from several species. They showed that oxyhemoglobin from certain vertebrates crystallized in more than one form. Although it is known that crystal form is influenced by several factors other than primary structure, these observations in the light of later studies (see the references following) can be considered evidence for inherited differences in the structure of proteins.

In 1923–24, Heidelberger and Landsteiner [339] and Hektoen and Schulhof [341] reported that the antigenic specificity of the hemoglobins varied in accord with the evolutionary relationships of the animals from which the hemoglobins were obtained. That is to say, hemoglobins from closely related species showed considerable cross-reaction to a common antibody, whereas hemoglobins from more distantly related species had less cross-reactivity. Examples of this phenomenon with other proteins are known,[83] including antigenic specificity of various human hemoglobins.[128] There is little doubt that the immunochemical methods remain one of the most valuable tools for judging the similarity of proteins in related species.

Amino-acid analysis provided an additional means of determing species differences. In the earliest studies, only certain of the amino acids could be determined with reasonable precision. Vickery and White,[894] in 1933, reported that the cysteine and noncysteine sulfur content was different in hemoglobins from the horse, sheep, and dog. Block,[67] on the other hand, reported in 1934 that a constant respective ratio of 1:3:8:9 existed between the iron, arginine, histidine, and lysine content in these same hemoglobins. In 1937 Bergmann and Niemann [53] found essentially the same ratio in bovine hemoglobin. Brand and his co-workers [94, 96] later demonstrated differences in the leucine, isoleucine, and methionine content of various species of hemoglobin. This method has become more useful since the recent development of very precise procedures for amino-acid analysis.[550, 552] Species differences have been demonstrated by this means in other proteins, such as serum albumins [95] and the plant-seed globulins.[812, 813, 814]

In 1938, Landsteiner, Longsworth, and van der Scheer [479] reported studies on the differentiation of closely related hemoglobins by the technique of moving-boundary electrophoresis. Several of the hemoglobins they examined were shown to possess different electrophoretic mobilities under standard conditions. This is just one of the many studies of the kind that was made soon after the moving-boundary technique was developed by Tiselius.[875] By 1940, species variations had been demonstrated by this method among several proteins, including the ovalbumins from closely related birds,[479] several invertebrate, respiratory proteins,[876] and the plasma proteins from certain species.[875]   It is clear that electrophoresis has remained one of the most useful techniques for judging differences in closely related proteins.   This is best demonstrated by the fact that, some ten years after the studies of Landsteiner *et al.*, electrophoresis showed, for the first time, a clear difference between normal and sickle-cell hemoglobin.[646]

Thus, as early as 1940, several methods of judging inherited differences in protein structure were available, including antigenic specificity, amino-acid composition, and electrophoretic behavior.   With these methods alone, it was possible to show that the hemoglobins of different species have a great deal of similarity in addition to their common function in respiration.   On the other hand, these methods also revealed subtle differences that undoubtedly represent genetically controlled alterations in structure.

By 1950, many of the modern techniques for determining the sequence of amino acids in proteins had been developed.   These methods greatly strengthened the available means of comparing closely related proteins.   As an example, Porter and Sanger [672] showed in 1948 that the kind and number of amino end-groups of various hemoglobins differed considerably, although there was no correlation between the kind of end groups and the phylogenetic classification of the species.

The most satisfactory picture of species differences has been made in the last few years with those proteins for which the entire amino-acid sequence is established.   This is true for some of the pituitary hormones,[504] although a better example is provided by insulin.   Sanger and his colleagues [332] showed that the insulins from five vertebrate species differed only by one or two residues in either of three positions within the intrachain disulfide loop in the *A*-chain.   In every case only neutral amino acids were involved, for example, isoleucine replaced by valine, alanine by threonine, and glycine by serine.   It is interesting to note that these differences are not great and, in this sense, are analogous to those variations subsequently shown to exist in the abnormal human hemoglobins.   In light of the majority of the earlier studies on species differences, more extensive changes than these might have been expected.   Because these changes are small, it might be assumed that the genetic mechanism controlling the structure of insulin could be very similar, even in rather unrelated species such as the horse and whale.

These detailed, structural studies provided the first indication of the **exact** hereditary differences that can occur in the structure of a protein.   However, the interpretation of these studies is limited in one important respect, viz., it is impossible to correlate the structural differences with a definite pattern of inheritance of the protein.   Clearly, this limitation can be overcome only by studies on the variation of a protein within a species.   The only systematic study of this kind has been made with the human hemoglobins.   In this sense, this group of proteins is unique at this time in chemical genetics.

Structural Variation in the Human Hemoglobins.*—Several critical experimental observations were responsible for establishing the human hemoglobins as a model system for studying the genetic control of the structure of proteins. Neel,[608] in 1949, conclusively demonstrated that sickle-cell anemia is inherited in a strictly Mendelian manner. In the same year Pauling and co-workers [646] showed that the hemoglobin from individuals with sickle-cell disease has an electrophoretic mobility different from that of normal adult hemoglobin. Subsequently, several electrophoretically abnormal hemoglobins were discovered, some of which were associated with hereditary anemias.[409] At this point, however, it was impossible to decide whether the abnormal hemoglobins differed from normal by one or more specific amino acids in their primary structure or whether they differed in their three-dimensional configuration. Hemoglobin S was shown to have fewer titratable carboxyl groups [409] and a much lower solubility in the reduced state,[659] although a variety of methods, including amino-acid analysis,[380, 781] amino- and carboxyl-end-group determinations,[376] and measurements of the sulfhydryl content,[367] failed to provide additional clues as to the structural differences. The first demonstration of a precise chemical difference between adult and sickle-cell hemoglobins was made by Ingram in 1956.[400] He showed that two glutamyl residues per mole of hemoglobin *A* were absent in hemoglobin *S* and were replaced by valine. This result led to the discovery that other abnormal hemoglobins differ from hemoglobin *A* by at least one specific amino acid.

As a consequence of these studies, several hypotheses were proposed to explain the structural differences between normal and abnormal hemoglobins. The most generally accepted working hypothesis [588] suggests that the gene can be represented as a molecule of deoxyribonucleic acid and that the linear sequence of base pairs in this double-stranded, helical molecule determines the linear sequence of amino acids in the primary structure of the protein controlled by the gene, regardless of the exact mechanism of biosynthesis of the protein. From this, it is concluded that the presence of a different sequence of amino acids in an abnormal hemoglobin is the result of a mutation in the gene which controls the amino-acid sequence in the hemoglobin. The mutation is viewed as an alteration in the nature or sequence of certain base pairs in the deoxyribonucleic acid. However, it should be emphasized that this is only a working hypothesis and many detailed studies will be required in order to obtain a clearer picture. Whether or not this hypothesis is correct, it is reasonable to assume that important relationships between hemoglobin and the genes which control hemoglobin structure can be gained if the abnormal hemoglobins are characterized. Before this is possible, two major problems need solution.

---

* The nomenclature of the human hemoglobins used is that suggested by several investigators in this field.[404] Adult hemoglobin is hemoglobin *A*, fetal hemoglobin is hemoglobin *F*, sickle-cell hemoglobin is hemoglobin *S*, and so on. Several hemoglobins that are obviously different have been named with the letter *G*. Classification of all these varieties of hemoglobin has not been achieved at this time.

The first problem is the identification and localization of the aberrant amino-acid residues in the known abnormal hemoglobins. This can be accomplished by the techniques developed for the analysis of the sequence of amino acids in a protein. Of course, the logical conclusion to this problem is elucidation of the complete primary structure of hemoglobin. Whereas the complete structure will not be available for a few years, recent advances in our knowledge of globin allow at least a partial localization of those areas in the molecule which are potentially abnormal.

The second problem is to determine the behavior of the genes controlling the structure of hemoglobin. A solution to this problem is dependent on studies of families in which two or more of the genes responsible for abnormalities in the structure of hemoglobin are present together. Only a small number of such families have been described, and few of these have been studied by detailed chemical methods. However, some knowledge of the genetic control of certain parts of the hemoglobin molecule has been obtained.

METHODS FOR IDENTIFICATION OF ABNORMAL HEMOGLOBINS

A complete understanding of the abnormal human hemoglobins is not possible without knowledge of the structure of the normal protein. Consequently, in the following discussion, considerable attention will be given to the information that has been obtained with normal hemoglobin. Admittedly, this cannot be a comprehensive discussion of all aspects of hemoglobin chemistry, but it can give an adequate idea of important structural features. The excellent reviews by Schroeder [779] and Itano [409] should be consulted for more detailed information.

Moving-boundary and Zone Electrophoresis.—The application of electrophoretic methods to the identification of the abnormal hemoglobins [406, 469] has been discussed on several occasions, thus no attempt will be made to provide here the details of these procedures. However, since electrophoresis has been the most widely used technique for identification of abnormal hemoglobins, some comment on its utility should be made. The moving-boundary procedure of electrophoresis remains the most exact method and gives precise measurement of electrophoretic mobilities and isoelectric points. Satisfactory separation of some components with very similar mobilities also is achieved. However, the widespread use of zone electrophoresis on filter paper, starch gel, or starch block preparations, has permitted, in recent years, the rapid identification of several, previously unrecognized hemoglobins. These methods are generally not as precise as the moving-boundary procedures, but they have great utility, not only because of simplicity of operation and the small amount of hemoglobin required for a single analysis, but also because they can be used for screening a large number of hemoglobin samples in a short period of time. However, it should be emphasized that additional methods are needed for distinguishing between certain abnormal hemoglobins. For example, by electrophoretic analysis alone, hemoglobins *G,* *P,* and *Q,*[774, 775] as well as D_α, D_β, D_γ, and *S* are electrophoretically difficult to differentiate,[52] whereas there is little doubt that they are different hemoglobins as judged by either solubility analysis or comparison of tryptic digests of these hemoglobins (see following).

One advantage of zone-electrophoretic techniques is the ease with which the separated hemoglobins can be eluted from the supporting medium. This provides useful quantities of material for further analysis, although the hemoglobin obtained in this manner might be heterogeneous. It is clear by several kinds of analysis [177, 470, 678, 679] that hemoglobin obtained from hemolysates of red blood cells is heterogeneous. Minor heme containing proteins in a hemolysate, with the exception of hemoglobin A_2,[470] are not resolved by electrophoretic methods. The extent of the heterogeneity of hemoglobin can best be described by other methods, in particular, chromatography.

Chromatography.—Ion-exchange chromatography is one of the more promising techniques for demonstrating the heterogeneity of hemoglobin as well as for differentiating among the abnormal hemoglobins. The chromatographic procedures of Huisman, Prins, and co-workers [381, 678, 679] and Morrison and Cook [566] distinguish between certain abnormal hemoglobins as well as adult and fetal hemoglobins. However, the column-chromatographic systems described by Allen, Schroeder, and Balog [5] and Clegg and Schroeder [143] best demonstrate the heterogeneity of normal adult and fetal hemoglobins and also provide useful methods for separating abnormal hemoglobins. Their procedures involve use of the carboxylic acid resin, *IRC*–50, at 5° C. in conjunction with sodium-phosphate buffers between *p*H 6.8 and 7.2. Under these conditions, minor hemoglobin components in hemolysates are separated. Figure 27 shows a chromatogram of the hemoglobin in hemolysates of red blood cells from individuals with different types of hemoglobin. Whereas hemoglobin-*A* (figure 27) emerges at about two to three times the holdup volume of the column, hemoglobin-*S* or hemoglobin-*G* (figures 27B and 27C), can be removed as the sharp peaks shown only by warming the column to 25° C. The minor components, which emerge at column volume, normally comprise 10 to 15 per cent of the total hemoglobin in normal red blood cells.

As much as 25 to 30 per cent of the total hemoglobin is composed of minor components in red blood cells of individuals with certain hemoglobinopathies. The minor hemoglobin components shown in figure 27 are obviously heterogeneous. This is demonstrated by the pattern in figure 28, which was obtained when adult hemoglobin is chromatographed under slightly different conditions. At least eight different hemoglobins are seen in this chromatogram. These are not the result of the chromatographic treatment, because rechromatography of any of the several components does not give rise to measurable amounts of the other components.

Although hemoglobin had been demonstrated to be heterogeneous by several investigators,[177, 678] the results of Schroeder and his colleagues provide the first clear picture of this heterogeneity. Their methods now make available useful quantities of the minor hemoglobins and should allow chemical analysis of the

Fig. 27. COLUMN CHROMATOGRAPHY OF VARIOUS TYPES OF HEMOGLOBIN ON *IRC*–50.[858]

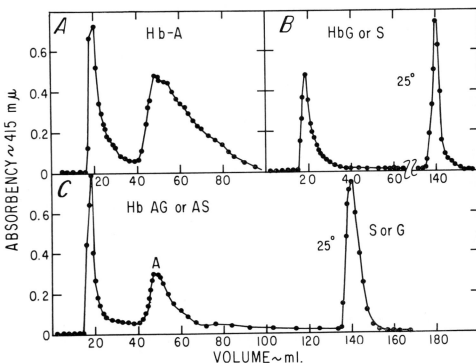

A. The resolution of minor hemoglobin components (peak at 20 ml.) of hemoglobin-*A* from the major component (peak at 50 ml.).

B. Same as Figure 27A except with hemoglobin *G* or *S*.

C. The same as Figure 27A, except with hemoglobin *AG* or *AS*. The conditions were in all cases, essentially those of Allen *et al.*[5] with phosphate buffer at *p*H 7.2 (developer no. 2) as the eluant. All columns (1 × 35 cm.) were maintained at 5° except where indicated.

different hemoglobin components. Several questions concerning the origin and function of the minor components cannot be answered at this time.[5, 143] For example, do they represent precursors of hemoglobin or are they degradation products of the aged cell; or on the other hand, are they genetically determined? At least one of the minor components, designated as A_{IIIb} in figure 28, which is identical to hemoglobin A_2 of Kunkel and Wallenius,[470] has recently been shown by Stretton and Ingram [846] to possess some amino-acid sequences different from those found in similar areas of the molecule of adult hemoglobin (A_{II} of figure 28). An independent genetic control of this component has been suggested.[122]

It should be emphasized that future genetic or chemical studies of the human hemoglobins must take into account this heterogeneity. No longer should one be content with experimental observations made on an impure protein.

Fig. 28.   COLUMN CHROMATOGRAPHY OF HEMOGLOBIN *A* ON *IRC*–50.[858]

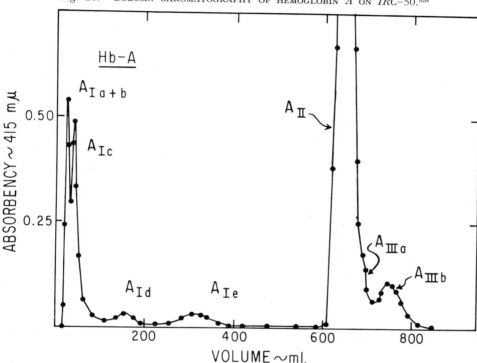

Chromatography was performed under conditions essentially like those of Clegg and Schroeder,[143] with the aid of developer no. 5.

Amino-acid Composition.—This method has provided little indication of the nature of the structural differences in several abnormal, human hemo-globins.[380, 781, 827] The results obtained to this date are also complicated by the fact that the hemoglobins that have been analyzed are, in every case, chroma-tographically heterogeneous. However, it is helpful to our understanding of hemoglobin structure to consider briefly the composition of some of the hemo-globins. Table 48 presents the composition of hemoglobins *A*, *A₂*, and *F* and the α- and β-chains of hemoglobin *A*. Discrepancies exist among the published data, consequently these values were chosen arbitrarily to be presented because they were obtained by methods of analysis which use one of the very precise ion-exchange procedures of Moore and Stein.[550, 552]

The hemoglobin *A*, the composition of which is given in table 48, was purified by zone electrophoresis on starch so that it was free of hemoglobin A_2 but did contain the A_1 components shown in figure 28. One of the most striking features of the composition of this hemoglobin is the absence of isoleucine. Brand and Grantham [94] had shown earlier by microbiological methods of analysis that human hemoglobin contained no isoleucine, although more recent studies

Table 48

AMINO-ACID COMPOSITION OF HUMAN HEMOGLOBINS

Amino acid	Hemoglobin (g./100 protein)				
	Adult [50]	α-chain [59]	β-chain [59]	A_2 [54]	F [50]
Aspartic acid	9.64	9.35	9.84	10.89	10.02
Threonine	5.13	5.78	4.61	4.23	5.69
Serine	4.05	6.03	2.82	6.03	5.42
Glutamic acid	6.55	4.42	8.77	7.62	7.38
Proline	5.02	4.83	4.82	4.05	3.78
Glycine	4.32	2.93	4.42	4.18	4.52
Alanine	9.15	10.60	7.37	7.58	8.80
Valine	10.36	8.37	12.22	7.37	9.47
Methionine	1.25	1.63	0.75	1.38	1.57
Leucine	13.94	13.35	14.62	11.74	13.75
Isoleucine	0.0	—	—	1.11	1.49
Tyrosine	3.05	2.84	2.94	2.82	2.54
Phenylalanine	7.33	6.67	7.85	6.60	7.27
Lysine	0.28	9.75	9.50	8.60	10.0
Histidine	8.32	9.45	8.17	5.76	7.28
Arginine	2.82	3.20	3.38	3.50	3.11
Cysteine	0.73–1.10	0.74	1.32	—	—
Tryptophan	2.03	1.27	2.48	—	—

revealed a small quantity of this amino acid. Allen *et al.*[5] have shown also that chromatographically purified hemoglobin contains essentially no isoleucine. Rossi-Fanelli *et al.*[732] have reported the absence of isoleucine in crystalline hemoglobin, although these results could not be confirmed.[5] Other aspects of the composition are noteworthy. Aside from the ubiquitous aspartic and glutamic acids, hemoglobin has a very high content of the aliphatic amino acids, valine, leucine, and alanine, in addition to a high histidine content. The sulfur content of hemoglobin is rather low. Five to six residues of both methionine and cysteine [147] are found per mole (*MW* 68,000). Several studies [147, 367] indicate that no cystine is present in the native protein.

Rossi-Fanelli *et al.*[733] recently reported the amino-acid composition of hemoglobin A_2. In contrast to hemoglobin *A*, this hemoglobin contains isoleucine. One would suspect that the presence of isoleucine in recrystallized hemoglobin could be caused by contamination with hemoglobin A_2. Unfortunately, this analysis is incomplete in that neither tryptophan nor cysteine were determined. The other differences in composition are consistent with the work of Stretton and Ingram,[846] who have shown some differences between the primary amino-acid sequence of this molecule and adult hemoglobin. It is possible that some of the differences could be the result of contamination of hemoglobin A_2 with catalase. It is known that this enzyme migrates electrophoretically with hemoglobin A_2 under the conditions used for isolation of the material that was analyzed.

The amino-acid composition of fetal hemoglobin differs from that of adult hemoglobin. Like hemoglobin A_2, significant quantities of isoleucine are present. The other differences undoubtedly reflect the variation seen in the primary structure of adult and fetal hemoglobins.[385, 439]

It is now established that hemoglobin is composed of two kinds of polypeptide chain.[717, 718] Although this will be discussed later, it should be noted that the average composition of these chains is in close agreement with the composition of intact hemoglobin.[350]

Amino-terminal Groups. The Polypeptide Chains of Human Adult Hemoglobin.—The examination of amino-terminal groups in the various human hemoglobins has not revealed any differences except in the case of fetal and adult protein.[337, 379, 530, 761, 778] However, this method is essential for precise sequence analysis (see following) and was very important in assessing the number of polypeptide chains in hemoglobin. Thus a few comments on results obtained by end-group analysis are warranted.

The quantitative application of Sanger's dinitrophenylation procedures to the amino-terminal residues of adult human hemoglobin originally yielded values from 3 to 5 valyl residues per mole.[337, 378, 530, 761, 778] From these studies it was apparent that hemoglobin was composed of a multiple number of polypeptide chains. However, the exact kind and number of chains was elucidated only recently by Rhinesmith, Schroeder, and Pauling.[716, 717] In a careful study of the acid-hydrolytic products of dinitrophenyl globin, these workers noted that almost 2 moles of dinitrophenyl valylleucine were quantitatively released per mole of globin after 15 minutes hydrolysis, whereas continued hydrolysis gave rise to an additional 2 moles of dinitrophenyl valine. From these observations, as well as from the stoichiometry of the release of the dinitrophenyl valylleucine and dinitrophenyl valine, it was concluded that hemoglobin has four polypeptide chains. Two of the chains have the amino-terminal sequence, valylleucine (designated α-chains), and two have amino-terminal valine (designated β-chains). Studies by Hilse and Braunitzer [355] yielded the same conclusions concerning the nature of the chains in human hemoglobin. More recently the sequence at the amino terminus of the β-chains was shown to be valylhistidylleucine.[718] A stepwise, amino-terminal analysis [257] of both chains has been made by Shelton and

Schroeder.[797]   The first six residues of the α-chains and the first eight residues of the β-chains were shown by these methods to be in the following sequence:

α-chains   Val. Leu. Ser. Pro. Ala. Asp.

β-chains   Val. His. Leu. Thr. Pro. Glu. Glu. Lys.

After it was established that two kinds of polypeptides were in human hemoglobin, procedures were soon developed for their isolation. Hill and Craig[350] separated globin by countercurrent distribution into two distinct components, one of which possessed amino-terminal valylleucine (α-chain) and the other amino-terminal valylhistidylleucine (β-chain). The clean separation achieved by this method allowed a quantitative amino-acid analysis of each chain. The results of these analyses are given in table 48. Although some discrepancy has existed in the literature concerning the number of thiol groups per mole of hemoglobin,[147, 367] the analytical data support the view that there are six thiol groups per mole. One cysteine residue is present in each of the two α-chains and two cysteine residues are present in each of the β-chains.

Ingram[404] has described the application of the chromatographic system of Wilson and Smith[933] for separation of the chains of human hemoglobin. In this procedure, globin is fractionated at pH 1.9 on a chromatographic column of *IRC*–50 with a linear gradient of urea between 2 and 8M. Chromatography by this method does not yield as pure peptide preparations as those obtained by countercurrent distribution. However, this method, because of its simplicity, is especially useful for obtaining the two chains in quantities that are sufficiently pure for certain kinds of study. Hilse and Braunitzer[933] have reported similar procedures for separation of these chains.

It is interesting to note that human fetal hemoglobin can also be separated into its component polypeptide chains by the chromatographic procedures just described. It would appear from the evidence available at this time that fetal hemoglobin contains two α-chains per mole[429, 782] that are presumably identical to the α-chains of adult hemoglobin. Two additional chains are also present which differ from both the α- and β-chain as judged by either amino-terminal analysis[782] or analysis of the tryptic peptides derived from this polypeptide.[385] These chains have been designated γ-chains and have the amino-terminal sequence, glycylhistidylphenylalanine.[797]

Some caution should be observed in accepting this rather simple interpretation of fetal globin. Two distinct fetal components are observed on chromatography of cord blood. The component present in minor amounts appears to possess properties somewhat different than those of the major fetal hemoglobin.

The final identification of the kinds and number of polypeptide chains has resulted in a convenient nomenclature for the various hemoglobins. The symbols, α and β, are used to designate the kind of chains. The superscripts, for example, α^A or β^S, designate the source of the chain and the subscripts denote the number of chains per mole. Thus, hemoglobin A is written as $\alpha_2^A \beta_2^A$, hemoglobin F as $\alpha_2{}^A\gamma_2{}^F$, and hemoglobin A_2 as $\alpha_2{}^A\beta_2{}^{A_2}$.* The abnormal hemoglobins are written in terms of the chains which are aberrant (see following) ; thus hemoglobin S, which has aberrant β-chains, is written, $\alpha_2{}^A\beta_2{}^S$, whereas hemoglobin I, which has normal β-chains but abnormal α-chains, is written, $\alpha_2{}^I\beta_2{}^A$. Hemoglobin H which has only β-chains [428] is written as $\beta_4{}^A$.

Fig. 29. "FINGERPRINTS" OF HUMAN SICKLE-CELL AND NORMAL HEMOGLOBINS.

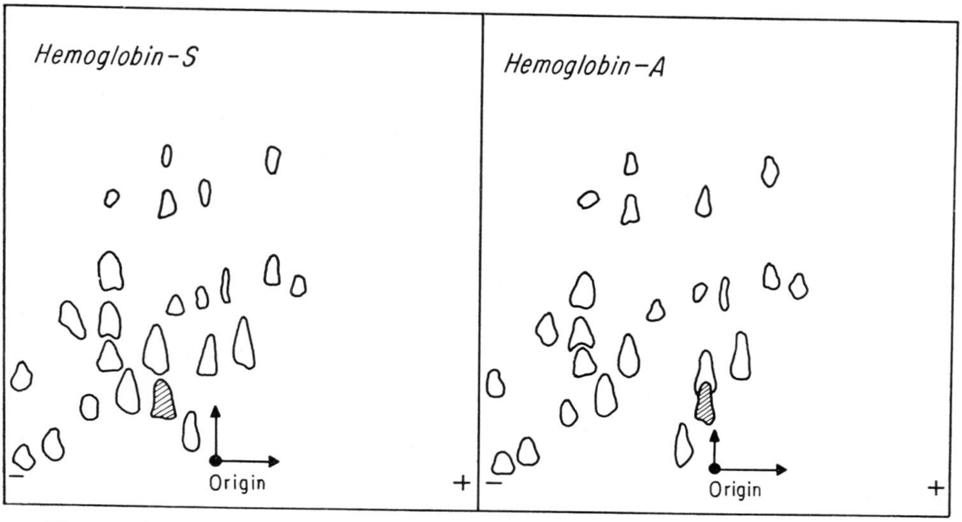

Electrophoresis was performed at pH 6.5 in pyridine-acetate buffer [352] for 3 hours at 15 v/cm. Descending chromatography in n-butanal-acetic acid-water solvent (200: 30:75). The shaded areas refer to those peptides absent in the digest of the other hemoglobin.

Analysis of Tryptic Peptides: The Hemoglobin "Fingerprint."—Ingram [400] pointed out that proteins which are too large for detailed amino-acid-sequence analysis, might be partially characterized in terms of the specific peptides that are obtained from a tryptic digest of the protein. Indeed, it was assumed that a replacement of a single amino-acid residue for another might be detected on examination of the tryptic peptides. When this approach was used to compare normal and sickle-cell hemoglobin, Ingram found that the tryptic peptides were identical with a single exception. Each hydrolysate contained one peptide not present in the other hydrolysate. The peptide mixture was resolved by a two-dimensional combination of paper electrophoresis and paper chromatography. The resulting chromatogram was called a "fingerprint" of the hemoglobin. Typical "fingerprints" for normal and sickle-cell hemoglobin are shown in figure 29.

---

* The demonstration [846] that hemoglobin A_2 contains β-chains which differ from those of hemoglobin A by at least four different amino acids, has resulted in the suggestion that these chains be called δ chains. Thus hemoglobin A_2 would be written $\alpha_2{}^A\delta_2{}^{A_2}$.

It is clear that those peptides, which are shaded, are absent in the digest of the other hemoglobin.

The detailed methods for analyzing hemoglobin by these procedures have been published [398]; however, some comment about this very important technique is warranted. One basic problem lies in the tryptic digestion. Ordinarily, digestion is obtained by incubating, at 40° C., heat-denatured hemoglobin with trypsin (2 per cent by weight of the hemoglobin) at pH 8.0. Under these conditions, digestion is essentially complete after 90 minutes as judged by measurement of hydrolysis of peptide bonds.[398] At the end of the digestion, some peptides are not soluble. These can be removed by centrifugation along with additional insoluble peptides after the digest is brought to pH 6.5. The supernatant solution contains those peptides shown in figure 29. The insoluble residue is analyzed separately. Thus, this fractionation of the tryptic digest prevents analysis of the whole molecule of hemoglobin by a single procedure. Those abnormal sequences that have been described have been found in the soluble tryptic peptides; however, the difficulty of examining the insoluble peptides leads one to suspect that aberrant sequences in this fraction might be difficult to find. The necessity of analyzing this fraction is apparent when one considers that it represents about one-third of the hemoglobin molecule.[779] Hunt and Ingram [389] have compared the insoluble tryptic residues of normal and sickle-cell hemoglobins and judged them to be identical. Their method of characterization involves splitting heme from the insoluble residue with the aid of HCl-acetone and then digesting the nearly colorless residue with chymotrypsin under essentially the same conditions as were described above for trypsin digestion. The chymotryptic peptides were then examined on paper by electrophoretic or chromatographic methods. These methods revealed peptides which appeared to be identical whether derived from hemoglobin A or S.

In my experience this technique is neither as "clean" nor as reproducible as analysis of the soluble tryptic peptides. Improved procedures for examining the tryptic residue should be found. Indeed, successful methods for examining large tryptic peptides, such as these, would have wide application, since many of the proteins, the structure of which is currently being determined, yield insoluble tryptic peptides.[351]

It is interesting to note that tryptic digestion of either the α- or the β-chains of both normal and sickle-cell hemoglobin results in an insoluble tryptic residue at pH 6.5. Thus the analysis shown in figure 29 is deficient in parts of both the α- and β-chains. In this regard, it is noteworthy that Konigsberg and Hill [461] have recently described the isolation and characterization of tryptic peptides from the α-chain. The entire amino-acid composition of these chains is accounted for by these peptides. Their choice of methods for peptide isolation wisely ignored the fact that certain peptides are insoluble at pH 6.5. Consequently, of the several peptides isolated, one was found to contain forty-two amino-acid residues, including a large amount of the aliphatic amino acids. This peptide might be expected to represent the pH 6.5-insoluble peptide of α-chains.

A second problem in Ingram's methods lies in the analytical procedures yielding the hemoglobin "fingerprint." It is obvious that structural differences that involve replacement of a charged amino acid with an uncharged amino acid have a good chance of being detected, whereas it is not apparent that replacements involving one neutral amino acid for another can be detected. The lack of assurance that all kinds of amino-acid replacements can be observed seriously limits this method.

In spite of the problems of the combined electrophoretic-chromatographic method, there is little doubt that some of the significant advances made in the elucidation of chemical differences in the hemoglobins are the result of using this technique. Indeed, its usefulness in protein analysis is reflected by the attention that has been given to modifications of the original design of the method. Katz, Dreyer, and Anfinsen,[440] in particular, have described detailed methods for obtaining "fingerprints" of tryptic digests. They show that certain tryptic digests can be analyzed best on paper by resolving the peptide mixture chromatographically and then electrophoretically; just the opposite of the Ingram technique. This could be useful for some tryptic-peptide mixtures, but little advantage can be seen for obtaining hemoglobin "fingerprints" by this method. Considerable variation in the conditions of electrophoresis and chromatography still provide good "fingerprints." A wide variety of voltage gradients can be used for the electrophoretic separation. In addition, it seems to be unimportant whether ascending or descending chromatographic resolution is used as the second dimension in this technique.

Paper Electrophoresis of Tryptic Peptides.—Paper electrophoresis also has proved useful for identification of chemical differences in the hemoglobins. With the aid of a system for cooling the paper during the electrophoresis, it is possible to use rather high voltages to obtain a rapid undimensional separation of tryptic peptides. Figure 30 shows typical patterns obtained after high-voltage separation of the tryptic peptides of hemoglobins A, S, and G.[858] A single basic peptide in the digests of S and G is absent in A, whereas a neutral peptide, present in the A digest, is absent in both S and G. These are the same peptides observed in the hemoglobin "fingerprints" shown in figure 29.

Of particular use in analyzing the peptides on paper are spray reagents that detect specific amino acids in the peptides. Peptides which contain arginine,[1] histidine,[68] tyrosine,[1] and trypotphan [816] can be located in this manner. Under certain circumstances this is useful for detecting some amino-acid interchanges. For example, differences in the D hemoglobins [52] were observed with the aid of these reagents.

Isolation and Characterization of Peptides.—A detailed discussion of this subject has been given in several reviews.[499, 551, 780] Those procedures described for the isolation of peptides from enzymic or acid hydrolysates could be applicable to hydrolysates of hemoglobin. In many cases, however, peptides have been isolated from tryptic digests of hemoglobin by separating the peptides by high-voltage electrophoresis (figure 30), or by use of the two-dimensional electrophoretic-chromatography procedure (figure 29). Most of the peptides, the sequences of which are shown in table 49, were isolated in this manner. When these methods are used for preparative purposes, the peptides are first resolved and then located on the paper by spraying with a dilute ninhydrin reagent.[499] The ninhy-

Fig. 30. Undimensional electrophoretic separation on paper of soluble tryptic peptides derived from hemoglobin A, S, and G.

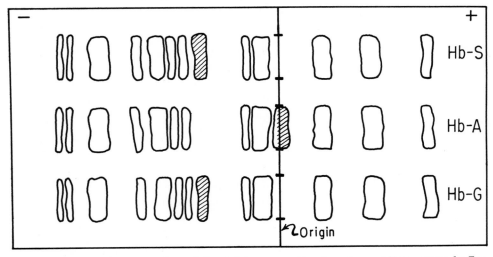

Electrophoresis was performed for 1.5 hours at 45 v/cm. in pyridine-acetate buffer, pH 6.4. The paper was cooled throughout the electrophoresis.[352]

drin-reactive spots are cut from the paper, extracted with acetone in order to remove excess ninhydrin and then eluted with an aqueous solvent to remove the peptide. Peptides resolved by high-voltage electrophoresis can be located with the aid of narrow strips cut from the sides of the paper. These strips are sprayed with ninhydrin, developed, and used as markers for locating those areas of the unsprayed paper containing the peptides of interest.

Isolation of large quantities of peptide by these methods is laborious and low yields invariably are encountered. For these reasons, a systematic isolation of large quantities of all of the peptides in a single hydrolysate of hemoglobin should be made by other methods, such as ion-exchange chromatography, partition chromatography, or countercurrent distribution.[901] For example, Konigsberg and Hill [461] in their studies with α-chains have systematically used fractional dialysis,[155] countercurrent distribution,[901] and separation by gel filtration [669, 670] for the isolation of rather large amounts of fifteen pure peptides. The composition of the peptides accounts for all the amino acids in the α-chain. Braunitzer [98, 355, 507] and his co-workers have also reported the isolation and partial characterization of twenty-one soluble, tryptic peptides derived from both α- and β-chains. Several methods of separation were used, including chromatography on Dowex-50 or Dowex-1. From the Dowex-1 system, twelve unique peptides from each chain were identified. Un-

Table 49

PRIMARY STRUCTURAL DIFFERENCES IN THE HUMAN HEMOGLOBINS

Hemoglobin	Sequence	Polypeptide chain	References
A [1]	Val. His. Leu. Thr. Pro. *Glu. Glu.* Lys.	β	349, 352, 399
S	Val. His. Leu. Thr. Pro. *Val.* Glu. Lys.	β	352, 399, 404
C [2]	Val. His. Leu. Thr. Pro. *Lys.* Glu. Lys.	β	386, 404
G	Val. His. Leu. Thr. Pro. Glu. *Gly.* Lys.	β	349, 352
A	Val. Asp. Val. Asp. Glu. Val. Gly. Gly. *Glu.* Ala. Leu. Gly. Arg	β	389
E	Val. Asp. Val. Asp. Glu. Val. Gly. Gly. *Lys.* Ala. Leu. Gly. Arg	β	389
$D\beta$	Defect in this peptide (no. 23)	β	52, 404
A	Ala (Val, Try, Gly) *Lys.* Val (His, Leu, Tyr, Gly₃, Glu₃, Ala₄) Arg	α	582, 412, 581
I	Ala (Val, Try, Gly) *Glu.* Val (His, Leu, Tyr, Gly₃, Glu₃, Ala₄) Arg	α	582, 412, 678
$D\alpha$	Defect in this peptide (no. 26)	α	52, 404
G_I Hopkin's-2 P K	All these hemoglobins have aberrant α-chains	α	798, 411, 179, 179
A	$\alpha_2{}^A\beta_2{}^A$	—	802, 333, 129
H	$\beta_4{}^A$	—	385
F	$\alpha_2{}^A\gamma_2{}^F$	—	390
"Bart's"	$\gamma_4{}^F$	—	

[1] The sequences for A, S, and C are slightly different than first reported.[402]
[2] The amino acids that are italicized are those which are altered in the indicated hemoglobins.

fortunately, certain discrepancies exist between the data of these workers and those of Konigsberg and Hill.[461] However, it is clear that the complete structure of hemoglobin must be established before it will be possible to obtain a thorough understanding of the abnormal hemoglobins. These studies represent the first important step toward this end.

Several peptide sequences in the human hemoglobins are known. Those that have been published thus far are shown in table 49. These sequences were established in conjunction with studies concerning the nature of the chemical abnormalities in certain hemoglobins and do not include those described by Braunitzer *et al.*[98, 355, 507] or Konigsberg and Hill.[461]

It is impossible to discuss in detail methods for establishing sequences of amino acids in peptides. Details of the procedures for sequence analysis can be found by consulting the several excellent reviews on this subject.[257, 503, 757] These methods, however, are extremely fundamental for determining the kinds of aberrant sequences which exist as the result of gene mutations, and certainly the final concepts of the nature of how a gene controls the structure in a protein will only be as good as the methods of sequence analysis. Consequently, certain fundamental principles of sequence analysis, which are often ignored, should be mentioned.

Fig. 31. PAPER-ELECTROPHORETIC SEPARATION OF THE SOLUBLE, TRYPTIC PEPTIDES DERIVED FROM HEMOGLOBINS A AND G AND THE α- AND β-CHAINS OF THESE HEMOGLOBINS.[352]

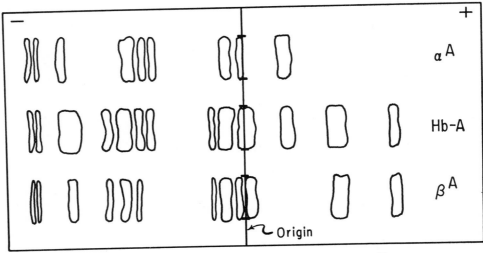

The conditions for separation are given in Figure 30.

Preliminary to any sequence analysis, it is imperative to establish, as unequivocally as possible, the purity of the peptide under study. This can be achieved by many methods, including various types of chromatography, end-group analysis, and so on. Then, when purity is established, it is necessary to determine, by quantitative methods,[550, 552] the precise amino-acid composition of the peptide. This will serve as another criterion of purity, but will also define the amino acids that must be dealt with in the sequence analysis. It is particularly important to know the composition of the peptide in that choice of techniques, for detailed sequence determination will depend on the amino acids present in the peptide.

Determination of the Aberrant Polypeptide Chain.—The direct method for locating the aberrant chain involves isolation of the individual polypeptide chains followed by analysis of the tryptic peptides derived from these chains. Obviously the chain that is aberrant would contain a tryptic peptide not normally found. An example of this, with hemoglobin G,[858] is shown in figure 31.

Another method, which is particularly useful, is based on the independent studies of Itano and Singer [413, 802] and Hasserodt and Vinograd.[333] They showed that hemoglobin will asymmetrically dissociate near pH 5 as follows:

$$\alpha_2^A\beta_2^A \rightleftharpoons \alpha_2^A + \beta_2^A,$$

whereas reassociation occurs at neutral pH. Consequently, when a mixture of hemoglobin A and an abnormal hemoglobin is brought to a dissociating pH and then to neutrality, hybrid molecules will be formed. The hybrids will behave as either the normal or abnormal hemoglobin and can be separated by electrophoretic or chromatographic methods. Considerable information can be obtained if it is possible to label one of the hemoglobins which is hybridized. The most useful label is with C^{14}-leucine, which can be incorporated biosynthetically into hemoglobin A.[896] After hybridization with C^{14}-leucine-hemoglobin A, it is possible to determine the amount of radioactivity in each of the chains of the isolated hybrids (by counting the C^{14} in the leucine of the dinitrophenylpeptides derived from the dinitrophenyl hybrids) and thus localize the aberrant chain. Consequently, sickle-cell hemoglobin [896] has been shown to form hybrids with C^{14}-leucine-hemoglobin A as follows:

$$\alpha_2^{A*}\beta_2^{A*} + \alpha_2^S\beta_2^S \rightleftharpoons \alpha_2^{A*}\beta_2^S + \alpha_2^S\beta_2^{A*}.$$

After hybridization, the hemogoblin S that can be isolated has radioactivity only in the α-chains. This result clearly shows that hemoglobin S contains abnormal β-chains and normal α-chains and should therefore be written as $\alpha_2^A\beta_2^S$.

CONCLUSIONS

The recent interest in the problem of the genetic control of protein structure has added a great deal of impetus to studies on the primary structural analysis of hemoglobin. Although these studies are far from complete, several structurally-different forms of human hemoglobin have been recognized. Normal adult hemoglobin $(\alpha_2^A\beta_2^A)$, which is composed of two α- and two β-chains, may be considered the norm from which all other human hemoglobins differ. At least nine hemoglobins (figure 28) in addition to hemoglobin A have been found in individuals who are homozygous for hemoglobin A in the classical genetic sense. The two fetal variants have two α-chains that are identical to those of major adult hemoglobin and two γ-chains.[129, 385, 782] The γ-chains are similar to β-chains as judged by comparing the tryptic peptides of fetal and adult hemoglobins.[385, 439] Differences have not been described in the primary structure of the two forms of fetal hemoglobin. Only one of the remaining seven variants, hemoglobin A_2, has been studied in detail. This hemoglobin [846] also contains two α-chains, but appears to possess two additional chains that are different from any found in adult or fetal hemoglobin. This hemoglobin, designated $\alpha_2^A\beta_2^{A_2}$, is present in only small amounts in red blood cells and together with the other minor hemoglobins comprise about 10 to 15 per cent of the total circulating hemoglobin. The remaining minor hemoglobins have not been characterized.

In contrast to these hemoglobins are those resulting from a mutation of the genes responsible for formation of hemoglobin A. These are the abnormal hemoglobins which are designated by their component chains in the same manner as hemoglobin A but with different superscripts to indicate the aberrant chain, for example, $\alpha_2^I \beta_2^A, \alpha_2^A \beta_2^S$, et cetera. Whereas only a few of the abnormal hemoglobins have been shown to be inherited in a strictly Mendelian manner, it is probable that they are formed as the result of a single gene mutation.[599] It is not known whether a series of abnormal hemoglobins analogous to the minor, A hemoglobins are present in individuals who carry mutant genes. Apparently they are found, but detailed studies will be needed to establish this point.[858]

In contrast to both these groups of hemoglobins are those composed of only one kind of polypeptide chain. Thus far, two of these hemoglobins are known, "Bart's"[390] and hemoglobin H,[129] which have the chain formulas γ_4^A and β_4^A, respectively. These have only been found in individuals who have thalassemia,[590] and their genetic basis is not fully understood. It would seem that they might arise as the result of insufficient α-chain synthesis, although this has not been demonstrated. Other hemoglobins composed solely of other kinds of chains have not been found, for example, α_4^A.

Thus, several variants of human hemoglobin are known. Some contain more than one primary structural difference, for example, fetal and A_2 hemoglobins, whereas others appear to contain only a single alteration, such as hemoglobins S, C, D, E, and so on. In addition are hemoglobins Bart's and H, which are composed solely of one kind of chain.

It has not been possible to study the primary and secondary structures of all types of human hemoglobin. However, on the basis of studies on hemoglobin A, it can be assumed that each hemoglobin, regardless of the differences found in its globin moiety, is composed of 4 polypeptide chains and 4 heme prosthetic groups. The various chains are held together by noncovalent forces to form the three-dimensional globular protein. These forces probably involve hydrogen bonds, ionic or salt linkages, hydrophobic forces, and others. The rather high content of leucine, valine, and alanine (see table 48) in hemoglobin suggests that hydrophobic interactions may play a large role in interchain attraction and folding. Little can be said about the secondary or tertiary structure of human hemoglobin, although it is noteworthy that Perutz and co-workers [660] have recently shown by X-ray analysis that equine oxyhemoglobin crystals have a polypeptide-chain fold similar to that described by Kendrew *et al.*[442] for sperm-whale myoglobin. It is not known whether this chain fold also occurs in human-hemoglobin crystals.

With these structural features in mind, it is interesting to consider certain genetic aspects of the formation of hemoglobin. On the basis of the structural differences in abnormal hemoglobins (table 49), apparently there is a tendency for the hemoglobin to be altered in only certain parts of the molecule, a situation reminiscent of the species differences in insulin. On the basis of existing hypotheses, this could reflect instability in certain parts of a gene (sequences in a deoxyribonucleic-acid molecule). This situation also could occur if alterations in certain parts of the gene produced changes that prevented the hemoglobin from functioning as a respiratory protein. Obviously such a hemoglobin in all probability never would be formed. In this light, no reason exists for belief in a greater mutability in one part of a gene than another.

One problem of great interest concerns the number of genes controlling the structure of hemoglobin. In 1957, Schwartz et al.[787] suggested two nonallelic genes must control the structure of hemoglobin. Whereas the evidence that led these workers to this conclusion now needs new interpretation, it is clear that their suggestion is indeed true. Smith and Torbert [815] also arrived at this conclusion in their studies on hemoglobin S and Hopkins-2 hemoglobin. Itano and Robinson [410, 411] have provided the best evidence for this nonallelism in studies on the hemoglobins in individuals who carry genes for both hemoglobin S and Hopkin's-2 hemoglobin. Because Hopkin's-2 hemoglobin has an aberrant α-chain [410] and hemoglobin S an aberrant β-chain,[404] an individual heterozygous for both genes would possess four different hemoglobin variants if the genes controlling these hemoglobins segregated independently. On examining the hemoglobins from such individuals, four species were found:

$$\alpha_{\frac{A}{2}} \beta_{\frac{A}{2}}, \quad \alpha_{\frac{H_o}{2}} \beta_{\frac{A}{2}}, \quad \alpha_{\frac{A}{2}} \beta_{\frac{S}{2}}, \quad \alpha_{\frac{H_o}{2}} \beta_{\frac{S}{2}}.$$

Of particular interest is the problem of whether more than one gene controls the structure of a single chain of hemoglobin. Most workers in this field have accepted the "one gene-one enzyme" concept of Horowitz and Fling [369] which, in essence, is consistent with the concept of genetic control of the structure of a protein. Indeed, this is understandable in that no evidence to the contrary is provided by analysis of human hemoglobins. However, in a recent study, Hill, Schwartz, and Swenson [352] showed that hemoglobin G contains an aberrant β-chain. The seventh residue from the amino terminus of this chain was found to be a glycyl instead of the glutamyl residue present in hemoglobin A (see table 49). This finding is in contrast to that found in hemoglobin S, where the sixth residue from the amino terminus of the β-chain is a valyl instead of a glutamyl residue. Because hemoglobin G was obtained from a family that also had the gene for hemoglobin S, and because analysis of a pedigree of the family indicated the genes for hemoglobins S and G were nonallelic, this seemed to indicate that two nonallelic genes controlled the structure of β-chains. This exception to current ideas was resolved when the pedigree was reexamined by the detailed chemical methods used to judge the nature of the abnormality in hemoglobin G. Clearly, one critical individual had no hemoglobin G as judged by these methods, although by electrophoretic analysis a component was present which migrated similarly. The electrophoretic component identified as hemoglobin G in the critical individual seems to be, in all probability, a mixture of hemoglobins found normally in minor amount. These components comprise 30 per cent of the total hemoglobin in this individual and appear to migrate electrophoretically at *pH* 6.5 like hemoglobin G. Additional studies will be needed to explain the nature of these minor-hemoglobins components; however, it can be assumed that the inheritance of hemoglobins G and S is consistent with the present ideas about the nature of the genetic control of hemoglobin structure. This study also provides an excellent example of the difficulties which can be encountered in chemical genetics when the most precise chemical methods are not used.

DISCUSSION

DR. BURDETTE: Dr. Jones, will you open the discussion, please?

DR. JONES: I would like to congratulate Dr. Hill on his paper and compliment him, Dr. Herbert Schwartz, and their collaborators on their study of the enigma of hemoglobin G. They deserve special credit for the careful application of the methods of protein chemistry which they have made to this genetic problem. They have recognized that studies of the primary structure of proteins are chemical in nature and yield genetic information only when properly applied. Their conclusion concerning the allelism of the hemoglobin S and hemoglobin G mutations has not been based primarily upon the observation that alterations of the amino-acid sequences of these two hemoglobins occur in adjacent residues of the beta chain. Rather, it has been based upon a critical study of the various hemoglobins present in the genetically informative members of this interesting family. Although this approach may seem too conservative, many protein chemists and geneticists believe that only through such detailed studies can valid conclusions of a genetic nature be made from investigations of chemical structure alone.

During the past few years a great deal of new information has been obtained about the relationship between the chemical structure and the genetic control of hemoglobins. Many methods have contributed to this advance. Several of these methods are currently being used in the laboratories at the California Institute of Technology. Brief mention of some of the applications of these procedures may be of interest. My discussion is not intended to be general, but will be limited mainly to work recently completed in our laboratories, much of which has not yet been published.

Starch-grain electrophoresis and column chromatography with Amberlite *IRC*–50 are useful both for qualitative identification and preparative isolation of many different hemoglobin proteins. Schnek and Schroeder [777] have compared the various components that can be isolated from normal adult hemoglobin by these two techniques.   A summary of some of their results is given in figure 32. The resolution of hemoglobin into components A_1, A_2, and A_3 by starch-block electrophoresis is represented in the middle portion of the figure (a).   Chroma-

Fig. 32.   COMPARISON OF HEMOGLOBIN COMPONENTS FROM STARCH-BLOCK ELECTROPHORESIS AND CHROMATOGRAPHY WITH *IRC*–50.

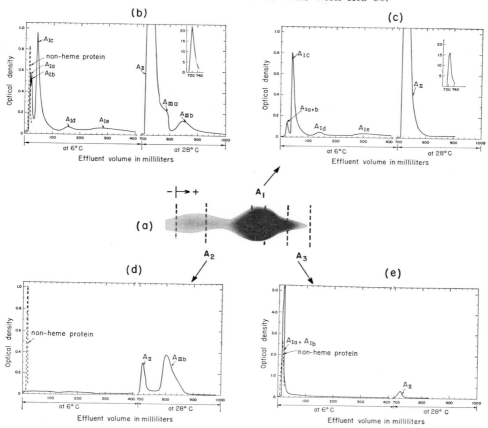

(a) Starch-block electrophoresis. (b) Chromatography with *IRC*–50. (c), (d), and (e) Chromatography of starch-block components, A_1, A_2, and A_3 respectively.

tography of the same hemoglobin with Amberlite *IRC*–50 is illustrated in the upper left corner (b).   Chromatography of components A_1, A_2, and A_3 first isolated by electrophoresis is also illustrated as shown.   It is apparent that the electrophoretic component A_1 is still an heterogeneous mixture when studied further by chromatography.

The hemoglobin components isolated by chromatography with *IRC*–50 from normal adult individuals have been found to be constant in kind and amount.

Studies of hemoglobin components from individuals with various abnormal hemoglobins, including hemoglobins *C, D, E, H, L, Q,* and *S* have revealed many other minor hemoglobin components. Several of the minor components associated with these abnormal hemoglobins are chromatographically similar to some of the normal minor components, although others are quite different. Many of the latter group of abnormal minor components reflect, at least in part, an alteration in structure analogous to that of their corresponding major hemoglobin component.[427]

The separation of two or more hemoglobins by electrophoresis or chromatography is primarily dependent upon differences in electrical charge. Therefore, two hemoglobins differing in primary structure may not be distinguishable by these techniques unless differences in charge exist. It is possible that genetic and biosynthetic variants of the normal adult hemoglobin exist which have not yet been detected by these charge-dependent procedures. However, small differences in chromatographic behavior of two similarly-charged hemoglobins can often be detected by the introduction of small amounts of an appropriate radioactive hemoglobin. Thus, when hemoglobin D^{848} is chromatographed after the addition of a small amount of hemoglobin *S* containing C^{14} leucine, one finds that the maxima of the radioactivity and optical-density curves do not coincide.[427] This tracer study indicates that hemoglobin *D* migrates slightly faster than hemoglobin *S,* in spite of the observation that mixtures of hemoglobins *S* and *D* result in only one maximum in the optical-density curves. Hemoglobins *S, D,* and mixtures of *S* and *D* are readily differentiated by this tracer technique. Procedures of this type may also be useful in the study of other hemoglobins having similar electrophoretic and chromatographic properties.

Hemoglobinopathies are generally difficult to diagnose at birth because of the presence of large amounts of fetal hemoglobin. Matsuda and co-workers [532] have found that column chromatography can be utilized to detect and estimate the "adult" forms of hemoglobins in the human fetus at as early an age as three months gestation. Hemoglobins *S* and *C* as well as hemoglobin *A* can readily be detected and differentiated from one another by these procedures. Several cases of homozygous sickle-cell disease have been detected at birth and followed during their first year of life by this method.[427]

Studies of the kinetics of denaturation of hemoglobins in alkaline solutions are valuable for the classification of hemoglobins into several groups. Weliky, Jones, and Pauling [917] have demonstrated that some components of human hemoglobin can be grouped into four general classes having rates of denaturation in the following order: tetra beta (major components of hemoglobin *H*); faster than normal, adult hemoglobin; faster than tetra gamma (a minor component present in some hemoglobin-*H* individuals which is also identical to hemoglobin Bart's); and faster than fetal. Several minor components isolated from umbilical-cord blood, from normal adults, and from adults with hemoglobinopathies have been found to have denaturation rates similar to the major, fetal-hemoglobin component, but to differ from it in chromatographic behavior.

From the examination of the ultra-violet-absorption spectra in the region of 289μ, it is possible to group hemoglobins according to their relative tryptophan content. Thus, fetal hemoglobin can be shown to contain more tryptophan than the adult hemoglobins, although tetra gamma hemoglobin contains almost twice the amount of tryptophan when compared to fetal. The various minor components can be grouped into those with tryptophan content like adult and those like fetal hemoglobin.[427]

While the methods of electrophoresis, chromatography, alkali denaturation, and ultra-violet spectroscopy are valuable for the empiric classification of hemoglobins, more critical methods of chemical characterization are necessary and available. Three of these are the determination of the N-terminal-amino-acid residue or peptide sequence by the *DNP*-method,[717] molecular-subunit hybridization [896] or chain-recombination procedures,[411] and the examination of peptide patterns resulting from partial, enzymatic hydrolysis of hemoglobins.[398] Employing these three procedures in combination with the determination of molecular weights, the gross structure of several hemoglobins have been determined to be the following: $A = \alpha_2\beta_2$, $L = \alpha_2^L\beta_2$, $F = \alpha_2\gamma_2$, major $H = \beta_4$, minor H or Bart's $= \gamma_4$, and $Q = \alpha_2^Q\beta_2$; in which α and β are the N-terminal val-leu and val-his-leu chains respectively of normal adult hemoglobin and γ is the N-terminal, glycine chain of fetal hemoglobin. The superscripts indicate alterations in primary structure which are characteristic in a particular hemoglobin. Recently, similar studies have been completed on the hemoglobin component of Kunkel's A_2 fraction.[776] This hemoglobin has been found to contain two chains, N-terminal in val-leu, which are identical to the α chain of the major component of A hemoglobin, and two chains, N-terminal in val-his-leu, which are similar but not identical to the β-chains of the major component of A hemoglobin. This conclusion has been confirmed independently by the recent primary-structure work of Stretton and Ingram.[846]

Questions concerning the chemical and genetic evolution of proteins are becoming increasingly interesting and approachable by new techniques. A comparative study of the structure of animal hemoglobins has been started by Zuckerkandl, Jones, and Pauling [958] which employs the method of gross examination of the peptide patterns resulting from tryptic hydrolysis of various hemoglobins. Figures 33 and 34 represent some of the peptide patterns that have been obtained. A strong correlation has become apparent between the fine structure of hemoglobin and the phylogenetic relationships of the animals studied. Peptide patterns from tryptic hydrolysis of piscine hemoglobins have few features in common with those of human hemoglobin, although the hemoglobin patterns of the hog and cow have many peptides that resemble those of the human. Patterns of gorilla and chimpanzee are nearly indistinguishable in appearance from the human,

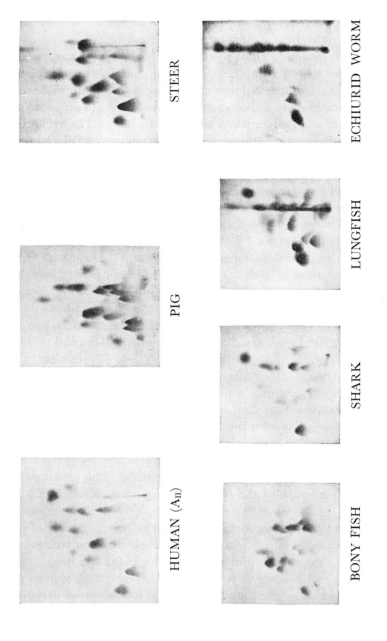

Fig. 33. PEPTIC PATTERNS OF TRYPTIC HYDROLYSATES OF VARIOUS ANIMAL HEMOGLOBINS.

STEER

ECHIURID WORM

PIG

LUNGFISH

SHARK

HUMAN (A$_{II}$)

BONY FISH

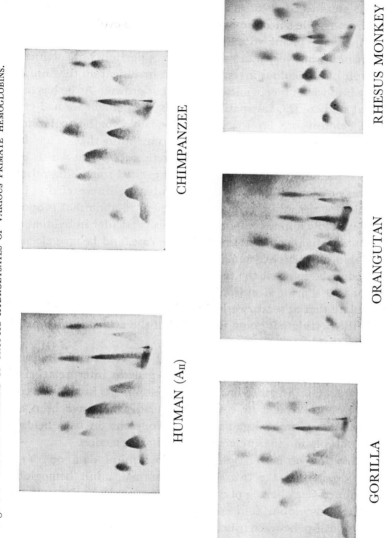

Fig. 34.   Peptide patterns of tryptic hydrolysates of various primate hemoglobins.

RHESUS MONKEY

CHIMPANZEE

HUMAN (A$_{\mathrm{II}}$)

ORANGUTAN

GORILLA

whereas other primate patterns show a few definite differences. It may be concluded that amino-acid substitutions have occurred at a number of sites in the hemoglobin molecule since the onset of vertebrate evolution.

DR. BURDETTE: Thank you, Dr. Jones. Dr. Neel also wishes to comment.

DR. NEEL: I want very much to take this opportunity to congratulate Dr. Hill on a real epic of meticulous investigation, an investigation that illustrates the full complexity into which biochemical genetics has grown. As he has brought out, in the original study of this fascinating pedigree we thought we had encountered an individual heterozygous for the genes responsible for hemoglobin G, hemoglobin S, and thalassemia.[787] The grounds for concluding that he had hemoglobin G was the apparent presence of an electrophoretic component which behaved as hemoglobin G, plus the solubility data that suggested something other than hemoglobin S was present It now appears that what we interpreted as hemoglobin G is a collection of the minor components present in abnormal proportions. That is, as a group, these minor components in this particular individual are present to a greater extent than would normally be the case, and behave electrophoretically as hemoglobin G. This accounts then for the evidence that seems to have led us astray. Remembering that the techniques of "fingerprinting" were not generally available at that time, it is an ironic commentary that the study of this family led us to postulate that at least two loci are involved in hemoglobin production. It looks as if the postulate is going to stand, but the evidence on which it was based has now collapsed.

DR. MOTULSKY: I am a great admirer of Dr. Hill's and Dr. Jones' work and certainly share their caution not to push genetic interpretations beyond the available data. On the other hand, the developments in the chemistry of hemoglobin have occurred at a time when great advances have been made in fine-structure genetics of microörganisms. Developments in both fields have yielded powerful concepts for the understanding of genic action.

Consideration of both family and biochemical data on the hemoglobins leave little doubt that the two chains of normal, adult hemoglobin (and the gamma chain of fetal hemoglobin as well) are under independent genetic control.[387] The hemoglobin data are suggestive that the current dogma of a one-to-one relationship between nucleotides in *DNA* and amino-acid sequence in proteins probably is correct. No doubt, absolute proof of this concept will have to come from microbial genetics, unless production of hemoglobin and somatic crossing-over can be achieved in human erythropoetic tissue cultures.

Figure 35 demonstrates the current concept of the genetic control of hemoglobin production. There are three polypeptide chains, the alpha, beta, and gamma chains. Each of these is under control of a pair of independent genes. Normal adult hemoglobin is $\alpha_2\beta_2$. Normal fetal hemoglobin is $\alpha_2\gamma_2$. Normally, the $Hb\gamma$ gene is almost completely suppressed in adult life. If an individual is an heterozygote for a beta-chain mutation, such as a sickle-trait carrier, he makes both normal ($\alpha_2\beta_2$) and sickling hemoglobin ($\alpha_2\beta_2{}^S$). Polypeptide

Fig. 35. GENETIC CONTROL OF HEMOGLOBIN PRODUCTION IN MAN.

Hb LOCI

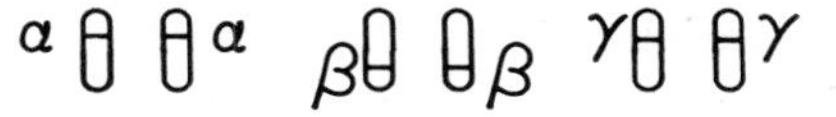

$$\alpha\boxed{}\;\boxed{}^{\alpha}\quad {}_{\beta}\boxed{}\;\boxed{}_{\beta}\quad {}^{\gamma}\boxed{}\;\boxed{}^{\gamma}$$

	GENE NOTATION	GENE PRODUCT(S)
Normal man	$Hb\alpha^+/Hb\alpha^+$ $Hb\beta^+/Hb\beta^+$ $(Hb\gamma^+/Hb\gamma^+)$	$\alpha_2^+\beta_2^+$ (Hb A) $\left[\alpha_2^+\gamma_2^+ \text{ (Hb F)}\right]$
Sickle trait carrier	$Hb\alpha^+/Hb\alpha^+$ $Hb\beta^+/Hb\beta^S$	$\alpha_2^+\beta_2^+$(Hb A)* $\alpha_2^+\beta_2^S$(Hb S)
Sickle cell anemia	$Hb\alpha^+/Hb\alpha^+$ $Hb\beta^S/Hb\beta^S$	$\alpha_2^+\beta_2^{\ S}$ (Hb S)
Hb I carrier	$Hb\alpha^I/Hb\alpha^+$ $Hb\beta^+/Hb\beta^+$	$\alpha_2^+\beta_2^+$(Hb A)** $\alpha_2^I\beta_2^+$(Hb I)
Double heterozygote S & I	$Hb\alpha^I/Hb\alpha^+$ $Hb\beta^+/Hb\beta^S$	$\alpha_2^I\beta_2^+$ (Hb I) $\alpha_2^+\beta_2^S$ (Hb S) $\alpha_2^+\beta_2^+$ (Hb A) $\alpha_2^I\beta_2^S$ (Hybrid S/I)

$*$ not $\alpha_2^+\beta^+\beta^S$
$**$ not $\alpha^+\alpha^I\beta_2^+$

Note that three independent genic pairs are involved, each synthesizing a different polypeptide chain. The symbol "+" refers to normal.

chains may be presumed to dimerize following biosynthesis since the hemoglobin chains only occur in duplicate, and single chains such as $\alpha_2\beta\beta^S$ do not occur. A homozygote for the sickling mutation lacks normal $Hb\text{-}\beta$ genes and, therefore, only makes the abnormal sickling hemoglobin $(\alpha_2\beta_2^S)$. The capacity for making normal fetal hemoglobin in such an individual is not affected by the mutation, and fetal hemoglobin $(\alpha_2\gamma_2)$ is often produced in increased quantities in cases of sickle-cell anemia. Heterozygotes for an alpha-chain mutation, such as hemoglobin *I*, makes two hemoglobins: hemoglobin *A* $(\alpha_2\beta_2)$ and hemoglobin $I(\alpha_2^I\beta_2)$. Double heterozygotes for the beta-chain mutations, such as hemoglobin C and S and Hb G and S (see following), only make these two abnormal hemoglobins, since they lack the normal β-chain to make hemoglobin A $(\alpha_2\beta_2)$.

The most instructive examples have been individuals heterozygotic for both an alpha- and a beta-chain mutation. Such individuals have four different hemoglobin genes: a normal and an abnormal Hb-α gene as well as a normal and abnormal Hb-β gene. If the genetic control of the two genes is indeed independent, such persons should make four types of hemoglobin: normal hemoglobin (Hb A or $\alpha_2\beta_2$), abnormal hemoglobin with the abnormal alpha gene (e.g., Hb I or $\alpha_2{}^I\beta_2$), abnormal hemoglobin with the abnormal beta gene (e.g., Hb S or $\alpha_2\beta_2{}^S$), and a new type of hybrid molecule having both abnormal alpha- and beta-chains (e.g., Hb S/I or $\alpha_2{}^I\beta_2{}^S$). Several such cases exist.[22, 407] In one instance, only three hemoglobins were found initially.[407] Because of similarities in electrophoretic charge, the double, hybrid molecule overlapped the normal hemoglobin on electrophoresis. Its true nature could be identified after isolation.

The demonstration that the amino-acid alteration of hemoglobin G affects the beta-chain made it difficult to understand the data of Dr. Schwartz's pedigree as published earlier.[787] An additional body of data in relation to thalassemia needs to be cited for full understanding of this pedigree. It is now apparent that the type of thalassemia associated with a high Hb-A_2 component seen frequently in patients of Mediterranean origin is allelic or closely linked to the beta-hemoglobin locus.[123] The action of this particular thalassemia gene leads to a decreased rate of beta-chain formation.* Dr. Schwartz tells me that the type of thalassemia in this family is associated with a high A_2 hemoglobin. Since hemoglobins S and G have close electrophoretic mobilities, I had always doubted the presence of hemoglobin G in individual III–1 and early this year drew up a pedigree of this family to fit the current concept of hemoglobin genetics, as previously outlined (figure 36). I find that Dr. Ingram's reasoning in this problem proceeded along similar lines.[405]

Dr. Hill's failure to find hemoglobin G in individual *III*–1 helps considerably in strengthening the interpretation outlined on the pedigree chart. Individuals *II*–5 and *II*–7 are sisters who have G-thalassemia. Since this thalassemia suppresses beta-chain formation, they have mostly hemoglobin G in their cells. Individual *II*–5 married a sickle-trait carrier, *II*–4. Their child, *III*–4, has Hb G–S "disease," inheriting the sickling gene from one parent and the Hb G gene from the other parent. Since both the Hb-S and Hb-G genes cause β-chain alterations, no Hb A is present in this individual. Individual *III*–1 inherited the thalassemia gene from one parent, the sickling gene from the other parent, and passed on the thalassemia gene to his child, *IV*–1. The absence of normal hemoglobin (Hb A) in *III*–1 with sickle-cell-thalassemia disease is due to the presence of the thalas-

---

* It is immaterial whether Ingram's hypothesis [405] of nonelectrophoretic beta-chain mutation (drastically altering the beta chain in such a manner that it cannot be formed in normal amounts) is accepted rather than the view that beta-chain formation is suppressed in an unknown manner.

semia gene. The rather high amount of the minor component present in this patient will need further study. One more point needs resolution in this pedigree. Individuals *I–3* is stated to be normal by blood films. I understand that his type of hemoglobin has not been determined yet. This individual certainly does

SCHWARTZ ET AL. PEDIGREE

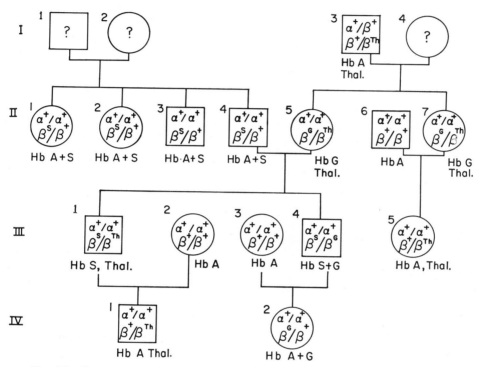

Fig. 36. PEDIGREE OF THE FAMILY REPORTED BY SCHWARTZ *et al.*[787] REDRAWN TO FIT CURRENT CONCEPTS OF HEMOGLOBIN PRODUCTION AND THALASSEMIA.

not have hemoglobin *G*, but he should have thalassemia, since his two daughters, *II–5* and *II–7*, carry the thalassemia gene. The unknown mother, *I–4*, could not have given both the hemoglobin *G* and the thalassemia chromosomes to her daughters, since these two genes should be allelic or closely linked in this family as indicated above. The Hb-A_2 value of patient *I–3* will be of great interest. It is well known that thalassemia sometimes is difficult to diagnose from blood smears alone. The presence of thalassemia, therefore, cannot be ruled out from a normal blood film.

There are no other areas in genetics where the detailed structure of the "primary" genic product is as well known as in the human hemoglobins. It is, therefore, comforting that the initially difficult and rather contradictory set of data contained in this pedigree can be fitted into the framework of all other genetic and biochemical data in this field at the present time.

DR. BURDETTE: Dr. Hill, will you close the discussion?

DR. HILL: I wish to comment on Dr. Motulsky's remarks first. I think that some of his suggestions are reasonable and fit into the framework of the hypotheses that are currently popular for explaining the genetic control of the structure of a protein. It might be wise, however, to use more caution in proposals such as these. The work I have just presented shows what troubles can be encountered in making genetic proposals based on incomplete chemical information. The assumption that thalassemia causes an alteration in the structure of one of the chains of hemoglobin is not founded on experimental facts. Ingram and Stretton [405] have considered the advantages of looking at thalassemia from this standpoint, although much evidence favors the view that thalassemia might result from alterations in the control or rate of hemoglobin formation.[408] [In future discussions, Dr. Motulsky should change the genotype of the individual (pedigree number *I*–3) in the first generation of the *G–S* family. In spite of the fact that his ideas would fit best if *I*–3 has thalassemia minor, there is no evidence for this at the present time.[787]]

The other point I would like to make concerns Dr. Neel's remarks. Obviously I did not make it clear that the minor components in individual *III*–1 comprise about 30 per cent of his total hemoglobin. As Dr. Neel pointed out, this high concentration of minor components gives rise to erroneous electrophoretic and solubility analyses. It can be mentioned that normal individuals have no more than 10 to 15 per cent of these components,[143] whereas individuals with either sickle disease or sickle trait have much higher concentrations. Individuals in the *G–S* family with thalassemia also have high levels of minor components. I believe this agrees with Dr. Jones' unpublished findings. Future work on hemoglobin from either a genetic or chemical view must account for the minor hemoglobin components. Questions concerning the genesis or function of these hemoglobins remain almost totally unanswered.

The last point I would like to emphasize concerns our methods for evaluating the structure of hemoglobin. The fact that we have found a single defect in the genetically-determined, abnormal hemoglobins is consistent with the "one gene-one enzyme" hypothesis and fits with ideas that a single mutation should result in alteration of a single amino acid in a protein.[143] However, it must be remembered that our methods of "fingerprinting" will, in all probability, not allow one to detect whether an amino acid has been replaced by one of like charge. In addition, it should be remembered that only two-thirds of the hemoglobin molecule is represented by the soluble tryptic peptides seen on "fingerprints." About one third of the molecule remains insoluble after tryptic digestion. The methods currently used to examine this insoluble residue are not very satisfactory. Thus it appears that, with present methods, we cannot judge precisely whether more than one amino-acid residue in the abnormal hemoglobins is altered.

William C. Boyd, Ph.D.

BLOOD GROUPS *and*
SOLUBLE ANTIGENS

THE BLOOD GROUPS

ABO Blood Groups.—The study of human genetics has consistently lagged behind that of lower forms. The classical Landsteiner blood groups were the first example of Mendelizing characteristics demonstrated in man, and even today few other normal hereditary characteristics have been as well studied as have the various blood-group systems. Van Dungren and Hirszfeld demonstrated in 1910 [900] that the agglutinogens *A* and *B* behave as simple, Mendelian dominants and proposed a theory of their inheritance as two independent pairs of genetic factors. This theory was corrected by the mathematician, Bernstein, who, looking for some human trait to illustrate the triple alleles that had been found in various lower organisms, demonstrated statistically that the *A, B,* and *O* genes formed such a series in man.[56] Hirszfeld, to his credit, urbanely accepted this correction.

In testing for blood groups, routine use is made of anti-*A* and anti-*B* reagents (agglutinins). The nature of these substances is discussed later. Genotypes *AA* and *AO* both behave in this test as group *A,* and *BB* and *BO* behave as group *B.* Group *O* is diagnosed by its failure to give a positive reaction with either anti-*A* or anti-*B.* If the gene *O* caused the production of a characteristic antigen, as do genes *A* and *B,* it ought to be possible, by using a suitable anti-*O,* to differentiate by laboratory test between *AA* and *AO* and between *BB* and *BO,* unless of course the gene *O* were completely recessive to *A* and *B.* It is commonly stated that *O* is completely recessive, and it is certainly true that no claims of serologic differentiation between the homozygotes and heterozygotes of groups *A* and *B* have ever been confirmed.

335

If gene *O* were recessive, but nevertheless produced a characteristic antigen when present in double dose, then blood group *O* ought to have characteristic serologic properties. Actually it is not difficult to obtain agglutinins from a variety of sources which agglutinate group-*O* erythrocytes, do not agglutinate bloods of group *B*, and do not agglutinate most group *A* and *AB* bloods. But it is obviously unjustifiable to call these reagents anti-*O*, for they do agglutinate bloods belonging to the subgroups A_2 and A_2B, although, in all bloods of the latter type and in some of the former, no *O* gene is present to cause the production of an *O* antigen. To take account of this fact and to emphasize the heteroge-

Table 50

GENETICS OF SUBGROUPS OF BLOOD GROUPS *A* AND *AB*,
ACCORDING TO THOMSEN, FRIEDENREICH, AND WORSAAE [873, 874]

Genotype	Phenotype	Reaction with			
		Anti-*A*	Anti-A_1	Anti-*H*	Anti-*B*
OO	*O*	−	−	+	−
A_1A_1	A_1	+	+	−	−
A_1O	A_1	+	+	−	−
A_1A_2	A_1	+	+	−	−
A_2A_2	A_2	+	−	+	−
A_2O	A_2	+	−	+	−
A_1B	A_1B	+	+	−	+
A_2B	A_2B	+	−	+	+

netic * nature of these agglutinins, Morgan suggested that they should be called anti-*H*, and this usage has been adopted except perhaps by a few of the most old-fashioned workers. It is the considered opinion of Watkins and Morgan [911] that no true anti-*O* serum has yet been identified beyond doubt. So far as we know now, therefore, it is possible that the gene *O* does not cause the production of any antigen solely characteristics of it.

The two varieties of the A antigens, A_1 and A_2, already referred to, probably correspond to two *A* alleles, for such a genetic theory explains nearly all the family data at present available, if it is assumed that the A_2 allele is dominant over *O* but recessive to A_1 (table 50). Rare alleles, A_3 and possibly A_4, seem to exist, and genes modifying the expression of *A* and *B*,[701] so that the genetics of the *ABO* blood group system has lost its pristine simplicity. According to Race [697] it is, in fact, the most complex of the blood-group systems.

Secretors and Nonsecretors.—The antigens of the *ABO* blood-group system are not confined to the erythrocytes. Instead, they may occur in practically all

---

* Heterogenetic antibodies are those acting against antigens from organisms not closely related taxonomically. The classical example is the Forssman antibodies to common antigenic components of guinea-pig organs and sheep erythrocytes. These heterogenetic Forssman antigens are found in animals, and even plants, of a variety of species.

tissues and fluids of the body, with the probable exception of the central nervous system. They may occur in two forms: water soluble and lipid soluble (that is, soluble in lipid solvents such as alcohol-ether and chloroform). All individuals apparently have the lipid-soluble form in their tissues, in conformity with their blood group.[85] However, the water-soluble form is found in only about 85 per cent of European individuals. Such persons are called secretors and those in whose tissues and body fluids water-soluble antigens corresponding to their blood group are not found are called nonsecretors.[770, 771] The ability to secrete the A and B antigens in water-soluble form is inherited, being controlled by a pair of genes S and s, S being dominant over s.

The presence of the blood-group antigens in the saliva of group-A and group-B secretors can be detected by the inhibiting action they have on anti-A and anti-B agglutinins. The technic is described below. If group-specific precipitins are available, soluble antigens A and B may be tested for directly.[82, 93] Group-O secretors are detected by the presence of the H antigen in their saliva.[767] The saliva of all secretors contains enough of the H antigen to make it possible to diagnose secretors by the inhibition technic, merely by the use of an anti-H reagent such as Ulex extract,[89] although this is easiest with group O, since saliva of group-O secretors contains the most H.

MNS Blood Groups. P System.—In 1927 Landsteiner and Levine,[478] using absorbed immune serums from rabbits, established the existence of three new blood antigens in man, M, N, and P. Further study showed that M and N are inherited as alleles on a pair of chromosomes different from the pair carrying the ABO alleles and are consequently not linked in inheritance with ABO. The factor, P, belongs to a different system. The discovery that the blood factor, Tj^a, is part of the P system has revealed that this system is in fact strikingly similar to the A_1A_2BO system.[758]

The inheritance of the M, N, blood groups is quite simple so long as attention is restricted to the original discoveries. There are just three genotypes, MM, MN, and NN, corresponding to blood types M, MN, and N. In this system, dominance is not observed. However, varieties of the M and N antigens exist, and in 1947 another blood factor, S, was found [909] which forms a part of the MN system.[700, 759] This means that, in most populations, not less than four alleles, M, MS, N, and NS are present, enabling us to make, when both anti-S and anti-s sera are available, the distinction of 9 phenotypes, namely, Ms, MS, MSs, MNs, MNS, MNSs, Ns, NS, and NSs.

Two other antigens which form part of the MN system, called Hunter and Henshaw from the donors in whom they were first found, are not too common in Africans and are virtually unknown in persons of European descent. A Vw antigen, also rare, forms part of the MN system.

Some Negro bloods lack both S and s. They are assumed to represent a homozygote, S^uS^u.[701]

Rh Blood Groups.—In 1939, Levine and Stetson [498] observed a case of erythroblastosis fetalis which they correctly ascribed to sensitization of the mother to a fetal-blood antigen inherited by the fetus from the father. It is now known that the new blood factor was in fact *Rh*, but Levine and Stetson did not propose any name, and it was not until Landsteiner and Wiener in 1940 reported the discovery that a rabbit serum against rhesus erythrocytes detected a new factor in human blood that the term, *Rh*, was introduced.[480] The new factor might not have attracted any more attention than others previously reported had not Wiener and Peters [926] showed that certain transfusion reactions were due to sensitization to *Rh*, and especially had not Levine *et al.*[495] demonstrated that *Rh* incompatibility between mother and child could be the cause of erythroblastosis fetalis.

Investigation soon showed that the *Rh* factor was antigenically and genetically complex. This work was carried on mainly by Wiener and Levine in this country and by the British workers now at the Lister Institute, although others contributed. As new facts were discovered, Wiener correspondingly modified his notation but retained the basic *Rh* symbol. He based his theory of inheritance upon the idea of multiple alleles, in analogy to the *ABO* series. In England, R. A. Fisher [696] noticed the antithetical relation of the 2 antiserums that Wiener called anti-*Rh'* and anti-*Hr'* (*Hr* was proposed by Levine as the reverse of *Rh*) and suggested that similar serums antithetical to anti-Rh_0 and anti-*Rh"* would be found. Fisher suggested the symbols *C, c, D, d, E,* and *e* for the blood antigens detected by these pairs of serums, the small letters corresponding to the *Hr* factors of Wiener and Levine. In this scheme there was therefore a 1 to 1 correspondence between genes and agglutinogens. Fisher furthermore suggested that the *Rh* antigens were inherited, not as a series of alleles at a single locus, but as a series of pairs of alleles each at 1 of 3 adjacent, closely linked loci on one chromosome. The correspondence of the 2 notations, in their present form, for the 8 genes or chromosomes originally recognized is listed in table 51 in the order given by Fisher and used here.

The anti-*e* serum predicted by the theory of Fisher was soon discovered, but anti-*d* has probably not been found. In the meantime, variants of the *C, D, E,* and other antigens have been found, symbolized as C^w, c^v, D^u and so on by the British workers, and as new variations of *Rh* (including Gothic letters, bars, and double bars) by Wiener. The British workers suggested that the real order of the loci on the *Rh* chromosome is *DCE*.

There has been an increasing tendency everywhere to prefer the British notation, and in some parts of the world it is the only one used. The reasons for this do not include a demonstration that the theory of inheritance underlying the British notation is correct. Wiener [925] has offered some forceful arguments against this theory. For practical purposes, there is no difference between Wiener's series of alleles and the closely or completely linked loci postulated by the British.

The same findings may be predicted from both theories and both are based on facts.* The swing to the British notation is rather to be traced to advantages it has as a notation and for teaching and medicolegal purposes. The desirable goal of a uniform notation has not been attained, but hopefully it may be reached within the next decade. Ford [247] has recently proposed a uniform notation for all the blood groups, and there is much to recommend his proposal.

The frequencies of the *Rh* antigens vary widely in different populations. One of the most striking differences is that the *Rh*-negative type (*cde*) does not occur in Asian and Pacific populations or in American Indians. The *Rh*-negative

Table 51

WIENER AND FISHER-RACE NOTATION FOR
Rh GENES OR CHROMOSOMES [925, 696]

Wiener	Fisher-Race
r	*cde*
r'	*Cde*
R⁰	*cDe*
r''	*cdE*
rʸ	*CdE*
R'	*CDe*
R²	*cDE*
Rᶻ	*CDE*

type has its highest frequency in the Basques, a relict population in certain regions of France and Spain. The Basques speak a non-Indo-European language and are known to represent the remnant of an earlier European population once probably dispersed over a much wider area, even extending into North Africa. The *D* antigen is most frequent in African populations and is so much more common there that it could almost be called an African antigen.

The Lewis Blood Groups.—An antibody, anti-*Leᵃ*, which agglutinates the red cells of 22 per cent of Europeans, was discovered by Mourant in 1946.[580] Further discoveries have considerably complicated the system. An antigen, *Leᵇ*, exists, and it was proposed that the gene, *Leᵇ*, is allelic to *Leᵃ*,[12] but the subject is still not entirely elucidated.[700]

The Lewis blood groups will be discussed somewhat more in detail than the

* The prevailing feeling of the outstanding *Rh* workers of today seems to be that the *Rh* complex can be called a gene locus, and the various antigenic complexes inherited as units can be called genes. The same workers, however, feel that the *Rh* "gene" is in reality a linear strip of *DNA*, of considerable complexity, belonging to some one of the 23 pairs of human chromosomes. It is thought that in this strip of *DNA* there are fully as many loci of possible mutations as envisaged by the Fisher-Race theory and probably many more. The methods currently available to human geneticists do not permit the sharp resolution of adjacent but not identical sites which is possible in bacterial genetics.

other new systems, primarily because they are connected with the secreting system. According to Race and Sanger [700] there seems to be general agreement that:

$Le(a+b-)$ red cells belong to nonsecretors,
$Le(a-b+)$ red cells belong to secretors,
$Le(a-b-)$ red cells usually, but not always, belong to secretors.

Lewis antigens also occur in saliva and plasma; the saliva of nonsecretors contains more Le^a than does the saliva of secretors. Le^b substance is present in the saliva of persons whose red cells are $Le(a-b+)$.

Table 52

BLOOD-GROUP ANTIGENS IN MAN [701]

System	Antigens detected by	
	positive reaction with specific antibody	positive reaction with one antibody, negative with another
A_1A_2BO	A_1, B, H	A_2, A_3, Ax
MNSs	M, N, S, s, Mi^a, Vw, Hu, He, M^a, Vr	M_2, N_2M^c
P	P_1, P^k	P_2
Rh	D, C, c, C^w, C^x, E, e, E^w, f, V	D^u, C^u, c^v, E^u
Lutheran	Lu^a, Lu^b	
Kell	K, k, Kp^a	
Lewis	Le^a, Le^b	
Duffy	Fy^a, Fy^b	
Kidd	Jk^a, Jk^b	
Diego	Di^a	
Sutter	Js^a	
"Private"	Levay, Becker, *Ven.*, Wr^a, Be^a, Rm, By	
"Public"	*Vel*, Yt^a, 1	

Le^a substance occurs in the serum only of persons of Lewis group $Le(a+b-)$; Le^b sometimes is found in the serum of persons of group $Le(a-b+)$. Grubb [296] suggests that the Lewis antigens are primarily a feature of the body fluids rather than the red cells. This is supported by observations that cells lacking Le^a or Le^b will take up these antigens if suspended in serum or plasma containing them.

Other Blood Groups.—Once laboratories were set up to examine routinely human sera which showed atypical agglutination reactions, the discovery of other human blood groups followed rapidly, and it is doubtful that all have yet been reported. They have generally been named for the donor in whose blood the antigen or the antibody was first identified and have names such as Lutheran, Duffy, Kell, and Kidd. A blood factor possibly different from any of these has been found with the aid of a plant agglutinin from peanuts,[87, 88, 89] but the number of blood-group systems is now so large that it is not possible to state, when a new blood factor is announced, whether it will eventually be found to form part of one of the older systems or not. A list of 59 antigens that have been identified on the human red cell is given in table 52, taken from Race and Sanger.[701]

REAGENTS USED IN BLOOD GROUPING

The reagents used in blood-grouping research are mainly agglutinins, although group-specific precipitins have been described [91, 93] which might have practical application under certain conditions. These agglutinins may be classified as follows: (1) naturally occurring human antibodies, (2) immune human antibodies, (3) immune animal antibodies, (4) heterogenetic animal antibodies, (5) plant agglutinins and precipitins (lectins). Each class will be treated only briefly and generally.

1. NATURALLY OCCURRING HUMAN ANTIBODIES

The prime example of human antibodies occurring naturally are the anti-*A* and anti-*B* normally found in the plasma or serum of individuals of blood groups *B* and *A* respectively. The plasma of group-*O* individuals contains both anti-*A* and anti-*B*, *AB* plasma contains neither. For a long time serum of group-*B* and group-*A* persons, chosen pretty much at random, was used in the routine determination of the *ABO* blood groups, but it was found that not all individuals possess agglutinins sufficiently potent for this purpose, and modern practice inclines to the use of serum from "stimulated" donors injected with *A* and *B* substances, which is classified under immune antibodies.

Origin of Anti-A and Anti-B.—The anti-*A* and anti-*B* of normal human plasma are exceptions to the rule that antibodies to blood-group antigens do not usually appear in a person's blood without some history of unusual antigenic stimulus. In fact, as Race and Sanger point out, when anti-*A* or anti-*B*, which would be expected according to Landsteiner's rules, are absent, there is usually some special explanation. It is natural to ask why these agglutins appear with such regularity, and there have been two main theories.

One theory of the origin of the isoagglutinins, anti-*A* and anti-*B*, is that they are a result of the action of the *ABO* genes just as are the *ABH* antigens. This theory has been supported by Furuhata [265] and the Wurmsers.[423] Whether or not this theory seems inherently plausible depends partly on which theory of antibody formation is adopted.

Another theory suggests that anti-*A* and anti-*B* are immune antibodies, as are most others, being formed in response to antigens (in food, bacteria, and animal parasites) which are chemically similar to the *A* and *B* antigens of man.[924] It is known that a number of such related antigens exist; worthy of mention is the cross-reaction of blood-group-*A* antigen and type-14 pneumococcus.

If the second theory of isoagglutinin formation is correct, one wonders why natural isoagglutinins for other human blood antigens, such as *M, N,* and *Rh,* are so seldom encountered. One reason might be the lower antigenicity of these other agglutinogens; there is definite evidence for this. Another reason might be that these other human agglutinogens are more unusual and perhaps more "specialized" in their structure than are A and B, so that closely related antigens are less common in lower organisms.

Other Natural Agglutinins.—Other naturally occurring agglutinins are less regularly present in normal human blood. Persons of group A_2 or A_2B, for example, may have anti-A_1 in their plasma, and persons of group A_1 or A_1B may have anti-H, but, in other individuals of these groups, such antibodies are lacking. An anti-P_1, originally designated by Landsteiner as "extra agglutinin 1," occurs in some bloods of P_2 individuals; in fact, if it is tested for by a sufficiently sensitive technique, it is found in most P_2 sera.[342] Rarely, agglutinins for other blood antigens, M, N, Rh, et cetera, are found in apparently normal individuals. As a rule these rare, naturally occurring antibodies are not strong enough for routine testing purposes.

2. IMMUNE–HUMAN ANTIBODIES

Immune-human hemagglutinins may be divided into two classes: unintentionally produced (for example, as the result of sensitization by repeated transfusion or pregnancy) and deliberately produced (by injection of volunteers with incompatible blood or blood-group substance).

As a rule, the deliberate production of hemagglutinins is not satisfactory as a means of producing routine reagents. For example, although anti-Rh antibodies can be produced in this way, they are generally weak and tend to be predominantly of the blocking type. In other cases, no antibody at all results. I am S-negative and have taken repeated transfusions of S-positive blood in an attempt to produce immune anti-S, with no perceptible results. Others have had similar experiences.

Immune anti-A and anti-B agglutinins form a notable exception to this rule. About half the volunteers injected with commercial A and B blood-group substances respond with a considerable rise in the titer of their isoagglutinins.[80] The new antibodies display some qualitative differences from normal isoagglutinins and may be distinguished as isoimmune antibodies. The technique of their production is simple, and has been described in a number of places. In my laboratory, I use isoimmune anti-A and anti-B almost exclusively.

Unintentionally produced isoantibodies are very important as laboratory reagents. In the case of Rh, they provide the only practical reagents, for, although the work of Landsteiner and Wiener [480] was carried out with an immune-rabbit serum and guinea-pig serums were used for a time, it has not proved possible routinely to produce anti-Rh sera in these animals or in any other laboratory animal tried even with the use of adjuvants.[793] Unintentionally produced, human antibodies are also the sole reagents for the detection of nearly all the other blood-group factors aside from *ABO* and *M* and *N*. Good, human, immune anti-*M* has also been found.[700] These unintentionally produced antibodies generally result from sensitization due to pregnancy, but some have been found in the plasma of individuals sensitized by repeated transfusion.

3. IMMUNE–ANIMAL ANTIBODIES

Anti-*A* and anti-*B* can be produced in rabbits,[82, 368, 476] although not all animals respond, and have been found useful in research.[84, 85] Since the resulting sera have to be absorbed with the proper type of human erythrocytes before they are ready for use, they are more difficult and expensive to prepare than isoimmune anti-*A* and anti-*B* (see preceding). Aside from the rare, naturally occurring or isoimmune, human anti-*M* and anti-*N*, immune agglutinins from the rabbit are the only source of routine reagents, although anti-*M* and anti-*N* of plant origin are coming into use (see following). The preparation of absorbed anti-*M* and anti-*N* from rabbits has been described in a number of places.[82, 535, 699, 769]

With these exceptions, animals have not generally been found to respond well to the human blood-group antigens. If the sera of rabbits injected with the intention of producing anti-*A*, anti-*B*, anti-*M*, or anti-*N* are absorbed with the bloods positive for these but negative for other antigens, it is rare to find any immune agglutinins remaining.

4. HETEROGENETIC ANIMAL "ANTIBODIES"

The serum of goats and chickens injected with *Shiga* bacilli have sometimes proved good sources of anti-*H* agglutinin.[201, 297, 359] This is evidently a true heterogenetic antibody, and its reactivity is probably accounted for by chemical similiarities between the *H* substance and certain *Shiga* antigens.

The serum of an animal of one species often contains agglutinins for the red cells of other species. Whether these should be called antibodies or not is a question,[477] but some of them have been of limited usefulness in laboratory testing.[431] Anti-*P* agglutinins sufficiently strong for some investigations have been produced by absorption of serum from the horse and cow.

Anti-*H* agglutinins are found in the serum of a number of species, especially in the serum of some eels.[700] Their place in routine testing has been largely taken by anti-*H* from certain seeds (see following). A natural anti-*M* occurs in horse serum.[496]

5. PLANT AGGLUTININS (LECTINS)

It has been known for a long time that certain seeds contain substances that agglutinate human and other red cells, and Landsteiner [477] even pointed out that some of these exhibited some degree of species specificity. It was not until 1945, however, that I observed that lima beans contain an agglutinin specific for the blood-group-*A* antigen. This observation, referred to briefly in 1947,[81] was not published until 1949.[92] In the meantime, Renkonen [714] had published his independent discovery of blood-group specificity in plant extracts, mentioning *Vicia cracca* as specific for *A* and *Laburnum alpinum, Cytisus sessilifolius,* and *Lotus tetragonolobatus* as specific for *H*.

A number of laboratories are now engaged in the study of these interesting substances, and reviews have been published by Krüpe,[465] Mäkelä,[522] and Bird.[62] Seeds of a number of plants contain anti-*A;* that of one of these, *Dolichos biflorus,* reacts so much more strongly with A_1 than with A_2 as to be virtually specific for A_1.[63] It has therefore come to be widely used in the diagnosis of the subgroups of

Table 53

DIAGNOSIS OF SUBGROUPS WITH LECTINS [87, 88, 89]

Subgroup	Reaction with extract of	
	Dolichos	*Ulex*
A_1	4	0
A_2	0	3
A_1B	$3\frac{1}{2}$	0
A_2B	0	3

groups *A* and *AB*.[91] An anti-*N* has been found in *Vicia graminea* [635] and more recently in *Bauhinia purpurea*.[86, 522] An anti-*M* is available commercially.

No good anti-*B* lectin is routinely available. Extracts of *Sophora japonica* agglutinate blood of group *B* more strongly than group *A* but react too strongly with *A* to be satisfactory as a laboratory anti-B reagent.[466] *Euonymous europeus* extracts contain anti-*B* and anti-*H* activity.[772] *Marasmius oreades,* which sometimes furnishes a satisfactory, though weak, anti-*B,* is a small mushroom not grown commercially.[204] The best anti-*B* is said to be that from *Bandeiraea simplicifolia,*[524] but the samples of this plant I have tested personally have been disappointing.

Because of the absence of an anti-*B,* lectins are not used routinely in the determination of the *ABO* blood groups, in spite of the fact that satisfactory anti-*A* is available from several plants. It has been found, however, that the anti-A_1 of *Dolichos biflorus* and the anti-*H* of *Ulex europeus* make an ideal combination of reagents for the routine determination of the subgroups of *A* and *AB*.[91] The way in which this is done is shown in table 53. Anti-A lectins, especially the one from lima beans, have had a number of applications in special experiments where a large amount of specific anti-*A* agglutinin is needed.[23] They have been used by Morgan and Watkins [560] to show that the blood-group antigen of group-*AB* individuals is not a mixture of *A* and *B* substances, but a unique molecule containing both *A* and *B* specificities. Testing for *H* substance in saliva, by inhibition of the anti-*H* of *Ulex,* is discussed later (table 55).

The anti-N of *Vicia graminea* is actually better than the absorbed anti-N as usually prepared from immune serum from rabbits and would doubtless be used routinely if more of the tiny seeds of this tropical plant were available. It has already proven valuable in clearing up the MN system in chimpanzees.[497] The anti-N of *Bauhinia* is not quite as good but may nevertheless come into routine use, as the seeds are available in many parts of the world.[86]

Another example of the practical use of lectins is provided by the work of Levine, Celano, Lange, and Berliner.[496] Using the anti-N of *Vicia graminea,* these workers observed that equine erythrocytes contain the N antigen, a fact that absorbed, rabbit sera had failed to reveal because of species-specific agglutinins remaining in the absorbed sera. The finding of N in the erythrocytes of the horse led Levine *et al.* to look for natural anti-M in horse serum. Having found it, they were led to predict that the horse should be a good producer of immune anti-M, and this prediction was verified, thus providing a new and abundant source of immune anti-M.

BLOOD GROUPING TECHNIQUE

Whole books can be, and have been, written on the subject of technique; spatial limitations permit no more than an outline of the basic typical procedures.

Taking Blood.—Sufficient red cells for a complete blood-grouping test can be taken from an ear or finger into saline (0.9 per cent NaCl solution) or 3 per cent sodium-citrate solution. In either case, the cells are centrifuged and the supernatant fluid replaced by fresh saline. It is best to prick the ear or finger, first scrubbed with alcohol, with a sterile, detachable, lancet blade or one of the disposable metal points sold for this purpose. Such samples are taken fresh each day. If more than a few drops of blood is needed, it is taken by venipuncture and put into an equal amount of Alsever's solution or mixed with about $\frac{1}{4}$ its volume of 5 per cent citrate solution.

If sterile blood is to be sent through the mails, it may be taken into an equal volume of Alsever's solution and to this a drop of penicillin and streptomycin added to give final concentrations of about 30 units and 0.005 *mg./ml.* respectively. Blood so prepared keeps very well, especially when kept cold.

If it is desired to preserve cells a long time, they may be frozen in glycerin.[700] This technique may prove very useful in preserving cells from remote populations or populations in the process of dying or losing their ethnic identity for future analysis of interesting antigens as yet undiscovered.

Agglutination Tests.—There are a number of ways of conducting blood-grouping tests; I shall describe the technique used in my own laboratory. More detailed descriptions of these and other methods may be found in Schiff and Boyd,[769] Race and Sanger,[700] and Stratton and Renton.[845]

We do the tests in small test tubes, made of Pyrex, without a lip, measuring about 75 × 9 mm. and holding about 4 ml. For *Rh* testing, we think smaller tubes, 70 × 6 mm., holding about 1.3 ml, are better, but apparently few other laboratories use them. The tubes are kept in suitable wire or aluminum racks.

Enough washed erythrocytes are placed in saline to give a 1 to 2 per cent suspension. A drop of this suspension is placed in a small test tube, one tube for each factor it is desired to test for plus (sometimes) one for a control. In the test for each blood agglutinogen, it is generally desirable to include known positive and known negative bloods. The drops of cellular suspension are transferred with the aid of a rubber bulb and capillary pipette (Pasteur pipette). One drop of a suitable dilution of the antiserum or lectin is added to each tube containing a drop of the cells it is desired to test for a given agglutinogen. The tubes are mixed by shaking, centrifuged at low speed (about 1200 r.p.m.) for one minute, and read by shaking. *Rh* tests are incubated one hour at 37° C. before being mixed again, centrifuged, and read. In making the readings, it is convenient to look at the reflection of the bottom of the tube in a small mirror kept on the desk, thus minimizing neck strain.

All tubes containing tests for *A* are shaken simultaneously in the same rack; *B, M,* and *N* are treated similarly. In the case of *M* and *N,* it may be necessary to shake quite vigorously to make the known negative controls appear negative. Each *Rh* tube is shaken individually and very gently.

Some anti-*Rh* sera do not agglutinate red cells suspended in saline. These are called blocking antibodies or (less appropriately) incomplete antibodies. Such antibodies will sometimes agglutinate red cells suspended in 20 per cent solution of bovine albumin. Serum from the same donor as the cells and sometimes other media such as gelatin, dextran, and polyvinylpyrrolidone (*PVP*) also work sometimes. Other antibodies require disclosure of their activity by the addition to the antibody-coated cells of an antihuman-globulin serum from rabbits (Coombs serum). These techniques are described more in detail by Race [700] and Stratton and Renton.[845] Sometimes blocking antibodies will agglutinate cells if these have first been treated with trypsin, ficin, papain, bromalin, or other proteolytic enzyme.[665, 700] It is not understood how these reagents improve the agglutinating power of blocking antibodies.

Inhibition Technique.—In testing for soluble antigens, as in the segregation of individuals into secretors and nonsecretors by test of their saliva and in chemical studies on substances related chemically to the blood-group antigens, the inhibition technique is extremely valuable. Most of our information, in fact, about the chemical constitution of the blood-group antigens comes from such work. The inhibition test depends upon the fact that soluble antigens and the separable, specifically reactive portion of some antigens (haptens) will combine with antibody under conditions such that no visible reaction occurs. However, subsequent addition of the appropriate antigen (red cells or undegraded protein or carbohydrate) in amounts that would ordinarily give a visible reaction (agglutination or precipitation) may reveal the fact that the antibody has already combined with antigen or hapten and is, in consequence, unable to bring about the expected visible reaction. This is called inhibition.

In work with haptens and soluble antigens the tests can be carried out in two different ways: different dilutions of the antigen or hapten may be tested against a constant concentration of antibody, or different dilutions of the antibody may be tested against a constant concentration of antigen or hapten. In testing human saliva for blood-group antigens, it is customary to make serial dilutions 1:2, 1:4, et cetera, of the saliva and incubate these with a suitable serum, diluted to such a point that it is not too strong, but still capable of giving good agglutination with the test cells. Typical results are shown in table 54.

Table 54

INHIBITION OF ANTI-*A* AGGLUTININ BY GROUP-*A* SALIVA FROM SECRETOR
NUMBERS INDICATE DEGREE OF AGGLUTINATION OF TEST CELLS

	Dilution of Saliva, 1:									
Serum	2	4	8	16	32	64	128	256	512	1024
Group *B* (anti-*A*) [1]	0	0	0	0	0	0	0	1	2	4
Group *A* (anti-*B*) [2]	4	4	4	4	4	4	4	4	4	4

[1] Diluted 1:16
[2] Diluted 1:8

We incubate the mixture of serum and diluted saliva for one hour at 37° C., add cells, mix, centrifuge, and read; Race and Sanger [700] add the cells after a few minutes, allow the mixture to stand at room temperature for about an hour, then read without centrifuging.

In the classical procedure for diagnosing secretors, salivas of group *A* were tested with group-*B* (anti-*A*) agglutinin, salivas of group *B* with anti-*B* agglutinin, while group *AB* could be tested with either. Diagnosing individuals of group *O* was more difficult because good anti-*H* was hard to find. Eel serum, often used, was not entirely satisfactory. The introduction of the anti-*H* from *Ulex europeus* [89] has removed this difficulty. In fact, Boyd and Shapleigh [89] showed that the anti-*H* from *Ulex* could be used to test the saliva of individuals of any blood group, thus making it possible to use one reagent and one type of cell instead of three of each. Race and Sanger [700] have some doubt about the safety of using *Ulex* to test *A₁B* saliva and, in any case, think it wise to retest all *A₁*, *A₁B*, and *B* nonsecretor salivas with anti-*A* or anti-*B*. The sort of results to be expected with *Ulex* extracts are shown in table 55.

In immunochemical studies, including investigations into the chemical structure of the blood-group antigens, it is often desired to compare the relative in-

Table 55 *

DIAGNOSIS OF SECRETORS AND NONSECRETORS WITH ULEX EXTRACT [87, 88, 98]

Saliva from	Dilution of Saliva, 1:				
	2	4	8	16	32
O secretor	0	0	0	0	0
O nonsecretor	4	4	4	4	4
A_1 secretor	0	±	3	3	3
A_1 nonsecretor	4	4	4	4	4
A_2 secretor	0	0	±	3	3
A_2 nonsecretor	4	4	4	4	4
A_1B secretor	0	±	3	3	3
A_1B nonsecretor	4	4	4	4	4
A_2B secretor	0	0	0	±	3
A_2B nonsecretor	4	4	4	4	4

* Numbers indicate strength of agglutination obtained when *Ulex* extract previously incubated with indicated dilutions of saliva is tested against group-*O* erythrocytes.

hibitory power of different compounds for one and the same antibody. Similar solutions of such compounds, starting with equimolar concentrations if the molecular weights are known, are prepared and tested against suitably diluted antibody. After incubation, suitable test cells are added, and the mixtures centrifuged and read. In general, the results obtained with haptens X, X', X'', and Y (where X' is closely related chemically to X, X'' less closely related, and Y unrelated) may be expected to give results similar to those shown schematically in table 56.

Table 56

REACTIONS OF ANTI-X SERUM WITH X-ANTIGEN IN PRESENCE OF X (HAPTEN),
X' AND X'' (RELATED HAPTENS), AND Y (UNRELATED HAPTEN)

Hapten	Hapten Diluted					
	1:2	1:4	1:8	1:16	1:32	1:64
X	0	0	0	0	+	+±
X'	0	0	±	+	++	++
X''	0	±	+	++	++	++
Y	++	++	++	++	++	++

The relative effectiveness of different haptens as inhibitors could be shown by plotting concentrations that inhibit or by plotting the effective dilutions. Another way of comparing the inhibiting powers of different haptens is to use the same concentrations (on a molar basis) of haptens against successive dilutions of the antiserum. In this case, again using the hypothetical haptens, X, X', X'', and Y, the type of result shown in table 57 may be obtained.

Instead of trying to find the antibody concentration that is completely inhibited by a given concentration of hapten or the hapten concentration that will

Table 57

REACTION OF ANTI-X SERUM WITH X-ANTIGEN IN PRESENCE OF X HAPTEN, X' AND X'' (RELATED HAPTENS), AND Y (UNRELATED HAPTEN)

Hapten	Serum Diluted					
	1:2	1:4	1:8	1:16	1:32	1:64
X	+	0	0	0	0	0
X'	++	+	±	0	0	0
X''	++	++	+	+	±	0
Y	++	++	++	++	+	±

completely inhibit a given concentration of antibody, it is more accurate, if we are dealing with a precipitating antibody, to measure the amount of precipitate produced under the various conditions and to estimate the amount of hapten which gives 50 per cent inhibition.

SOURCES OF BLOOD–GROUP ANTIGENS FOR STUDY

The human erythrocyte is a complicated structure, and the blood-group antigens make up only a small part of its mass. Therefore, it is not surprising that attempts to determine the chemical structure of the blood-group antigens by analyses of material isolated from erythrocytes have not given information of much value. Not only is the starting material complex and the desired antigens only a small portion of it, but also the antigens seem to be bound in some way to the lipids, and possibly to the proteins, which are present on the surface of the red cell, making purification extremely difficult.[562] If it were not for the occurrence, on a much more generous scale, of the blood-group substances in water-soluble form in the saliva, gastric juice, ovarian cyst fluid, and meconium of secretors and the occurrence of closely related antigens in the stomach of hog and horse, little would be known today of the chemistry of the blood-group substances.

A number of methods of isolating and purifying the blood-group substances from such sources have been described, but the extraction with cold 90-per-cent phenol described by Morgan and King[559] has been used more than any other. This eliminates most of the accompanying nonspecific protein and other impurities. High-speed centrifugation and further fractionation from water and other solvents gives further purification.

As a result of such methods of purification, four blood-group substances have been obtained in amount sufficient for chemical study: A, B, H, and Le^a. The Le^a antigen (one of the antigens of the Lewis blood-group system) has been studied nearly as thoroughly as the ABH antigens, because it also occurs in water-soluble form in body fluids.

Analytical Results.—The results of chemical analyses of the ABH and Le^a blood-group antigens have been rather disappointing because they have not revealed any chemical differences that can be correlated with the differences in blood-group activity. At first glance, the four antigens seem to be very much alike. Each contains the same two sugar components (*L*-fucose and *D*-galactose) and the same amine-sugar components (*D*-glucosamine and *D*-galactosamine). They also contain the same eleven amino acids.[431, 562] The role of the amino acids is not clear; the specificity of the antigens seems to be determined mainly by the carbohydrate portions. However, Morgan believes that the blood-group an-

Table 58

TYPICAL ANALYTICAL FIGURES FOR PREPARATIONS OF
HUMAN BLOOD-GROUP SUBSTANCES [562]

	Nitrogen (per cent)	Fucose (per cent)	Acetyl (per cent)	Hexosamine (per cent)	Reducing Sugar (per cent)
A substance	5.4	19	9.0	29	54
H substance	5.3	18	8.6	28	50
Le^a substance	5.0	14	9.9	32	56
B substance	5.6	16	7.0	24	52
AB substance	5.6	17		26	54

tigens are not merely a loose combination of a macromolecular polysaccharide with protein, but consist of carbohydrate chains and peptide units bound together by primary chemical bonds. Typical analytical values for preparations of the specific substances are shown in table 58. The observed differences are within the range of variation found with different preparations of the same antigen.[562]

From the analytical results, the conclusion is inescapable that the specific serologic differences found between the *A, B,* and *H* antigens are due, not to dif-

Fig. 37. *L*-FUCOSE.

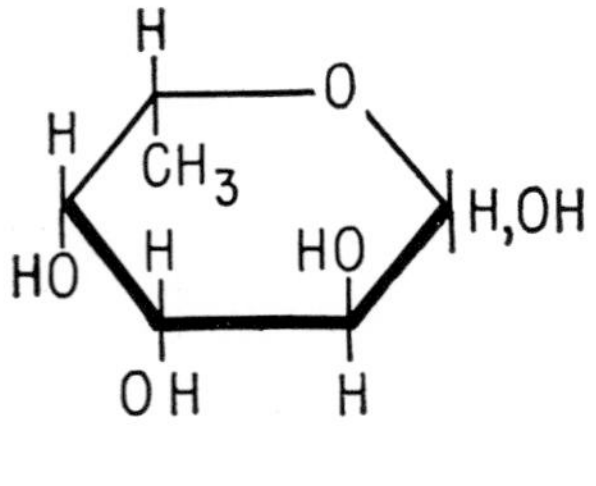

ferences in over-all composition, but to differences in the details of arrangement of the component parts. In fact, recent evidence suggests that only certain parts of the complex, polysaccharide molecules are responsible for the specific serologic properties. The first indication of this was obtained when it was shown that of all the sugars present in the *H* blood-group substance, only *L*-fucose (figure 37) specifically inhibited the agglutinating action of an anti-*H* from eel serum. Similar results were found with an anti-*H* of plant origin, an extract of the seeds of *Lotus tetragonolobus.*

It was also found that anti-*H* from either of these sources was inhibited more strongly by alpha-methyl-*L*-fucopyranoside than by the beta furanoside or by fucose alone. These results suggested that *L*-fucose was an important part of the *H* molecule; probably, in light of Landsteiner's findings with composite haptens, it is the terminal group of the specific part. The fact that the alpha-methyl-fucopyranoside inhibited better than the beta-methyl-pyranoside suggested that the fucose was connected by an alpha linkage to the next residue of the reactive portion of the *H* molecule.

The first information about the role of a particular sugar in the specificity of the *A* substance was obtained by tests on anti-*A* reagents of plant origin.[562] It was found that anti-*A* lectins were specifically inhibited by *N*-acetylgalactosamine (figure 38). This amino sugar did not inhibit most of the human anti-*A* reagents

Fig. 38. *N*-ACETYL-*D*-GALACTOSAMINE.

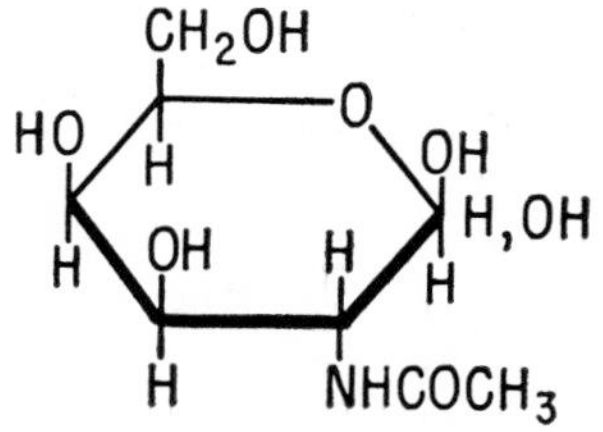

tested, but they were inhibited by the disaccharide, *O*-alpha-*N*-acetyl-*D*-galacto-sylaminoyl-(1,3)-*D*-galactose (figure 39). This suggests that this disaccharide

Fig. 39. *O*-α *N*-ACETYL-*D*-GALACTOSAMINOYL-(1,3)-*D*-GALACTOSE.

O-α-N-Acetyl-D-galactosaminoyl-
(1—3)-D-galactose

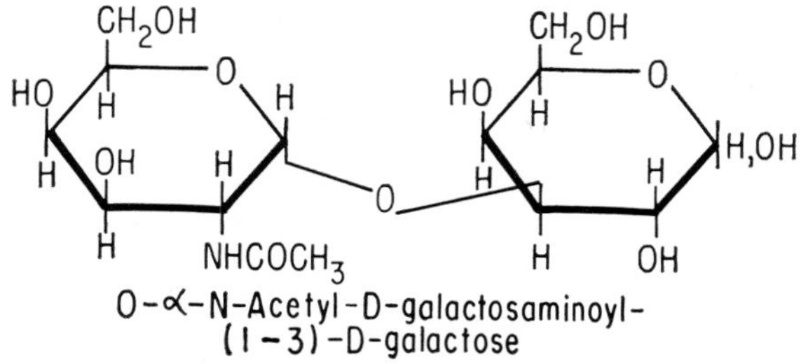

must be very similar to the terminal disaccharide portion of the specific part of the *A* substance or is possibly identical with it.[562]

Kabat and co-workers,[431] also using the inhibition technique, found similar evidence as to the structure of the specific part of the *B* antigen. Of the mono-saccharides present in the molecule, *D*-galactose is the best inhibitor of the anti-*B* antibodies. Kabat found that the galactose-containing disaccharide, melibiose,

Fig. 40.   MELIBIOSE, RAFFINOSE, AND STACHYOSE.

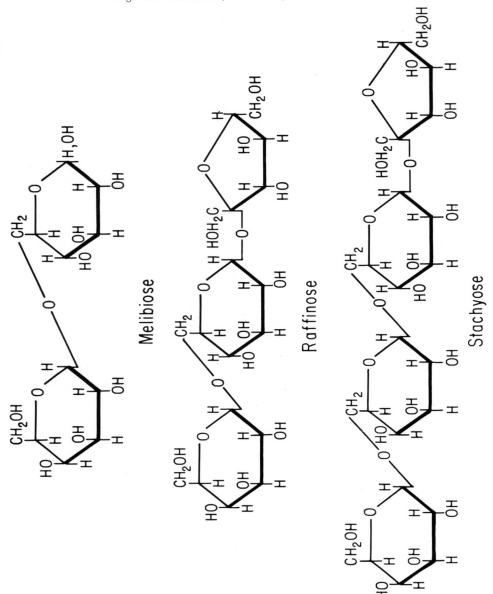

the trisaccharide, raffinose, and the tetrasaccharide, stachyose, (figure 40) inhibited even better than galactose alone (figure 41).   This is what would have
been expected if the specific part of the *B* antigen consists of a terminal non-
reducing galactose unit linked by an alpha linkage to another sugar unit.   That
the linkage is alpha is demonstrated by the fact that alpha-methyl galactoside inhibits better than galactose, but the beta galactoside does not inhibit as well
(figure 41).

Fig. 41.   RELATIVE INHIBITING POWER FOR ANTI-*B* OF VARIOUS SUGARS.

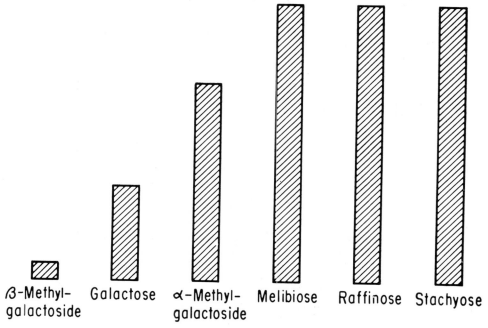

β-Methyl-   Galactose   α-Methyl-   Melibiose   Raffinose   Stachyose
galactoside   galactoside

From data of Kabat *et al.*[431]

Kabat was also able to draw some conclusions as to the nature of the sugar
residue next to galactose in the specific side chain of the *B* antigen.   It cannot be
glucose, because glucose is not a part of the *B* molecule.   Also it is not likely to
be another galactose.   If it were, stachyose, which contains a terminal galactose
bound by 1, 6 alpha linkage to another galactose, should be a better inhibitor
than melibiose and raffinose, when the sugar next to galactose is glucose; but stachyose is no better as an inhibitor than melibiose and raffinose.   Kabat believes that
this leaves *N*-acetylglucosamine as the only remaining possibility for the next-to-
terminal sugar in the specific side chain of *B*.   This information may eventually
see practical application.   Kabat suggests that the introduction of a number of
melibiosyl residues into a polysaccharide would endow it with substantial blood-
group-*B* activity.

How many other sugars are present in the active side chains of the A and B molecules is not known, but Kabat believes, on the basis of his determinations of the size of the reactive group of dextran, that the total is of the order of six. This would mean that the specific portions of the A and B molecules would, in the light of our present knowledge, look something like the structure shown in figure 42, where the symbol "x" designates a number of the order of four.

Fig. 42. SUGGESTED STRUCTURES OF TERMINAL PORTIONS OF ACTIVE SIDE CHAINS OF A AND B BLOOD-GROUP SUBSTANCES.

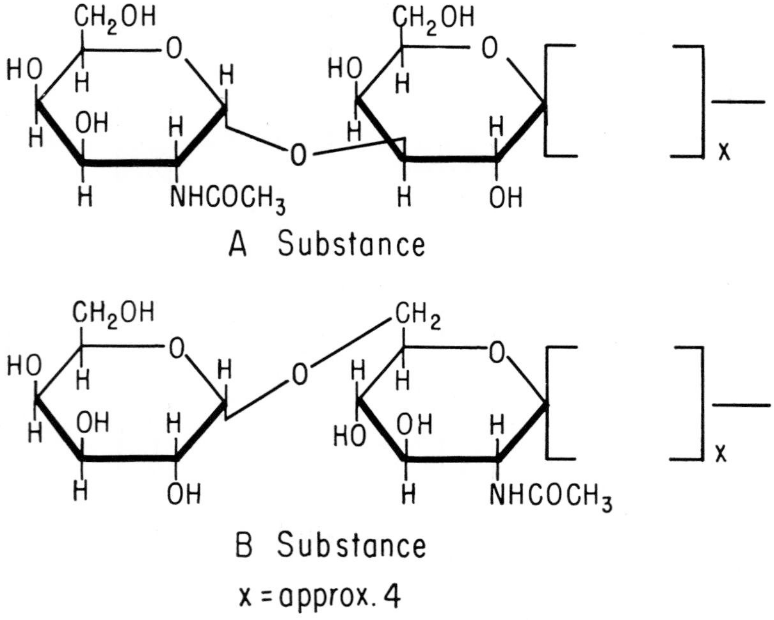

A Substance

B Substance

x = approx. 4

Additional and independent evidence for the part played by L-fucose, N-acetylgalactosamine, and D-galactose in H, A, and B specificity, respectively, was obtained by Watkins and Morgan [911] from the results of enzymatic inhibition by these sugars. It is known that an enzyme can be inhibited by an excess of one of the products of its action on its substrate, and an enzymatic preparation was available from *Trichomonas foetus* which destroyed the activity of A, B, and H substances. It was found, as expected, that the action of the enzymatic preparation on A substance was inhibited by N-acetylgalactosamine, the action on B substance by galactose, and the action on H substance by fucose.

There is, at the present time, no evidence concerning the identity of the monosaccharide unit adjacent to fucose in the specific part of the H substance, so the best picture of its structure is that shown in figure 43, where the symbol "x" designates a number of the order of five.

Fig. 43.   Suggested structure of terminal portion of active side chain of blood-group *H* substance.

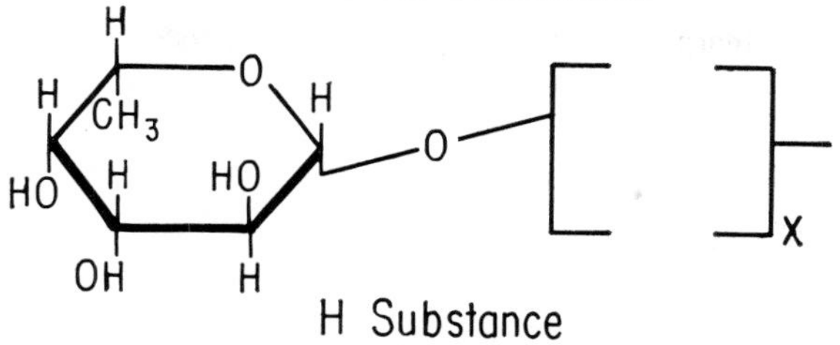

H Substance
X = Approx. 5

Watkins and Morgan [912] found that the destruction of the serologic activity of the *Le^a* antigen by the *Trichomonas* enzymes was inhibited by *L*-fucose, which suggested a role for this sugar in the specificity of the *Le^a* antigen.   However, the agglutination of *Le(a+)* red cells by human or rabbit, anti-Lewis sera was not detectably inhibited by *L*-fucose or by any other components of the blood-group substances.   Certain oligosaccharides containing fucose did inhibit, however, and

Fig. 44.   Suggested structure of terminal portion of active side chain of blood-group *Le^a* substance.[562]

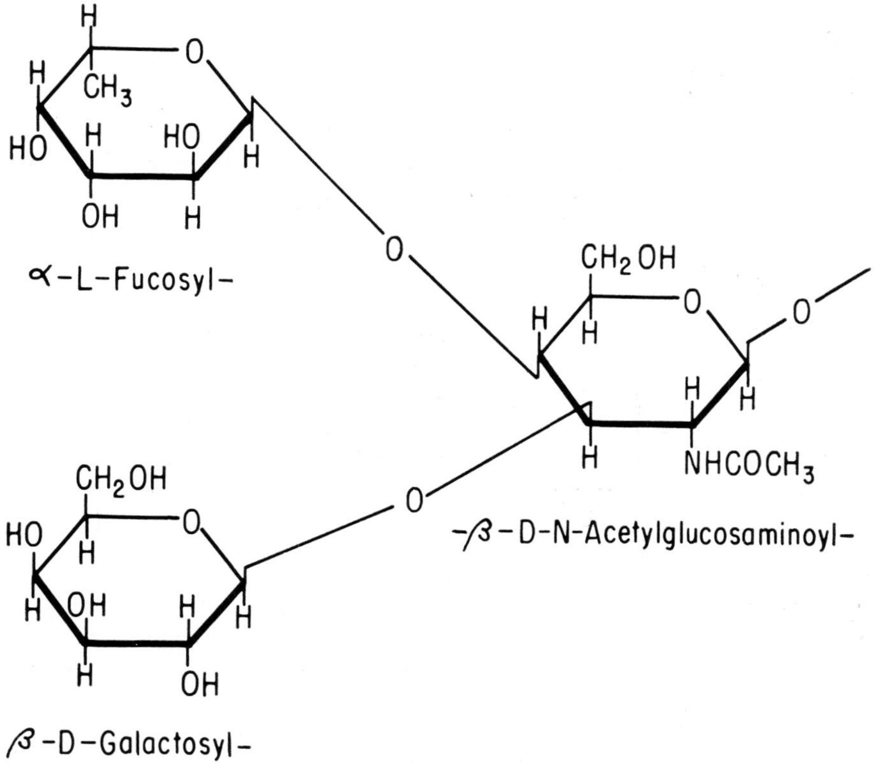

α–L–Fucosyl–

–β–D–N–Acetylglucosaminoyl–

β–D–Galactosyl–

showed that alpha-L-fucopyranosyl groupings were involved in *Le^a* specificity.[562] From a study of the inhibitory activity of various oligosaccharides, mostly isolated by Kuhn and his colleagues from human milk,[468] Morgan and Watkins suggest that the terminal portion of the specific part of the *Le^a* substance is a trisaccharide with the structure shown in figure 44.

Fig. 45.   SIMPLIFIED SCHEME OF ACTION OF GENES TO PRODUCE HUMAN
BLOOD-GROUP SUBSTANCES.

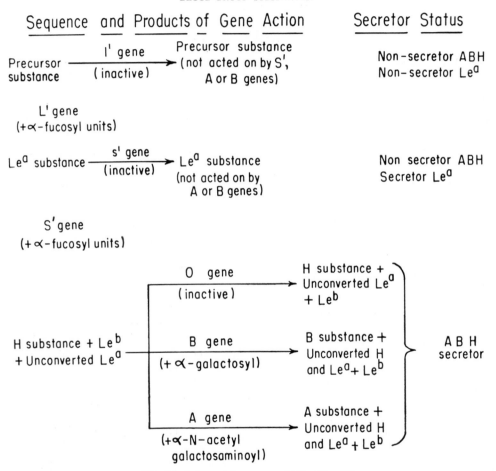

Modified from Morgan and Watkins.[562]

The detailed way in which the *ABO,* secretor, and Lewis genes coöperate to produce the various blood-group substances found in the body fluids of persons of different genotypes is still to be determined.   Watkins and Morgan [913] have proposed the following scheme as a first approximation.   Three independent genic systems, *L'* and *l', S'* and *s',* and the *ABO* genes, are supposed to be involved.   In

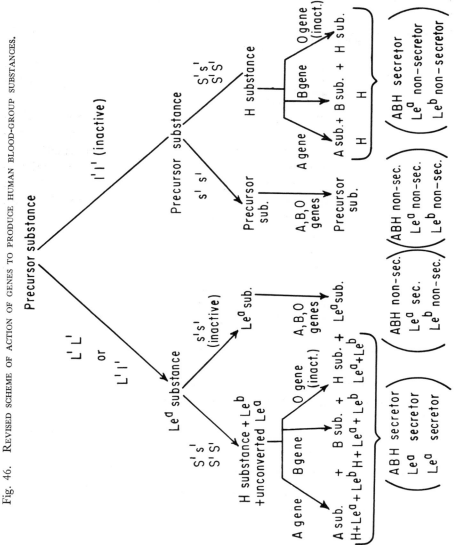

Fig. 46.   REVISED SCHEME OF ACTION OF GENES TO PRODUCE HUMAN BLOOD-GROUP SUBSTANCES.

Modified from Morgan and Watkins.[562]

various ways they act to modify the precursor substance, a mucopolysaccharide that Watkins and Morgan think is identical with the material found in the secretions of the individuals who secrete neither *ABH* substances nor *Le^a* and *Le^b* substance. The *L'* gene acts to add alpha-fucosyl units to this precursor substance, and the *S'* gene adds alpha-galactos aminoyl units (figure 45). The scheme in figure 45 is inadequate in some respects, and Watkins and Morgan suggest replacing it by the more complicated scheme shown in figure 46.

Other Human Red-Cell Receptors.—In addition to the red-cell receptors characteristic of the various blood groups, there are a number of receptors, some common to all human erythrocytes, about which a certain amount of information has been gained by a study of the inhibition of various lectins by various carbohydrates. Much of this information depends, directly or indirectly, on work with lectins; it will therefore be desirable first to discuss their specificities more in detail.

Specificity of Plant Agglutinins.—When blood-group-specific plant agglutinins were first discovered, it was natural to suppose, since their reaction with the blood-group antigens was thought to be merely a chemical accident, that their specificity would be less sharp than that of the normal isoagglutinins. However, the contrary has proved to be true, at least in some cases.

The specificity of some lectins is nearly absolute. The anti-*A* of human group-*B* plasma reacts with A_1 and A_2 erythrocytes and also reacts with the *J* substance of the blood of cattle, the *R* antigen of sheep, the *A* substance of hog, and the Forssman antigen.[768] The anti-*A* of *Dolichos biflorus,* on the other hand, is not specific for the antigens from any of these animal sources,[4] but detects a previously undescribed heterogenetic factor. It does react with both A_1 and A_2, but its affinity for A_2 is so much less than for A_1 (less than $\frac{1}{500}$) that it is virtually specific for A_1.

The anti-*A* of lima beans is somewhat less specific; its affinity for A_2 is higher, and, when very concentrated, it weakly agglutinates *B* cells also. The anti-*H* lectins are still less specific, since they may agglutinate A_1 and *B* cells when concentrated enough. This, however, may be because human erythrocytes of these groups contain some *H* antigen.

In addition to these relatively specific lectins, others are known which seem to react with more than one receptor on the red cell, such as *A* and *B* or *B* and *H*. Finally, those which agglutinate human cells of all groups should be mentioned. Even some of these, however, may have their own specificity.

Human isoagglutinins react not only with purified *A* and *B* substances but also with certain fragments into which these antigens can be split, as by hydrolysis. In the case of fragments too small to precipitate, the specificity must be demonstrated by the inhibition technique. It has been mentioned that the smallest fragment of the *A* antigen which specifically inhibited the reaction of human anti-*A* antibody with the *A* antigen was a disaccharide containing *N*-acetyl-galactosamine as a terminal group. The anti-*A* agglutinins from plants were not only inhibited by this disaccharide, but also by *N*-acetyl-galactosamine itself.[562] From this observation either of two opposed conclusions may be drawn: plant anti-*A* reagents are less specific than human anti-*A*, since they are inhibited by a simpler substance, or they are more specific, since they cross-react less.

In the case of anti-*H* agglutinins, the plant reagents have not proved any less specific than those of animal origin. The anti-*H* of eel serum is inhibited by

Fig. 47. STRUCTURES OF SUGARS INHIBITING ANTI-*H* OF *Lotus tetragonolobus*, SHOWING IDENTICAL CONFIGURATION AT CARBONS 3 AND 4.

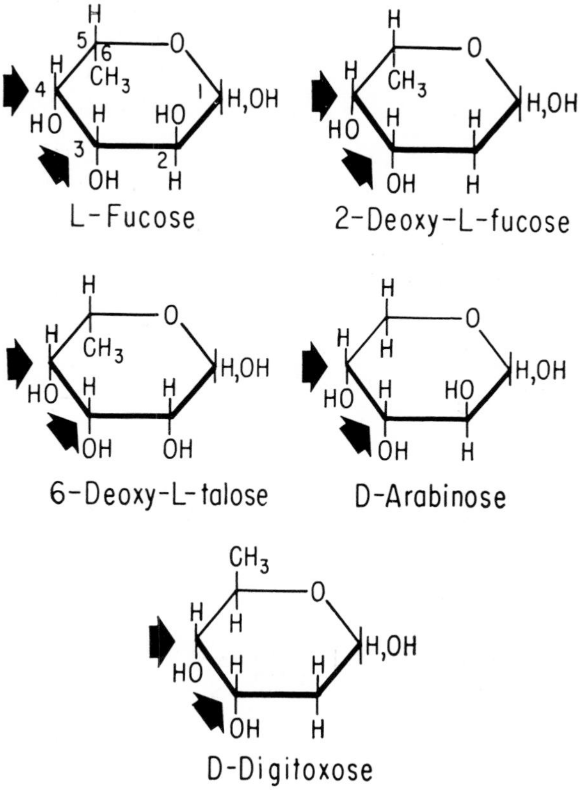

L-Fucose   2-Deoxy-L-fucose

6-Deoxy-L-talose   D-Arabinose

D-Digitoxose

Modified from Morgan and Watkins.[561]

L-fucose, and *L*-fucose inhibits the anti-*H* of *Lotus tetragonolobus* and *Ulex europeus*. *L*-fucose does not inhibit the anti-*H* of *Cytisus sessifolius* or of *Laburnum alpinum*, but salicin, a glucoside of *D*-glucose and saligenin, does.[63] If this means that the lectins of *Cytisus* and *Laburnum* are directed toward a part of the *H* antigen different from that recognized by the animal, anti-*H* reagents, it suggests that the specificity of the lectins is greater, not less, than that of animal agglutinins.

The anti-*H* of *Lotus tetragonolobus* is inhibited also by 2-deoxy-*L*-fucose, *L*-galactose, 6-deoxy-*L*-talose, *D*-arabinose, and *N*-acetyl-glucosamine. Morgan and Watkins [561] pointed out that, aside from the last, all the inhibiting sugars, when written in the pyranose form, have the same configuration at carbon atoms three

Fig. 48. STRUCTURES OF SUGARS INHIBITING ANTI-*A*, *B* AGGLUTININ OF *Sophora japonica* SHOWING IDENTICAL CONFIGURATION AT CARBONS 3 AND 4 OF TERMINAL SUGAR.

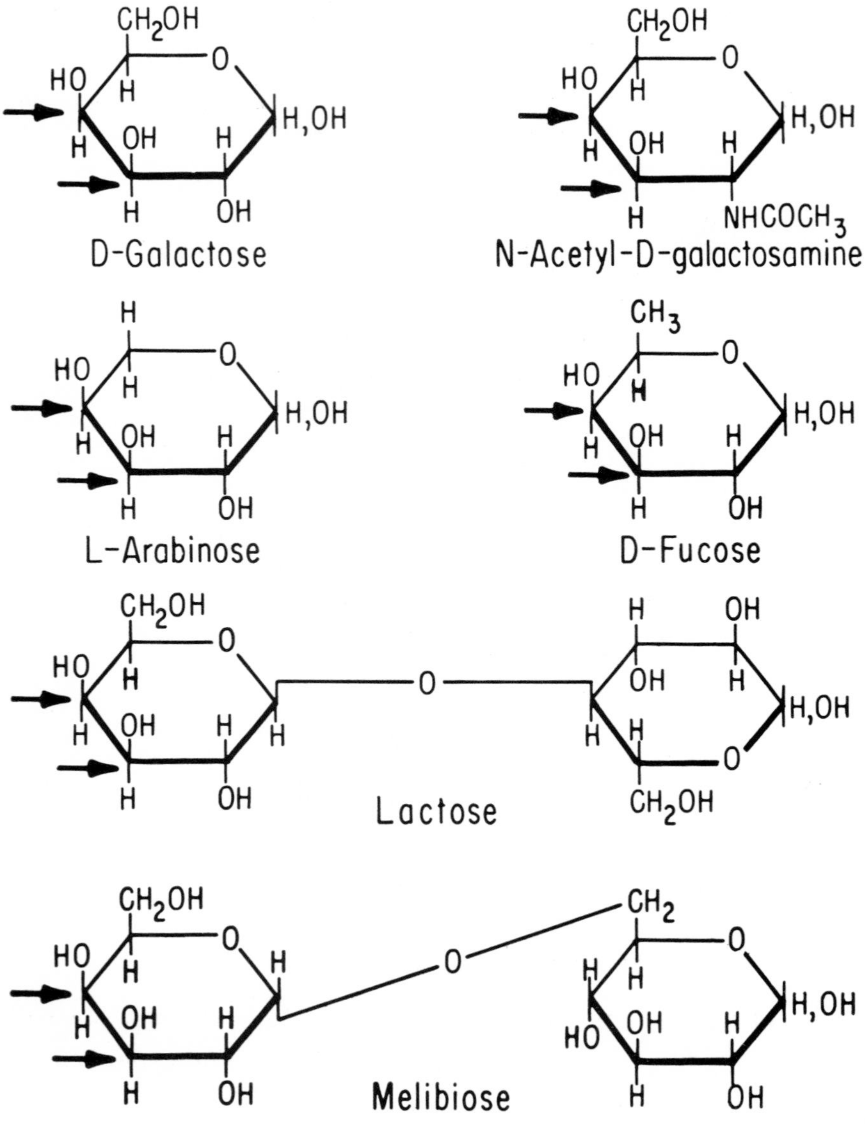

D-Galactose

N-Acetyl-D-galactosamine

L-Arabinose

D-Fucose

Lactose

Melibiose

Modified from Krüpe, 1956.[465]

and four (figure 47). In all of them, the hydroxyl groups are on the same side of the pyran ring pointing downward.

Krüpe [465] noticed that the sugars which inhibited the anti-$(A + B)$ agglutinin of *Sophora japonica* (*N*-acetyl-*D*-galactosamine, *D*-galactose, lactose, melibiose, *L*-arabinose, and *D*-fucose) all had the same configuration at carbons three and four. The opposite is found in the sugars that inhibit Lotus (figure 48). Mäkelä,[522] who made a much more extensive study of plant agglutinins and their inhibition reactions suggested that monosaccharides fall into four classes with re-

Fig. 49. Mäkelä's four classes of pyranoses, based on configurations of carbons 3 and 4.[522]

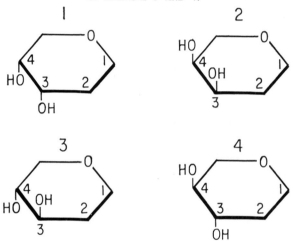

spect to their specific inhibiting activity for plant agglutinins, and that this is based on the configuration at carbons three and four (figure 49).

In some cases, the addition of an inhibiting sugar to a nonspecific plant agglutinin does not suppress all activity, but leaves the preparation able to agglutinate cells of certain blood groups, thus revealing a new specificity. Mäkelä [522] found that the agglutinin of *Bandeiraea simplicifolia* became *B*-specific when tested against cells suspended in 2 per cent glucose, and a suitable concentration of galactose makes the anti-$(A + B)$ agglutinin of *Calpurina aurea* *A*-specific.[62] Boyd, Everhart, and McMaster [86] found that some preparations of *Bauhinia purpurea* were nonspecific (really, specific for sugars of group two and derivatives), but could be made *N*-specific by the addition of *D*-galactose or other sugars of group two. Nonspecific plant agglutinins do not usually develop a new specificity when treated with inhibiting sugars in this way, however, so the three just mentioned seem to be exceptional.

Erythrocyte Receptors.—Inhibition studies have given some clues about the structure of the receptors in the human erythrocyte with which three different classes of lectins combine. First, one that is detected by extracts of *Bauhinia pur-*

purea [86] will be discussed.   Extracts of the seeds of this plant can be specific for the *N* antigen, and, if they are not *N*-specific, they can be made so by adding galactose or a similar sugar.   The anti-*N* specificity of *Bauhinia* extracts depends on the presence of sufficient amounts of one or more sugars, probably galactose or galactose derivatives; removal of these sugars by dialysis makes *Bauhinia purpurea* ex-

Fig. 50.   HYPOTHETICAL STRUCTURES OF TERMINAL PORTIONS OF ACTIVE SIDE CHAINS OF THREE GALACTOSE-CONTAINING RECEPTORS OF THE HUMAN RED CELL.

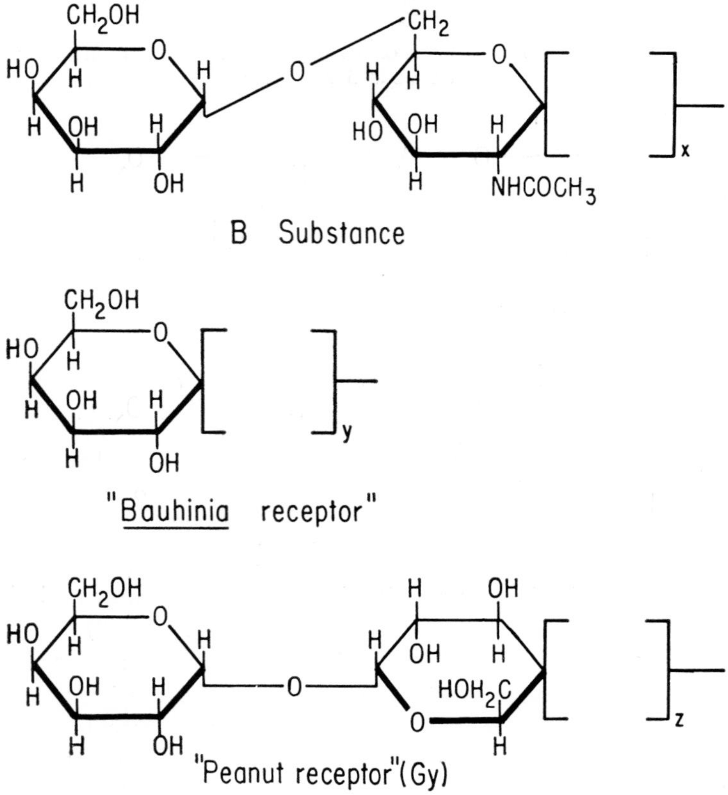

B   Substance

"Bauhinia   receptor"

"Peanut receptor"(Gy)

tracts nonspecific, that is, they then agglutinate human blood irrespective of blood groups.[86]

From the table in the paper cited, it is seen that the nonspecific activity of *Bauhinia* extracts is inhibited by sugars of Mäkelä's group 2.   It will be seen from the inhibiting power of the disaccharide, trisaccharide, and tetrasaccharide containing galactose that the *Bauhinia* receptor is probably an oligosaccharide containing at least one more unit beyond galactose; probably it contains several units, although, if it does, one can conclude that the next-to-terminal unit is not galactose (as in stachyose) for this sugar is not a very good inhibitor in that position. Although the *Bauhinia* receptor thus has some features in common with the *B* receptor, it is obviously not the same, since it occurs in all human erythrocytes.   As a matter of fact, B susbtance does not react with the *Bauhinia* agglutinin. It may well be that the structure I have called the *Bauhinia* receptor is the same as the one recognized by other group-2 lectins, for they are inhibited in the same way.[523] The others we have examined, however, do not contain any other specificity that remains after the group-2 specificity has been inhibited.

Another receptor has been detected with the aid of extracts of ordinary peanuts. This plant agglutinin is also inhibited by sugars of group 2 [87, 88, 89]; and the fact that galactose is the most effective, monosaccharide inhibitor suggests that galactose is the terminal group of this receptor also. The receptor consists of more than one sugar unit, however, as is obvious from the fact that two disaccharides, trehalose and lactose (the former not even containing galactose), inhibit better than galactose. Two other disaccharides containing only glucose (maltose and cellobiose) also inhibit well. The inhibitory power of the three, all-glucose disaccharides, and the fact that lactose contains glucose suggest that glucose may be the next-to-terminal group in the peanut receptor. This is supported by the observation that glucose itself has some inhibiting power. The effectiveness of trehalose may suggest that the peanut receptor contains glucose as the next-to-terminal unit, possibly linked to the terminal galactose by a 1 → 1 link (the type of linkage in trehalose). The peanut agglutinin, in spite of containing galactose as a terminal unit, is not inhibited by B substance and resembles the *Bauhinia* agglutinin in that respect. The two plant receptors are obviously different, since the peanut receptor, unlike the *Bauhinia* receptor, is not found on all human erythrocytes. Another sign of difference between the two receptors is the fact that the

Table 59 *

INHIBITION OF ANTI-*D* BY VARIOUS SUGARS

		Serum Diluted				
Sugar	Undiluted	1:2	1:4	1:8	1:16	1:32
(saline)	3	3.5	4	4	3	2
D-glucose (3)	4	3	3.5	3.5	1.5	1
L-glucose (4)	3	2.5	0	0	0	0
D-mannose (3)	3.5	3	3	3.5	2.5	0
L-mannose (4)	2.5	2	0	0	0	0
D-gulose (4)	0	1	0	0	0	0

*Strength of agglutination graded from 4 (all erythrocytes united into a single clump) to 0.5 (weak agglutination just visible to the naked eye); 0 indicates no agglutination. Figures in parentheses indicate group of the sugar according to Mäkelä.

peanut agglutinin is inhibited by some sugars (cellobiose, trehalose, melezitose, and maltose), which do not inhibit the *Bauhinia* agglutinin. Thus, at least three receptors contain galactose as a terminal unit, one of them present on all human red cells the others only on those of certain individuals. Their structure in the light of present scanty information is shown in figure 50.

It has already been mentioned that plant agglutinins inhibited by sugars of Mäkelä's group 3 obviously react with a red-cell receptor present on all human red cells. It would seem likely that the terminal unit in this receptor is a sugar of group 3. In view of the fact that Mäkelä [522, 523] found in several cases that mannose was the best inhibitor for such agglutinins, one could speculate that the terminal unit is mannose, but it is difficult to go further than this on the information now available.

The Rh Receptors.—Encouraged by the work with lectins, we have recently attempted to apply the specific-inhibition technique to a study of the *Rh* blood-group receptors. It was reported by Hackel, Smolker, and Fenske that anti-*Rh* sera are inhibited specifically by a number of ribonucleic-acid derivatives, suggesting that the *Rh* antigens are at least partly ribonucleotide in nature. We have found that human anti-*D* serum is inhibited also, weakly but apparently specifically, by unnatural sugars of Mäkelä's group 4, including *L*-lyxose, *L*-mannose, *L*-glucose, and *D*-glucose [88] (table 59). It is not inhibited by the natural enantiomers of these sugars or by any other sugars we have tried.

These results suggest that the *D* receptor may contain a sugar of group 4 as the terminal unit. This suggestion is supported by the observation that streptomycin, a natural glycoside of *N*-methyl-*L*-glucosamine, and rutinose [6–0(beta-*L*-rhamnosyl)-*D*-glucose] also inhibit. Streptomycin does not inhibit much better than *L*-mannose and *L*-glucose. However, rutinose, on a molar basis (the rutinose solution available at that time was only 0.14 as strong as the other sugar solutions studied) inhibits better than streptomycin or *L*-mannose, which suggests that the next-to-terminal unit in the *D* receptor is *D*-glucose or a similar sugar. The likelihood that the terminal unit of rutinose, *L*-rhamnose, is the terminal unit of the *D* receptor is diminished by the observation that rhamnose itself does not inhibit.

It is hardly necesary to mention that a knowledge of the chemical structure of the *D* antigen would have considerable practical value; it might make the preparation of good anti-*D* agglutinins possible, and injections of a nontoxic oligosaccharide with high *D* activity into pregnant women might possibly neutralize the anti-*D* of the mother's blood and prevent erythroblastosis in the infant.

The inhibition behavior of anti-*C* and anti-*E* is more complicated. *L*-glucose has some inhibitory effect on both of these agglutinins, but anti-*E* is also inhibited by *D*-glucose. Other sugars of group 4 do not always seem to inhibit anti-*C*, so I do not yet feel any confidence that the terminal unit is a sugar of this group. It seems possible that a sugar of group 3 is the terminal unit of the *E* receptor, considering the effectiveness of sugars of this group in inhibiting the anti-*E*. We also found inhibition by *L*-allose, a group-2 sugar, which is unexpected, and is not paralleled by inhibition by other group-2 sugars.

If the *D* receptor and possibly the *C* and *E* receptors as well contain a sugar of group 4, it is surprising, for sugars of this group have not previously been found in human tissues. It may be suspected, however, that many surprises remain in the field of the natural occurrence of sugars. In the field of protein chemistry, it is commonly assumed that only one enantiomer of each amino acid occurs in nature, but Oncley [634] has pointed out reasons for questioning this. It may well turn out that a number of the "unnatural" sugars actually occur in nature. As already mentioned, a derivative of *L*-glucose occurs in streptomycin, a di-deoxy-*L*-mannose has been isolated from *Parscaris equorum*,[251] and a di-deoxy-*L*-galactose plays an important role in the structure of the somatic antigen of group-*p Salmonella*.[921]

DISCUSSION

DR. BURDETTE: Dr. Owen will discuss Dr. Boyd's paper.

DR. OWEN: Dr. Boyd's coverage of the subject of methodology in the study of human blood groups and soluble antigens has been thorough and leaves little that need be added. I might remark, however, that Dr. Boyd's textbook, *Fundamentals of Immunology*,[81] has served through the years as a source of information about both the concepts and the methodology of immunology of unique usefulness to students of other disciplines, including genetics.

There are, of course, numerous other applications of immunology to studies of human genetics in addition to those dealing with blood groups and secretor substances. Of these, I wish to mention two. First, the *Gm* types of human globulin are recognized by a test system that involves determining whether a given human, gamma globulin will combine with the antibody-like material found in sera from certain patients with rheumatoid arthritis and thus block this material from agglutinating globulin-coated, test cells. The test cells are human red cells of group *O* that have been coated with particular "incomplete" *Rh* antibodies. References to this technique may be found in a paper by Steinberg, Giles, and Stauffer.[830]

Second, the inherited individual differences that lead to the rejection of incompatible transplants of tissues and organs provide a challenging field for study in the future. We know very little about the genetics of man in this regard, and it would doubtless be very worth while to learn more. It is possible, for example, that "matching" a prospective donor and recipient for major tissue antigens may make it possible to obtain prolonged survival of grafts of a sort that might be extended even farther by other developing approaches to the control of incompatibility reactions. In this area, however, methodology has been the limiting factor; we do not at present have acceptable techniques for defining human variation in this category of characteristics. The growing appreciation of inherited individuality in leucocyte antigens may prove relevant to transplantation antigens as well; in any case, the leucocyte variations accessible through such test techniques as leucocyte agglutination or cytotoxic determinations of various sorts provide further examples of applications of serologic methods [110, 475, 730] to human genetics.

APPENDIX I

Use of Computers in Biology and Medicine

Edward Novitski, Ph.D.

REFERENCES

Some applications of computers to research in the various branches of biology and medicine may be found in the following references.

GENERAL

Bibliography on the 11th Annual Conference on Electrical Techniques in Medicine and Biology. Minneapolis, Minn., p. 19, Nov., 1958.

Kirsch, R. A., L. Cahn, C. Ray, and G. H. Urban: Experiments in Processing Pictorial Information with a Digital Computer. Proc. Eastern Joint Computer Conf., Wash., D.C., Dec., 1957.

Ledley, R. S.: Digital Electronic Computers in Biomedical Science. 130: 1225, 1959.

Summaries of Current Work in the Division of Biological and Medical Research. Argonne National Laboratory, Chicago, May, 1958.

Weinrauch, H., and E. Pattishall: Air Force and Application of Computers to Medicine and Biology. Military Med., **122:** 178, 1958.

MEDICINE

Barlow, J. S.: Autocorrelation and Crosscorrelation Analysis in Electroencephalography. Presented at the 11th Annual Conference on Electrical Techniques in Medicine and Biology, Minneapolis, Minn., Nov., 1958.

Barlow, J. S., M. A. B. Brazier, and W. A. Rosenblith: The Application of Autocorrelation Analysis to Electroencephalography. Proc. Natl. Biophys. Conf., **1:** 622, 1959.

A BIZMAC Program for Medical Data Processing. Radio Corporation of America, New York, 1957.

Brazier, M. A. B., and J. S. Barlow: Some Applications of Correlation Analysis to Clinical Problems in Electroencephalography. Electroencephalog. and Clin. Neurophysiol., **8:** 1956.

Farley, B. G., L. S. Frishkopf, W. A. Clark, Jr., and J. T. Gilmore, Jr.: Computer Techniques for the Study of Patterns in the Electroencephalogram. M.I.T. Research Laboratory of Electronics Tech. Rept., no. 337, Nov., 1957.

Farley, B. G., L. S. Frishkopf, M. Freeman, C. Molnar, and T. Weiss: A Summary Description of EEG Data with the Aid of a Digital Computer. Presented at the 11th Annual Conference on Electrical Techniques in Medicine and Biology, Minneapolis, Minn., Nov., 1958.

Farrar, J. T.: Digital Computer Diagnosis of Gastrointestinal Disorders from Complex Intraluminal Pressure Record. Conference on Diagnostic Data Processing, Rockefeller Institute, New York, Jan., 1959.

Gertler, M. M., M. A. Woodbury, L. G. Gottsch, P. D. White, and H. A. Rusk: The Candidate for Coronary Heart Diseases. J. Amer. Med. Assoc., 170: 149, 1959.

Hawkes, A. K.: A Technique for Digital Computer Analysis of Electrocardiograms. Presented at the 11th Annual Conference on Electrical Techniques in Medicine and Biology, Minneapolis, Minn., Nov., 1958.

High-speed Computer to Analyze Human Heart-beat. Natl. Bur. Standards, (U.S.) Tech. News Bull., Sept., 1959.

Ledley, R. S.: Logical Aid to Systematic Medical Diagnosis. J. Operations Research Soc., Amer., *3*: no. 3, abstract, 1955.

Lipkin, M., and J. D. Hardy: Mechanical Correlation of Data in Differential Diagnosis of Hematological Diseases. J. Amer. Med. Assoc., **166**: 2, 1958.

McLannan, M. A., and E. G. Correll: Instrumentation for a Periodic Analysis of the EEG. Proc. IRE Conf., Houston, Tex., 1957.

Pattishall, E. G.: The Application of Computer Techniques to Medical Research. Report of Los Angeles Conference, Contract AF 18 (600)–118, Oct., 1957.

Pipberger, H. V., and E. D. Freis: A Project of the U.S. Veterans Administration for the Electronic Computation of Electrocardiographic Data. Presented at the 11th Annual Conference on Electrical Techniques in Medicine and Biology, Minneapolis, Minn., Nov., 1958.

Rao, R., R. Allard, and O. H. Schmitt: An Electronic Computer Program for Calculating Figures of Merit and Orthogonalization Constants of Electrocardiographic Lead Systems from Transfer Impedance Data. Presented at the 11th Annual Conference on Electrical Techniques in Medicine and Biology, Minneapolis, Minn., Nov., 1958.

Taback, L., E. Marden, H. L. Mason, and H. V. Pipberger: Digital Recording of Electrocardiographic Data for Analysis by a Digital Computer. IRE Trans. Med. Electronics, Sept., 1959.

Talbot, S. A.: Computers Applied to Ballistocardiography. Presented at the 11th Annual Conference on Electrical Techniques in Medicine and Biology, Minneapolis, Minn., Nov., 1958.

GENETICS

Barker, J. S. F.: Simulation of Genetic Systems by Automatic Digital Computers. III. Selection between Alleles at an Autosomal Locus. Australian J. Biol. Sci., **11**: 603, 1958. IV. Selection between Alleles at a Sex-linked Locus. Australian J. Biol. Sci., **11**: 613, 1958.

Fraser, A. S.: Simulation of Genetic Systems by Automatic Digital Computers. I. Introduction. Australian J. Biol. Sci., **10**: 484, 1957. II. Effects of Linkage on Rates of Advance under Selection. Australian J. Biol. Sci., **10**: 492, 1958.

Lewontin, R. C., and L. C. Dunn: The Evolutionary Dynamics of a Polymorphism in the House Mouse. Genetics, 1960 (in press).

Novitski, E., and E. R. Dempster: An Analysis of Data from Laboratory Populations of *Drosophila melanogaster*. Genetics, **43**: 470, 1958.

BIOCHEMISTRY

Chance, B., G. R. Williams, W. F. Holmes, and J. Higgins: Respiratory Enzymes in Oxidative Phosphorylation: V. A. Mechanism for Oxidative Phosphorylation. J. Biol. Chem., **217**: 439, 1955.

Garfinkel, D., J. J. Higgins, and J. D. Rutledge: A Univac Program for Representing the Kinetic Behavior of Chemical and Biological Systems. Presented at the 11th Annual Conference on Electrical Techniques in Medicine and Biology, Minneapolis, Minn., Nov., 1958.

Karreman, G., I. Isenberg, and A. Szent-Gyorgyi: On the Mechanism of Action of Chlorpromazine. Science, **130**: 1191, 1959.

Kraut, J.: The Shape of the Chymotrypsinogen Molecule from X-ray Diffraction. Biochem. et Biophys. Acta, **30**: 265, 1958.

Ledley, R. S.: Digital Computational Methods in Symbolic Logic with Examples in Biochemistry. Proc. Natl. Acad. Sci. (Wash.), **47**: 7, 1955.

Lewallen, C. G., M. Berman, and J. E. Rall: Studies of Iodoalbumin Metabolism. I. A Mathematical Approach to the Kinetics. J. Clin. Invest., **38**: 66, 1959.

Moore, R., R. H. Thompson, E. R. Buskirk, and G. D. Whedon: A Proposed Data-handling System for Human Metabolic Studies. Presented at the 11th Annual Conference on Electrical Techniques in Medicine and Biology, Minneapolis, Minn., Nov., 1958.

Pauling, L., and R. B. Corey: The Configuration of Polypeptide Chains in Proteins. Fortschr. Chem. Org. Naturstoffe, 11: 181, 1954.

Posner, A. S., A. Perloff, and A. F. Diorio: Refinement of the Hydroxyapatite Structure. Acta Cryst., 11: 308, 1958.

Rogoff, M., and T. Gallagher: Electronics in Biochemical Spectroscopy. Presented at the IRE National Convention. Medical Electronics Session, New York, 1958.

Steinrauf, L. K., J. Peterson, and L. H. Jensen: The Crystal Structure of 1-Cystine Hydrochloride. J. Amer. Chem. Soc., 80: 3835, 1958.

BIOPHYSICS

Berman, M., and R. L. Schoenfeld: Invariants in Experimental Data on Linear Kinetics and the Formulation of Models. J. Appl. Phys., 27: 1361, 1956.

Berman, M., and R. L. Schoenfeld: Information Content of Tracer Data with Respect to Steady-state Systems. Symposium on Information Theory in Biology. Pergamon, New York, 1958.

Oncley, J. L., ed.: Biophysical Science. A Study Program. Revs. Modern Phys., 31: 1, 1959.

PHYSIOLOGY

Bourghardt, S.: Computing Microphotometer for Cell Analyses, J. Sci. Inst., 32: 186, 1955.

Brownell, G. L., et al.: Electrical Analog for Analysis of Compartmental Biological Systems. Rev. Sci. Inst., 24: 704, 1953.

Clark, W. A., and B. G. Farley: Generalization of Pattern Recognition in a Self-organizing System. Proc. Western Joint Computer Conf., Los Angeles, p. 86, Mar., 1955.

Cole, K. S., H. A. Antosiewicz, and P. Rabinowitz: Automatic Computation of Nerve Excitation. J. Soc. Ind. Appl. Math., 3: 153, 1956. Correction. Ibid, 6: 196, 1958.

Communications Biophysics Group of the Research Laboratory of Electronics. W. M. Siebert, ed. Processing Neuroelectric Data Tech. Rept. no. 351, Technology Press, Cambridge, Mass., 1959.

Elias, P., and W. A. Bosenbilth: The Use and Misuse of Digital Models in the Study of the Nervous System. Presented at the 11th Annual Conference on Electrical Techniques in Medicine and Biology, Minneapolis, Minn., Nov., 1958.

Farley, B. G., and W. A. Clark: Simulation of Self-organizing Systems by Digital Computer. Trans. IRE Professional Group on Information Theory, Sept., 1954.

Hoffman, G., N. Metropolis, and V. Gardiner: Digital Computer Studies of Cell Multiplication by Monte Carlo Methods. J. Natl. Cancer Inst., 17: 175, 1956.

Kosten, F. H.: Nuclear Size Changes during Autolysis in Normal Mouse Liver, Kidney, and Adrenal Gland. Proc. Soc. Exp. Biol. and Med., 98: 275, 1958.

Pittendrigh, C. S.: The Oscillatory Nature of Physiological Processes. Presented at the 11th Annual Conference on Electrical Techniques in Medicine and Biology, Minneapolis, Minn., Nov., 1958.

Reed, R. W., and K. F. Gregory: Punch Card for Abstracts of Bacteriological Papers. Science, 118: 360, 1953.

Rochester, N., J. H. Holland, L. H. Haibt, and W. L. Duda: Tests on a Cell Assembly Theory of the Action of the Brain, Using a Large Digital Computer. IRE Trans. on Information Theory, p. 80, 1956.

Trank, J. W.: A Statistical Study of the Effects of Electric Fields on the Movements of Mammalian Sperm Cells. Presented at the 11th Annual Conference on Electrical Techniques in Medicine and Biology, Minneapolis, Minn., Nov., 1958.

Von Neumann, J.: The Computer and the Brain. Yale Univ. Press, New Haven, 1958.

Welkowitz, W.: Programming a Digital Computer for Cell Counting and Sizing. Rev. Sci. Inst., 25: 1202, 1954.

VITAL STATISTICS

Newcombe, H. B., J. M. Kennedy, S. J. Axford, and A. P. James: Automatic Linkage of Vital Records. Science, 130: 954, 1959.

Novitski, E., and A. W. Kimball: Birth Order, Parental Ages, and Sex of Offspring. Amer. J. Human Genetics, 10: 268, 1958.

VENDORS OF COMPUTERS

The following are names and addresses of companies manufacturing some of the more commonly used computers.

Bendix Computer Division, 5360 Arbor Vitae, Los Angeles, California.
Burroughs Corporation, 6071 Second Boulevard, Detroit, Michigan.
Control Data Corporation, Minneapolis, Minnesota.
International Business Machines Corp., Accounting Machine Division., 590 Madison Avenue, New York, New York.
Minneapolis Honeywell, 2954 Fourth Avenue, S., Minneapolis 8, Minnesota.
Philco Computer Division, 445 E. Tioga Street, Philadelphia, Pennsylvania.
Remington Rand Electronic Computer Division of Sperry Rand Corp., Fourth Avenue and 23rd, New York, New York.

APPENDIX II

A Proposed Standard System of Nomenclature for Human Mitotic Chromosomes * 76

The rapid growth of knowledge of human chromosomes in several laboratories, following advances in technical methods, has given rise to several systems
by which the chromosomes are named. This has led to confusion in the literature
and thus to the need for resolving the differences. Consequently, at the suggestion
of Dr. C. E. Ford, a small study group convened to attempt the formulation of
a common system of nomenclature. The meeting was arranged, through the
good offices of Dr. T. T. Puck, to be held at Denver, in the University of Colorado,
under the auspices of the Medical School. The meeting of this study group was
made possible by the support of the American Cancer Society, to whom grateful
appreciation is due. For practical reasons, it was decided to keep the group as
small as possible and to limit it to those hyman cytologists who had already published karyotypes.† In addition, three counselors were invited to guide the
group and to aid the discussions, and, if necessary, to arbitrate. Fortunately, the
last office did not prove necessary, and it was possible by mutual agreement to
arrive at a common system that has flexibility.

---

* Supplied by Dr. Theodore T. Puck

† In contemporary publications, the terms "karyotype" and "idiogram" have often
been used indiscriminately. The term, karotype should be applied to a systematized array of
the chromosomes of a single cell prepared either by drawing or by photography, with the
extension in meaning that the chromosomes of a single cell can typify the chromosomes of an
individual or even a species. The word idiogram would then be reserved for the diagram-
matic representation of a karyotype, which may be based on measurements of the chromosomes
in several or many cells.

The principle basis of the system was to be simplicity and freedom, as far as possible, from ambiguity and risks of confusion, especially with other forms of nomenclature in human genetics. It should also be capable of adjustment and expansion to meet the needs of new knowledge of human chromosomes. The system should be agreed to by the greatest possible proportion of cytologists working in the field, but the risk that a minority may be unable to accept the system as a whole should not be allowed to delay adoption by a majority.

It was agreed that the autosomes should be serially numbered, 1 to 22, as nearly as possible in descending order of length, consistent with operational conveniences of identification by other criteria. The sex chromosomes should be referred to as X and Y, rather than by number, which would be an additional and, ultimately, superfluous appellation.

It was generally agreed that the 22 autosomes can be classified into seven groups, between which distinction can readily be made. Within these groups, further identification of individual chromosomes can, in many cases, be made relatively easily. Within some groups, especially the group of chromosomes numbered 6 through 12 and also the X chromosomes, the distinctions between chromosomes are very difficult to make by criteria presently available. However, separating chromosomes 6 and the X from the remainder of this group presents lesser difficulties. It is believed that, with very favorable preparations, distinction can be made between most, if not all, chromosomes.

Autosomes are first to be ordered by placing the seven groups as nearly as possible in descending order of size. Within each group the chromosomes are arranged, for the most part, by size. It was desired specifically to avoid the implication that size relationships have been permanently decided in every instance, but it is hoped that the assignment of numbers will be permanently fixed. In those cases where distinction is at present doubtful, final definition of each chromosome can be left until further knowledge has accrued, though an attempt is made to provide a characterization of each. These principles make it possible to draw up a conspectus of the chromosomes, a table of their quantitative characteristics, and a table of the synonyms that authors have already published (tables 60, 61, and 62).

In table 61 showing the diagnostic characters of the chromosomes, three parameters are relied upon: (1) the length of each chromosome relative to the total length of a normal, X-containing, haploid set, that is, the sum of the lengths of the 22 autosomes and of the X chromosome, expressed per thousand; (2) the arm-ratio of the chromosomes expressed as the length of the longer arm relative to the shorter one; and (3) the centromeric index expressed as the ratio of the length of the shorter arm to the whole length of the chromosome. The two latter indices are, of course, related algebraically quite simply, but it is thought useful to present both here. In some chromosomes, the additional criterion of the presence of a satellite is available (table 60), but, in view of the apparent mor-

Table 60

CONSPECTUS OF HUMAN, MITOTIC CHROMOSOMES

Group 1–3	Large chromosomes with approximately median centromeres. The three chromosomes are readily distinguished from each other by size and centromere position.
Group 4–5	Large chromosomes with submedian centromeres. The two chromosomes are difficult to distinguish, but chromosome 4 is slightly longer.
Group 6–12	Chromosomes of medium size with submedian centromeres. The X chromosome resembles the longer chromosomes in this group, especially chromosome 6, from which it is difficult to distinguish. This large group presents major difficulty in identification of individual chromosomes.
Group 13–15	Chromosomes of medium size with nearly terminal centromeres (acrocentric chromosomes). Chromosome 13 has a prominent satellite on the short arm. Chromosome 14 has a small satellite on the short arm. No satellite has been detected on chromosome 15.
Group 16–18	Rather short chromosomes with approximately median (in chromosome 16) or submedian centromeres.
Group 19–20	Short chromosomes with approximately median centromeres.
Group 21–22	Very short, acrocentric chromosomes. Chromosome 21 has a satellite on its short arm. The Y chromosome belongs to this group.

phologic variation of satellites, they and their connecting strands are excluded in computing the indices.

Table 61 shows the range of measurements determined by various workers. Some of the variation expresses the uncertainty due to measurement of relatively small objects; but many of the discrepancies between different workers' observations are due to the measurement of chromosomes at different stages of mitosis and to the effect of different methods of pretreatment and preparation for microscopic study. The ranges shown, therefore, represent the maxima and minima of the means found by different workers using different techniques. However, within any one worker's observations, the variations are not so broad.

Two other matters of nomenclature arose. In the first place, it is considered that no separate nomenclature for the groups is needed. Any group to which it may be necessary to refer will be a sequence of those designated by Arabic numerals. Hence, any chromosome group may be referred to by the Arabic numerals of the extreme chromosomes of the group, joined together by a hyphen, for example, the group of the three longest chromosomes would be Group 1-3. This scheme has the merit of great flexibility. For instance, chromosomes X and 6 may be separated from the Group 6-12 whenever they can be distinguished.

Second, there is the problem raised by the abnormal chromosomes that are being encountered in the more recent studies. The discussion of their nomenclature came to no definite conclusion. Broadly, it was agreed, however, that any symbol used should avoid incorporating a specific interpretation which was not reasonably established. Arbitrary symbols, prefixed by a designation of the laboratory of origin, should usually be assigned to the abnormal chromosome.

Table 61

QUANTITATIVE CHARACTERISTICS OF THE HUMAN MITOTIC CHROMOSOMES †

	Tjio and Puck [881]			Chu and Giles [134]			Levan and Hsu [494]			Fraccaro and Lindsten *			Lejeune and Turpin [490]			Jacobs and Harnden *			Range		
	A	B	C	A	B	C	A	B	C	A	B	C	A	B	C	A	B	C	A	B	C
1	90	1.1	48	90	1.1	48	85	1.1	49	82	1.1	48	87	1.1	48	83	1.1	48	82–90	1.1	48–49
2	82	1.6	39	83	1.5	40	79	1.6	38	77	1.5	40	84	1.5	40	79	1.6	38	77–84	1.5–1.6	38–40
3	70	1.2	45	72	1.2	46	69	1.2	45	65	1.2	45	67	1.2	46	63	1.2	46	63–72	1.2	45–46
4	64	2.9	26	63	2.9	26	63	2.7	27	62	2.6	28	62	2.6	25	60	2.6	28	60–64	2.6–2.9	25–28
5	58	3.2	24	58	3.2	24	59	2.6	28	60	2.4	29	57	2.4	30	57	2.4	30	57–60	2.4–3.2	24–30
X	59	1.9	34	57	2.0	38	52	1.6	38	54	1.6	38	58	2.2	32	51	1.7	37	51–59	1.6–2.8	32–38
6	55	1.7	37	56	1.8	36	56	1.7	37	54	1.6	38	56	1.7	37	56	1.6	38	54–56	1.6–1.8	36–38
7	47	1.3	43	52	1.9	35	51	1.9	35	50	1.7	37	51	1.8	36	50	1.7	37	47–52	1.3–1.9	35–43
8	44	1.5	29	46	1.7	29	48	1.6	33	47	1.7	37	48	2.4	29	46	1.5	40	44–48	1.5–2.4	29–40
9	44	1.9	40	46	2.4	38	47	1.8	36	45	2.0	33	47	1.9	35	44	2.1	32	44–47	1.8–2.4	32–40
10	43	2.4	27	45	2.3	30	45	2.0	33	45	2.6	34	45	2.6	27	44	1.9	35	43–45	1.9–2.6	27–35
11	43	2.8	34	44	2.1	32	44	2.2	31	43	2.2	31	44	1.6	39	43	1.5	40	43–44	1.5–2.8	31–40
12	42	3.1	24	43	3.1	24	42	1.7	32	43	1.7	37	42	2.8	27	42	2.1	32	42–43	1.7–3.1	24–37
13	35	8.0	11	32	9.7	10	32	5.0	16	34	4.8	17	33	6.8	14	36	4.9	17	32–36	4.8–9.7	10–17
14	32	7.3	12	34	9.5	9	37	4.0	18	35	4.4	19	32	7.0	13	34	4.3	19	32–37	4.3–9.5	9–19
15	29	10.5	9	31	11.9	8	35	4.7	17	33	4.6	22	31	10.0	9	34	3.8	22	29–35	3.8–11.9	8–22
16	32	1.8	36	27	1.6	38	30	1.4	42	31	1.4	42	29	1.4	41	33	1.4	31	27–33	1.4–1.8	31–42
17	29	2.8	26	30	2.1	33	29	2.4	30	30	1.9	35	29	3.1	23	30	1.8	36	29–30	1.8–3.1	23–36
18	24	3.8	21	25	3.8	22	25	2.6	28	27	2.5	29	26	4.2	21	27	2.4	29	24–27	2.4–4.2	21–29
19	22	1.4	41	22	1.9	34	24	1.2	40	25	1.3	43	22	1.4	42	26	1.2	45	22–26	1.2–1.9	34–45
20	21	1.3	44	19	1.3	44	21	1.2	40	23	1.3	43	20	1.2	43	25	1.2	46	19–25	1.2–1.3	40–46
21	18	3.7	21	15	6.8	13	13	2.5	28	19	2.5	29	15	2.3	31	20	2.5	29	13–20	2.3–6.8	13–31
22	17	3.3	23	12	6.0	14	16	2.0	33	17	2.3	30	13	4.0	20	18	2.7	27	12–18	2.0–6.0	14–33
Y	19	∞	0	11	∞	0	18	4.9	17	22	2.9	26	18	∞	0	18	4.9	17	11–22	2.9–∞	0–26

† All measurements were made from cells of normal individuals, except those made by Fraccaro and Lindsten, which included cases of Turner's Syndrome. Column *A* is the relative length of each chromosome, *B* is the arm-ratio, and *C* the centromere index, as defined in the text.

    * Unpublished data.

In this connection, two further requisites for coördination of research were discussed. One is the storage of documentation for references, perhaps in a central depository, which is additional to material published. Also, cultures should be preserved, by the satisfactory methods now used, so they are available for reference, comparison, and exchange.

Some consideration was also given to the desirability of using a uniform system for presenting karyotoypes and idiograms, but because individual variation in taste is involved, rigidity of design was thought undesirable. However, it was recommended that the chromosomes should be arranged in numerical order, with the sex chromosomes near to but separated from the autosomes they resemble. Similar ones should be grouped together with their centromeres aligned.

Although choice between the different possible schemes of nomenclature is arbitrary, uniformity for ease of reference is essential. Hence, individual pref-

Table 62

SYNONYMY OF CHROMOSOMES AS PUBLISHED BY VARIOUS WORKERS

New chromo-some number	Tjio and Puck [881]	Chu and Giles [134]	Levan and Hsu [494]	Ford Jacobs and Lajtha [240]	Böök Fraccaro and Lindsten [75]	Lejeune, Turpin and Gautier [490]
1	1	1	1	1	1	$G1$
2	2	2	2	2	2	$G2$
3	3	3	3	3	3	$G3$
4	4	4	4	4	4	$G4$
5	5	5	5	5	5	$G5$
6	6	6	6	6 *	6	$M1$
7	7	7	7	(8)	7	$M2$
8	8	8	8	(9)	8	$Md1$
9	9	9	9	(11)	9	$M3$
10	10	10	10	10	10	$Md2$
11	11	11	11	(12)	11	$M4$
12	12	12	12	(13)	12	$Md3$
13	18	14	20	14	14	$T1$
14	19	15	18	15	15	$T2$
15	20	13	19	16	13	$T3$
16	13	17	15	19	16	$C1$
17	14	16	13	17	17	$P1$
18	15	18	14	18	18	$P2$
19	16	19	16	20	19	$C2$
20	17	20	17	21	20	$C3$
21	21	21	22	22	21	Vh
22	22	22	21	23	22	Vs
X	X	X	X	?(7)	X	X
Y	Y	Y	Y	Y	Y	Y

* In the published idiogram, the chromosomes of group 6–12 (including X) were indicated by discontinuous lines and left unnumbered owing to the uncertainty of discrimination at that time. For the purpose of this table, these chromosomes have been assigned the numbers shown in brackets, in serial order of length.

erences have been subordinated to the common good in reaching this agreement. This human-chromosome-study group therefore agrees to use this notation and recommends that any who prefer to use any other scheme should, at the same time, also refer to the standard system here proposed.

List of Signatories

Participants

Dr. J. A. Böök
University of Uppsala
The Institute for Medical Genetics
V. Agatan 24
Uppsala, Sweden

Dr. E. H. Y. Chu
Biology Division
Oak Ridge National Laboratory
P. O. Box Y
Oak Ridge, Tennessee, U.S.A.

Dr. C. E. Ford
Medical Research Council
Radiobiological Research Unit
Harwell, Didcot, Berks, England

Dr. M. Fraccaro
The Institute for Medical Genetics
V. Agatan 24
Uppsala, Sweden

Dr. D. G. Harnden
Medical Research Council
Group for Research on the General Effects of
   Radiation
Department of Radiotherapy
Western General Hospital
Crewe Road
Edinburgh 4, Scotland

Dr. T. C. Hsu
Section of Cytology
The University of Texas
M. D. Anderson Hospital and Tumor Inst.
Texas Medical Center
Houston 25, Texas, U.S.A.

Dr. D. A. Hungerford
The Institute for Cancer Research
7701 Burholme Avenue
Fox Chase
Philadelphia 11, Pennsylvania, U.S.A.

Dr. P. A. Jacobs
Medical Research Council
Group for Research on the General Effects of
   Radiation
Department of Radiotherapy
Western General Hospital
Crewe Road
Edinburgh 4, Scotland

Dr. J. Lejeune
Université de Paris
Chaire d'Hygiène et de Clinique de la
   Première Enfance
Hôpital Trousseau
158, Ab. du Général Michel-Bizot
Paris (12), France

Dr. A. Levan
Institute of Genetics
University of Lund
Lund, Sweden

Dr. S. Makino
Zoological Institute
Faculty of Science
Hokkaido University
Sapporo, Japan

Dr. Theodore T. Puck
Department of Biophysics
University of Colorado Medical Center
4200 East Ninth Avenue
Denver 20, Colorado, U.S.A.

Dr. A. Robinson
Department of Biophysics
University of Colorado Medical Center
4200 East Ninth Avenue
Denver 20, Colorado, U.S.A.

Dr. J. H. Tjio
National Institute of Arthritis and Metabolic
   Diseases
National Institutes of Health
Bethesda 14, Maryland, U.S.A.

Counselors

Dr. D. G. Catcheside, Chairman
Department of Microbiology
The University
Edgbaston
Birmingham 15, England

Dr. H. J. Muller
Department of Zoology
Indiana University
Bloomington, Indiana, U.S.A.

Dr. Curt Stern
Department of Zoology
University of California
Berkeley 4, California, U.S.A.

APPENDIX III

Conspectus of Cases of Chromosomal Abnormality in Man

C. E. Ford, D.Sc.

The following statement is believed to be reasonably complete to the end of May, 1960. All references to numbers of chromosomes are in accordance with the proposed standard system of nomenclature of human chromosomes agreed at Denver, Colorado, in April, 1960.[76] The presence of sex chromatin [39, 41] is indicated by the term "chromatin-positive" and its absence by the term "chromatin-negative." At present there is no published exception to the rule that chromatin-positive individuals have at least two X chromosomes and chromatin-negative individuals one X chromosome only. Recent evidence suggests that the sex-chromatin body represents a single heteropycnotic X chromosome.[632] Since the X chromosomes and autosome 6 are so similar, evidence of abnormality of the number of X chromosomes should be regarded as presumptive. The names "Klinefelter's syndrome" and "Turner's syndrome" are adopted here as being the most generally used of a number of clinical designations. In their extensive review, Grumbach and Barr [298] prefer the names seminiferous-tubule dysgenesis and gonadal dysgenesis respectively.

ABNORMALITIES OF THE SEX CHROMOSOMES

XXY, 47 chromosomes. Klinefelter's syndrome.—Cases have been reported by Jacobs and Strong,[421] Ford *et al.*,[242] Court-Brown *et al.*,[153] *Harnden*,[324] Crooke and Hayward,[158] Bergman *et al.*,[54] and Nowakowski *et al.*[629] Two were presumptive *XXY/XX* mosaics, and all were chromatin-positive. Chromatin-negative cases of the syndrome, which form a distinct subgroup on histologic evidence [221, 298] have an apparently normal male karyotype [153, 242] and are presumably etiologically distinct.[615] The frequency of the chromatin-positive syndrome in the general population may be one in a thousand [675] or higher.[549]

XO, 45 chromosomes. Turner's syndrome.—Chromatin-negative cases of Turner's syndrome with 45 chromosomes and a presumptive *XO* constitution have been described by Ford *et al.*,[242] Tjio *et al.*,[883] Fraccaro *et al.*,[254] Stewart,[843] Jacobs and Keay,[420] Court-Brown *et al.*,[153] and Harnden.[324] Recently, Bahner *et al.*[28] reported a woman with an *XO*-chromosomal constitution who had borne a normal son and who exhibited none of the clinical stigmata of Turner's syndrome, other than short stature. Chromatin-positive cases of Turner's syndrome also occur but have not been distinguished as a clinical subgroup. They account for about 20 per cent of all cases of the syndrome [298] and include examples with an apparently normal female karyotype [153] as well as *XO/XX* mosaics.[245]

XXX, 47 chromosomes.—Two phenotypic females, aged 34 and 21 respectively at the time of examination, have been found to have 47 chromosomes in their somatic cells, including, apparently, three X chromosomes.[416, 419] Both are below normal in mental performance. The older patient is recorded as having menstruated for some 5 years at puberty; the younger one is still actively menstruating. In both, a high proportion of cells with two sex-chromatin bodies were found in smears from the buccal mucosa. A third phenotypic female, many of whose buccal-mucosa cells contained two sex-chromatin bodies, has been found to be an *XXX/XO* mosaic.[419] She was 51 years old, had never menstruated, and is described as of low-normal intelligence. De Carli *et al.*[174] have found 47 chromosomes in cells cultured from a 19-year-old female, the extra chromosome being a member of the difficult medium-sized group (6–12 and X). The patient is said to have some mongoloid features; she menstruates regularly and is of poor intelligence. Although two sex-chromatin bodies were found in at least 8 per cent of buccal-mucosa cells, the authors favor the view that the case is another example of autosomal trisomy. The use of the term "super female" for the *XXX* condition [419] has been criticized by Stern [835] and Lennox.[491] Jacobs *et al.* have replied.[415]

XY, 46 chromosomes. Female phenotype.—Patients exhibiting a familial form of male pseudohermaphroditism, commonly known as testicular feminization, have an essentially female phenotype with generally good secondary sexual development but bear rudimentary testes, either abdominally or, more usually, in bilateral inguinal hernias. The condition is transmitted through the female line only, there is a marked preponderance of apparent females in the affected families, and the abnormal individuals are chromatin-negative. The chromosomes of five patients exhibiting this condition appear to be indistinguishable from those of normal males.[417, 693]

Harnden and Stewart [327] have found an apparently normal male karyotype in cultures established from a tall, eunochoidal female with primary amenorrhoea. The patient is chromatin-negative and is considered to be a case of pure gonadal dysgenesis,[362] a condition expressed in both chromatin-negative and chromatin-positive individuals and attributed to primary failure of gonadal development.[430]

XX, 46 chromosomes. True hermaphrodites.—Five examples of true hermaphroditism have now been studied.[21, 220, 325, 383, 527] All were chromatin-positive and all had an apparently normal female karyotype. No evidence of mosaicism of the sex chromosomes was obtained in any of these cases.

XX-deficiency, 46 chromosomes.—A female patient with primary amenorrhea, poor secondary sex development, and low-normal intelligence was found to have a consistent but abnormal karyotype in which only 15 chromosomes of the medium-sized group (6–12, X) were present. The total of 46 chromosomes was made up by an additional small chromosome similar to autosome 16. This was interpreted as a deficient X.[419] In a buccal-smear examination carried out before the chromosomal studies were made, the sex-chromatin body was recorded as being smaller than usual and only present in 7 per cent of cells.

ABNORMALITIES OF THE AUTOSOMES

Triplo-21, 47 chromosomes. Mongolism.—Reports of an extra, small, acrocentric chromosome in mongoloid imbeciles have been published by Lejeune *et al.*,[488] Ford *et al.*,[241] Jacobs *et al.*,[418] Böök *et al.*,[75] Lehmann and Forssman,[487] and Court-Brown *et al.*[153] The extra chromosome bears a satellite [254] and is autosome in the Denver system of numbering. Two cases of undoubted, clinical mongolism in which the karyotype consisted of 46 chromosomes only are apparent exceptions to the rule of trisomy. They can both be explained as translocation derivatives (see following).

Triplo-17–18, 47 chromosomes.—Edwards *et al.*[199] have found an extra autosome (considered to be a no. 17) in cultured cells from skin and muscle taken at autopsy from a child with multiple, congenital abnormalities. These included a large and misshapen head, small mouth, receding chin, and ears that were long, oblique and low-set, giving them a pixie-look. Mental development was retarded. Two more examples of this condition have been found by Pätau *et al.*[644] who thought the extra was a no. 18.

Triplo-13–15, 47 chromosomes.—Another autosomal trisomic syndrome has been identified by Pätau *et al.*[644] The extra chromosome is a member of the 13–15 group, that is, a long acrocentric. Two cases are mentioned; one is described in detail. Among the principle defects exhibited by this case are apparent anophthalmia and deafness, hare-lip and cleft palate, receding chin, and retardation of mental development.

Presumptive translocation derivatives.—The three cases mentioned below can all be interpreted as instances of duplication-deficiency unbalance arising from a (symptomless) reciprocal translocation present in one of the parents. It is well known that individuals heterozygous for a reciprocal translocation usually give rise to balanced and unbalanced gametes in approximately equal numbers, and that the unbalanced zygotes die.[821] Nevertheless, where the translocation is decidedly unequal, one of the two types of duplication-deficiency gamete would be deficient for a small chromosomal segment only and might be viable.

A boy with multiple skeletal abnormalities (polydysspondylie) has been described by Turpin *et al.*[889] Cells in tissue culture contained 45 chromosomes only. There were only 5 long, acrocentric chromosomes (group 13–15) and only 3 short, acrocentric chromosomes (group 21–22). An extra chromosome had a long arm about equal in length to the long arm of a chromosome of the 13–15 group and a short arm about equal to the long arm of a chromosome of the 21–22 group. Assuming a reciprocal translocation between members of the two groups, with breaks close to the centromeres in both the chromosomes involved, the child would be deficient for little more than the two short arms.

Polani *et al.*[667] have reported a typical mongol child whose bone-marrow cells contained 46 chromosomes instead of the expected 47, the case showing considerable formal, cytogenetic similarities to the last. There were 5 instead of 6 members of the 13–15 group and two normal autosomal pairs, 21 and 22. An extra chromosome very similar morphologically to the extra in the polydysspondylie case was interpreted to arise as a result of a reciprocal translocation between a long, acrocentric autosome (?15) and an autosome 21, with breaks close to the centromeres. According to this interpretation, the child would be deficient (simplex) for the short arm of autosome 15 and duplicated (triplex) for the long arm of autosome 21, giving a chromosomal segmental constitution effectively equivalent to that normally observed in mongolism.

Another mongol child with 46 chromosomes has been reported by Fraccaro *et al.*[253] The karyotype includes only one normal autosome 21, but there is an extra, small, metacentric chromosome that could be a product of reciprocal translocation between the two homologous members of autosome pair 21 (again with both breaks close to the centromeres), or alternatively, an isochromosome derived from the long arm of autosome 21. In either case the chromosomal segmental constitution would be nearly equivalent to the last case and to that characteristic of (cytogenetically) uncomplicated mongolism. The interpretation is rendered difficult by evidence pointing to trisomy of autosome 19 (or possibly 20) in the father of the child, who is a professional man and presumably free of clinical stigmata. Since the extra, small, metacentric chromosome in the mongol son is virtually indistinguishable from autosomes 19 and 20, the possibility must be envisaged that the son received two autosomes 19 (or 20) from his father. If so, he would be trisomic 19 (or 20) and monosomic 21, a condition which on *a priori* grounds would hardly be expected to be viable, let alone exhibit the clinical features of mongolism.

Triploid, 69 chromosomes.—Böök *et al.*[77] have published a preliminary report on a baby boy with a cerebral defect and other congenital abnormalities. Most of the cells in tissue culture contained 69 chromosomes, including three representatives of each autosome and an *XXY* group of sex chromosomes.

ADDENDUM

Additional Selected References

Anders, G., A. Prader, E. Hauschtek, K. Schärer, R. E. Siebenmann, and R. Heller: Multiples Sex Chromatin und Komplexes chromosomates Mosaik bei einem Knaben mit Idiotie und multiplen Milzbildungen. Helvetica Paediatrica Acta, *15;* 515, 1960.

Barr, M. L., and D. H. Carr: Sex Chromatin, Sex Chromosomes, and Sex Anomalies. Canad. Med. Assoc. J., *83;* 979, 1960.

Carr, D. H., M. L. Barr, E. R. Plunkett, M. M. Grumbach, A. Morishima, and E. H. Y. Chu: An XXXY Sex Chromosome Complex in Klinefelter Subjects with Duplicated Sex Chromatin. J. Clin. Endocrinol., *21;* 491, 1961.

Carr, D. H., M. L. Barr, and E. R. Plunkett: An XXXX Sex Chromosome Complex in Two Mentally Defective Females. Canad. Med. Assoc. J., *84;* 131, 1961.

Clarke, C., J. H. Edwards, and V. Smallpiece.: 21-Trisomy-Normal Mosaicism. Lancet, *1;* 1028, 1961.

Delhanty, J. D. A., J. R. Ellis, and P. T. Rowley: Triploid Cells in a Human Embryo. Lancet, *1;* 1286, 1961.

Ferguson-Smith, M. A., A. W. Johnston, and S. D. Handmaker: Primary Amentia and Microorchidism Associated with an XXXY Sex-chromosome Constitution. Lancet, *2;* 184, 1960.

Fraccaro, M., D. Ikkos, J. Lindsten, R. Luft, and K. Kaijser: A New Type of Chromosomal Abnormality in Gonadal Dysgenesis. Lancet, *2;* 1144, 1960.

Fraccaro, M., and J. Lindsten: A Child with 49 Chromosomes. Lancet, *2;* 1303, 1960.

Fraser, J. H., J. Campbell, R. C. MacGillivray, E. Boyd, and B. Lennox: The XXX Syndrome. Frequency among Mental Defectives and Fertility. Lancet, *2;* 626, 1960.

Hirschhorn, K., H. L. Cooper, and O. J. Miller: Gonadal Dysgenesis (Turner's Syndrome) with Male Phenotype and XO Chromosomal Constitution. Lancet, *2;* 1449, 1960.

Jacobs, P. A., D. G. Harden, K. E. Buckton, W. M. Court Brown, M. J. King, J. A. McBride, T. N. MacGregor, and N. MacLean: Cytogenetic Studies in Primary Amenorrhea. Lancet, *1;* 1183, 1961.

MacLean, N., D. G. Harden, and W. M. Court Brown: Abnormalities of Sex Chromosome Constitution in Newborn Babies. Lancet, *11;* 406, 1961.

Muldal S., and C. H. Ockey: The Double Male. A New Chromosome Constitution in Klinefelter's Syndrome. Lancet, *2;* 492, 1960.

Penrose, L. S., and J. Delhanty: Triploid Cell Cultures from a Macerated Fetus. Lancet, *1;* 1261, 1961.

Smith, D. W., K. Patau, and E. Therman: Autosomal Trisomy Syndromes. Lancet, *11;* 211, 1961.

Stewart, J. S. S., and A. Sanderson: Fertility and Oligophrenia in an Apparent Triplo-X Female. Lancet, *2;* 21, 1960.

BIBLIOGRAPHY

1. Acher, R., and C. Crocker: Réactions Colorées Spécifiques de L'arginine et de la Tyrosine Réalisées après Chromatographie sur Papier. Biochim. Biophys. Acta, **9:** 704, 1952.
2. Aldrich, R. A., A. G. Steinberg, and D. C. Campbell: Pedigree Demonstrating a Sex-linked Recessive Condition Characterized by Draining Ears, Eczematoid Dermatitis and Bloody Diarrhea. Pediatrics, **13:** 133, 1954.
3. Alee, W. C., A. E. Emerson, O. Park, T. Park, and K. P. Schmidt: Principles of Animal Ecology. W. B. Saunders Company, Philadelphia, p. 837, 1949.
4. Allan, W.: Heredity in Diabetes. Ann. Intern. Med., **6:** 1272, 1933.
5. Allen, D. W., W. A. Schroeder, and J. Balog: Observations on the Chromatographic Heterogeneity of Normal Adult and Fetal Human Hemoglobin: A Study of the Effects of Crystallization and Chromatography on the Heterogeneity and Isoleucine Content. J. Amer. Chem. Soc., **80:** 1628, 1958.
6. Allison, A. C.: Malaria in Carriers of the Sickle-cell Trait and in Newborn Children. Exp. Parasit., **6:** 418, 1957.
7. Allison, A. C.: Protection Afforded by Sickle-cell Trait Against Subtertian Malarial Infection. Brit. Med. J., **1:** 290, 1954.
8. Allison, A. C., and D. F. Clyde: Malaria in African Children with Deficient Glucose-6-Phosphate Dehydrogenase in Erythrocytes. Brit. Med. J. **1:** 1346, 1961.
9. Ames, B. N., and B. Garry: Coordinate Repression of the Synthesis of Four Histidine Biosynthetic Enzymes by Histidine. Proc. Nat. Acad. Sci. (Wash.), **45:** 1453, 1959.
10. Anderson, E. P., H. M. Kalckar, and K. J. Isselbacher: Defect in Uptake of Galactose-1-Phosphate into Liver Nucleotides in Congenital Galactosemia. Science, **125:** 114, 1957.
11. Anderson, R., R. R. Huestis, and A. G. Motulsky: Hereditary Spherocytosis in the Deer Mouse. Its Similarity to the Human Disease. Blood, **15:** 491, 1960.
12. Anderson, V. E., H. O. Goodman, and S. C. Reed: Variables Related to Human Breast Cancer. Univ. of Minnesota Press, Minneapolis, p. 172, 1958.
13. Andresen, P. H.: Blood Group System L. New Blood Groups L_2. Case of Epistasy within Blood Groups. Acta Path. Microbiol. Scand., **25:** 728, 1948.
14. Andrewartha, H. G., and L. C. Birch: The Distribution and Abundance of Animals. Univ. Chicago Press, Chicago, p. 782, 1954.
15. Annual Report of Notifiable Diseases. Dominion Bureau of Statistics, Catalog No. 82–201, Ottawa, Canada, 1958.
16. Anonymous Editorial: Human Chromosomal Abnormalities. Lancet, **2:** 448, 1959.
17. Anscombe, F. J.: Large-Sample Theory of Sequential Estimation. Proc. Camb. Phil. Soc., **48:** 600, 1952.
18. Anscombe, F. J.: Fixed Sample Size Analysis of Sequential Observations. Biometrics, **10:** 89, 1954.
19. Arias, I. M.: Gilbert's Disease. Bull. N.Y. Acad. Med., **35:** 450, 1959.
20. Aronson, M., and R. W. I. Kessel: New Method for Manipulation, Maintenance, and Cloning of Single Mammalian Cells *in vitro*. Science **131:** 1377, 1960.

383

21. Assis, L. M., D. R. Epps, C. Bottura, and I. Ferrari: Chromosomal Constitution and Nuclear Sex of a True Hermaphrodite. Lancet **2**: 129, 1960.
22. Atwater, J., I. R. Schwartz, and L. M. Tocantins: A Variety of Human Hemoglobin with Four Distinct Electrophoretic Components. Blood, **15**: 901, 1960.
23. Atwood, K. C., and S. L. Scheinberg: Somatic Variation in Human Erythrocyte Antigens. J. Cell Comp. Physiol., 52 Suppl., **1**: 97, 1958.
24. Auerbach, C.: A Possible Case of Delayed Mutation in Man. Ann. Hum. Genet. (Lond.), **20**: 266, 1956.
25. Aycock, W. L.: Constitutional Types and Susceptibility to Paralysis in Poliomyelitis. Amer. J. Med. Sci., **202**: 456, 1941.
26. Aycock, W. L.: Familial Susceptibility to Leprosy. Amer. J. Med. Sci., **201**: 450, 1941.
27. Baber, R. E., and E. A. Ross: Changes in the Size of American Families in One Generation. Univ. Wisc. Stud. in Soc. Sci. and Hist., Madison, **10**: 99, 1924.
28. Bahner, F., G. Schwarz, D. G. Harnden, P. A. Jacobs, H. A. Hienz, and K. Walter: A Fertile Female with *XO* Sex Chromosome Constitution. Lancet, **2**: 100, 1960.
29. Baikie, A. G., W. M. Court Brown, and P. A. Jacobs: Chromosome Studies in Leukemia (letter to the editor). Lancet, **1**: 168, 1960.
30. Baikie, A. G., W. M. Court Brown, and P. A. Jacobs: Chromosome Studies in Leukemia (letter to the editor). Lancet, **1**: 280, 1960.
31. Baikie, A. G., W. M. Court Brown, P. A. Jacobs, and J. S. Milne: Chromosome Studies in Human Leukemia. Lancet, **2**: 425, 1959.
32. Bakken, P., V. J. Evans, R. E. Stevenson, and W. R. Earle: Establishment of a Strain of Human Skin Cells on Chemically Defined Medium NCTC 109. Amer. J. Hyg., **73**: 96, 1961.
33. Baldridge, R. C., L. Borofsky, H. Baird III, F. Reichle, and D. Bullock: Relationship of Serum Phenylalanine Levels and Ability of Phenylketonurics to Hydroxylate Tryptophan. Proc. Soc. Exp. Biol. (N.Y.), **100**: 529, 1959.
34. Barker, J. S. F.: Simulation of Genetic Systems by Automatic Digital Computers. III. Selection between Alleles at an Autosomal Locus. Australian J. Biol. Sci., **11**: 603, 1958.
35. Barker, J. S. F.: IV. Selection between Alleles at a Sex-linked Locus. Austalian J. Biol. Sci., **11**: 613, 1958.
36. Barnes, D. W. H., C. E. Ford, S. M. Gray, and J. F. Loutit: Spontaneous and Induced Changes in Cell Populations in Heavily Irradiated Mice. Peaceful Uses of Atomic Energy. Proc. Second Internat. Conf., Geneva, 1958. Pergamon, London, **23**: 10, 1959.
37. Barness, L.: Personal communication.
38. Baron, D. N., C. E. Dent, H. Harris, E. W. Hart, and J. B. Jepson: Hereditary Pellagra-Like Skin Rash with Temporary Cerebellar Ataxia, Constant Renal Amino-Aciduria, and other Bizarre Biochemical Features. Lancet, **2**: 421, 1956.
39. Barr, M. L.: Sex Chromatin and Phenotype in Man. Science, **130**: 679, 1959.
40. Barr, M. L.: Sex Chromatin (letter to the editor). Science, **130**: 1302, 1959.
41. Barr, M. L., and E. G. Bertram: A Morphological Distinction between Neurones of the Male and Female, and the Behaviour of the Nucleolar Satellite during Accelerated Nucleoprotein Syntheses. Nature, **163**: 676, 1949.
42. Bauer, W., and E. Calkins: Diseases of Metabolism. G. G. Duncan, ed., W. B. Saunders Co., Philadelphia, p. 666, 1959.
43. Baxter, H.: Textbook of Surgery. H. F. Moseley, ed., Third edition, C. V. Mosby, St. Louis, Missouri, p. 237, 1959.
44. Bayreuther, K.: Chromosomes in Primary Neoplastic Growth. Nature, **186**: 6, 1960.
45. Becker, P. E.: Die Häufigkeit und Bedeutung der Mutationen beim Menschen. Ver. dtsch. Ges. inn. Med., **64**: 255, 1958.
46. Becker, P. E.: Neue Ergebnisse der Genetik du Muskeldystrophien. Act. Genet., **7**: 303, 1957.
47. Beeson, P. B., H. Rocha, and L. B. Guze: Experimental Pyelonephritis: Influence of Localized Injury in Different Parts of the Kidney on Susceptibility to Hematogenous Infection. Trans. Assoc. Amer. Physicians., **70**: 120, 1957.
48. Bender, M. A.: X-ray Induced Chromosome Aberrations in Normal Diploid Human Tissue Cultures. Science, **126**: 974, 1957.
49. Bennett, I. L., and E. W. Hook: Infectious Disease (Some Aspects of Salmonellosis). Ann. Rev. Med., **10**: 8, 1959.
50. Bennett, J. H., and J. Brandt: Some More Exact Tests of Significance for *O-A* Maternal-Fetal Incompatibility. Ann. Eugen., **18**: 302, 1954.

51. Benzer, S.: The Elementary Units of Heredity. A Symposium on the Chemical Basis of Heredity. W. D. McElroy and B. Glass, editors, Johns Hopkins Press, Baltimore, 70, 1957.
52. Benzer, S., V. M. Ingram, and H. Lehmann: Three Varieties of Human Hemoglobin D. Nature (Lond.), **182**: 852, 1958.
53. Bergman, M., and C. Niemann: On the Structure of Proteins: Cattle Hemoglobin, Egg Albumin, Cattle Fibrin, and Gelatin. J. Biol. Chem., **118**: 301, 1937.
54. Bergman, S., J. Reitalu, H. Nowakowski, and W. Lenz: The Chromosomes in Two Patients with Klinefelter Syndrome. Ann. Hum. Genet. (Lond.), **24**: 81, 1960.
55. Bernini, L., V. Carcassi, B. Latte, A. G. Motulsky, and M. Siniscalco: Genetica Della Predisposizione al Favisima dati Popolazionistici: Interazione con la Talassemia e la Malaria al Livella Popolazionistico. Rendiconti Accedemi Lincei., Roma (in press).
56. Bernstein, F.: Zusammenfassende Betrachtungen über die Erblichen Blutstrukturen des Menschen. Ztschr. f. Induktive, Abstamm., **37**: 237, 1925.
57. Berry, H. K., T. Dobzhansky, S. M. Gartler, H. Levene, and R. H. Osborne: Chromatographic Studies on Urinary Excretion Patterns in Monozygotic and Dizygotic Twins. I. Methods and Analysis. Amer. J. Hum. Genet., **7**: 93, 1955.
58. Beutler, E.: The Hemolytic Effect of Primaquine and Related Compounds: A Review. Blood, **14**: 103, 1959.
59. Beutler, E., R. J. Dern, and C. L. Flanagan: Effect of the Sickling-cell Trait on Resistance to Malaria. Brit. Med. J., **1**: 1189, 1955.
60. Bickel, H.: Die Nicht-Diabetischen Melliturien des Kindes. Modern Problems in Pediatrics, Vol. 6, S. Karger, Basel, 1960
61. Bickel, H., J. Gerrard, and E. M. Hickmans: The Influence of Phenylalanine Intake on the Chemistry and Behaviour of a Phenylketonuric Child. Acta Paediat., **43**: 64, 1954.
62. Bird, G. W.: Hemagglutinins in Seeds. Brit. Med. Bull., **15**: 165, 1959.
63. Bird, G. W.: Anti-*A* Hemagglutinins in Seeds. Current Sci., **20**: 198, 1951.
64. Bittner, J. J.: Genetic Concepts in Mammary Cancer in Mice. Ann. of N.Y. Acad. Sci., **71**: 943, 1958.
65. Blanc, W. A., and G. Baens: Ear Malformations, Abnormal Facies, and Genitourinary Tract Anomalies. Abstract, Amer. Ped. Soc., 1960.
66. Bleyer, A.: Indications that Mongoloid Imbecility Is a Gametic Mutation of Digressive Type. A. M. A. Amer. J. Dis. Child., **47**: 342, 1934.
67. Block, R. J.: The Basic Amino Acids of Three Crystalline Mammalian Hemoglobins. Further Evidence for a Basic Amino Acid Anlage of Tissue Proteins. J. Biol. Chem., **105**: 663, 1934.
68. Block R. J., E. L. Durrum, and G. Zweig: A Manual of Paper Chromatography and Paper Electrophoresis. N.Y. Acad. Press, p. 710, 1958.
69. Boas, N. F., and W. B. Ober: Hereditary Exophthalmic Goiter—Report of Eleven Cases in One Family. J. Clin. Endocr., **6**: 575, 1946.
70. Bockus, H. L.: Gastro-Enterology, Vol. III. W. B. Saunders Co., Philadelphia and London, p. 585, 1946.
71. Böhringer, H. R.: Statistik, Klinik und Genetik der Schweizerischen Retinoblastomfälle. Arch. Klaus-Stift. Vereb. Forsch., **31**: 1, 1956.
72. Bongiovanni, A. M.: *In vitro* Hydroxylation of Steroids by Whole Adrenal Homogenates of Beef, Normal Man, and Patients with the Adrenogenital Syndrome. J. Clin. Invest., **37**: 1342, 1958.
73. Böök, J. A.: Genetical Investigations in a North Swedish Population. The Offspring of First-cousin Marriages. Ann. Hum. Genet., **21**: 191, 1957.
74. Böök, J. A.: Fréquence de Mutation de la Chondrodystrophie et de L'épidermolyse Bulleuse dans une Population du sud de la Suède. J. Génét. Hum., **1**: 24, 1952.
75. Böök, J. A., M. Fraccaro, and J. Lindsten: Cytogenetical Observations in Mongolism. Acta Paediat., **48**: 453, 1959.
76. Böök, J. A., J. Lejeune, A. Levan, E. H. Y. Chu, C. E. Ford, M. Fraccaro, D. G. Harden, T. C. Hsu, D. A. Hungerford, P. A. Jacobs, S. Makino, T. T. Puck, A. Robinson (Secretary), T. H. Tjio, D. G. Catcheside, H. J. Muller, and C. Stern: A Proposed Standard System of Nomenclature of Human Mitotic Chromosomes. Lancet, **1**: 1063, 1959.
77. Böök, J. A., and B. Santesson: Malformation Syndrome in Man Associated with Triploidy (69 Chromosomes). Lancet, **1**: 858, 1960.
78. Bossak, E. T., C. I. Wang, and D. Adlersberg: Clinical Aspects of the Malabsorption Syndrome in The Malabsorption Syndrome. D. Adlersberg, ed., Grune and Stratton Publishers, New York, 112, 1960.

79. Bousser, J., and R. Neyde: La Neutropenie Familiale. Sang., **18**: 521, 1947.

80. Boyd, W. C.: Production of Potent Blood Grouping Sera. J. Lab. Clin. Med., **32**: 1275, 1947.

81. Boyd, W. C.: Fundamentals of Immunology, Second edition. Interscience Pub., New York, p. 503, 1947.

82. Boyd, W. C.: Production and Preservation of Specific Anti-sera for Blood-group Factors *A*, *B*, *M*, and *N*. J. Immunol., **37**: 65, 1939.

83. Boyd, W. C.: The Proteins: Chemistry, Biological Activity, and Methods. H. Neurath, and K. Bailey, editors, Academic Press Publ., New York, **2**: 755, 1953.

84. Boyd, W. C., and L. G. Boyd: Blood Grouping Tests on 300 Mummies. Notes on Precipitin Test. J. Immunol., **32**: 307, 1937.

85. Boyd, W. C., and L. G. Boyd: Blood Grouping in Forensic Medicine. J. Immunol., **33**: 159, 1937.

86. Boyd, W. C., D. L. Everhart, and M. H. McMaster: The Anti-*N* Lectin of *Bauhinia purpurea*. J. Immunol., **81**: 414, 1958.

87. Boyd, W. C., D. M. Green, D. M. Fujinaga, J. S. Drabik, and E. Waszczenko-Zacharczenko: A Blood Factor, Possibly New, Detected by Extracts of *Arachis hypogaea*. Vox Sang., **4**: 456, 1959.

88. Boyd, W. C., M. H. McMaster, and E. Waszczenko-Zacharczenko: Specific Inhibition of Anti-Rh Serum by "Unnatural" Sugars. Nature, **184**: 989, 1959.

89. Boyd, W. C., and E. Shapleigh: Separation of Individuals of Any Blood Group into Secretors and Non-Secretors by Use of a Plant Agglutinin (Lectin). Blood, **9**: 1195, 1954.

90. Boyd, W. C., and E. Shapleigh: Specific Precipitating Activity of Plant Agglutinins (Lectins). Science, **119**: 419, 1954.

91. Boyd, W. C., and E. Shapleigh: Diagnosis of Subgroups of Blood Groups A and AB by Use of Plant Agglutinins (Lectins). J. Lab. Clin. Med., **44**: 235, 1954.

92. Boyd, W. D., and R. M. Regurera: Hemagglutinating Substances for Human Cells in Various Plants. J. Immunol., **62**: 333, 1949.

93. Boyd, W. D., and E. R. Warshaver: Serological Relationship of Blood Group *A* and *B* Antigens; Production of Anti-*B* Precipitins. J. Immunol., **50**: 101, 1945.

94. Brand, E., and J. Grantham: Methionine and Isoleucine Content of Mammalian Hemoglobins. Its Significance for Nutritional and Metabolic Studies. J. Amer. Chem. Soc., **68**: 724, 1946.

95. Brand, E., B. Kassel, and T. J. Saidel: Amino-acid Composition of Plasma Proteins. J. Clin. Invest., **23**: 437, 1944.

96. Brand, E., F. J. Ryan, and E. M. Diskant: Leucine Content of Proteins and Foodstuffs. J. Amer. Chem. Soc., **67**: 1532, 1945.

97. Brauer, R. W., G. F. Leong, and R. L. Pessotti: Liver Cell Proliferation in X-Irradiated Rats after Single and Repetitive Partial Hepatectomy. Radiation Research, **12**: 451, 1960.

98. Braunitzer, G.: Vergleichende Untersuchungen zur Primärstruktur der Proteinkomponente einiger Hämoglobine. Z. Physiol. Chem., **312**: 72, 1958.

99. Bretthauer, R. K., R. G. Hansen, G. Donnell, and W. R. Bergren: A Procedure for Detecting Carriers of Galactosemia. Proc. Nat. Acad. Sci. (Wash.), **45**: 328, 1959.

100. Brink, R. A.: Paramutation at the R Locus in Maize. Cold Spring Harbor Symp., Quant. Biol., **23**: 379, 1958.

101. British Columbia Hospital Insurance Service. Seventh Annual Report, 1955, Victoria, B.C., 1956.

102. Bryant, J. C., V. J. Evans, E. L. Schilling, and W. R. Earle: Massive Fluid-Suspension Cultures of Certain Mammalian Tissue Cells. Exploratory Supplementation Experiments with Chemically Defined Medium NCTC 109. J. Nat. Cancer Inst., **24**: 859, 1960.

103. Bryant, J. C., E. L. Schilling, and W. R. Earle: Massive Fluid-suspension Cultures of Certain Mammalian Tissue Cells. I. General Characteristics of Growth and Trends of Population. J. Nat. Cancer Inst., **21**: 331, 1958.

104. Buckwalter, J. A.: Disease Associations of the *ABO* Blood Group. Acta Genet. Statist. Med., **6**: 561, 1956–1957.

105. Buckwalter, J. A.: Personal communication.

106. Buckwalter, J. A., A. E. Lark, and L. A. Knowler: A Study in Human Genetics: The *ABO* Blood Groups and Disease in South Africa. Arch. Int. Med., **107**: 558, 1961.

107. Buckwalter, J. A., R. T. Tidrick, L. A. Knowler, E. B. Wohlwend, D. E. Colter, J. H. Turner, L. Raterman, H. H. Gamber, G. J. Roller, C. B. Pollock, and V. K. Naifeh: The Iowa Blood Type Disease Research Project. J. Iowa St. Med. Soc., **48**: 76, 1958.

108. Burdick, A. B.: Dominance as Function of Within Organism Environment in Kernel-Row Number in Maize (*Zea mays L.*). Genetics, **36**: 652, 1951.
109. Burnet, F. M.: Biological Aspects of Infectious Diseases. Cambridge Univ. Press, p. 213 and 260, 1940.
110. Butler, J. J.: A Study of the Antigens of Normal Leucocytes. J. Lab. and Clin. Med., **55**: 110, 1960.
111. Campbell, M. F.: Clinical Pediatric Urology. Saunders Co., Philadelphia, p. 1113, 1951.
112. Canada Yearbook. Dominion Bureau of Statistics, Ottawa, Canada, 1957–1958.
113. Canadian Sickness Survey (Nos. 6, 9, and 10). Dominion Bureau of Statistics, and Department of National Health and Welfare, Ottawa, Canada, 1950–1951.
114. Carleton, R. A., W. H. Abelmann, and E. W. Hancock: Familial Occurrence of Congenital Heart Disease. New Eng. J. Med., **259**: 1237, 1958.
115. Carson, P. E., C. L. Flanagan, C. E. Ickes, and A. S. Alving: Enzymatic Deficiency in Primaquine-sensitive Erythrocytes. Science, **124**: 484, 1956.
116. Carter, C. O.: Personal communication.
117. Carter, C. O., and J. Wilkinson: Congenital Dislocation of the Hip. The Results of Conservative Treatment. J. of Bone and Joint Surg., **42B**: 669, 1960.
118. Carter, T. C.: The Estimation of the Total Genetical Map Length from Linkage Test Data. J. Genet., **53**: 21, 1955.
119. Carter, T. C., M. F. Lyon, and R. J. S. Phillips: Genetic Hazards of Ionizing Radiations. Nature, **182**: 409, 1958.
120. Case Records of the Massachusetts General Hospital, Case 39501. New Eng. J. Med., **249**: 990, 1953.
121. Census of Canada. Dominion Bureau of Statistics, Ottawa, Canada, 1951 and 1956.
122. Ceppellini, R.: Hemoglobin B_2. Ciba Foundation Symposium Biochemistry of Human Genetics. G. E. W. Wolstenholme and C. M. O'Connor, editors, Little Brown & Co., Boston, Mass., and Churchill, London, p. 134, 1959.
123. Ceppellini, R., L. C. Dunn, and H. G. Kunkel: Biochemistry of Human Genetics. Ciba Foundation Symposium. G. E. W. Wolstenholme and C. M. O'Connor, editors, Little Brown & Co., Boston, Mass., and Churchill, London, p. 136, 1959.
124. Cesar, A. B.: Familial Chronic Malignant Neutropenia. Bol. Soc. Cubana Pediatria, **15**: 900, 1943.
125. Chalmers, J. N. M., and S. D. Lawler: Data on Linkage in Man: Elliptocytosis and Blood Groups. 1. Families 1 and 2. Ann. Eugen. (Lond.), **17**: 267, 1953.
126. Chang, R. S.: Genetic Study of Human Cells *in vitro;* Carbohydrate Variants from Cultures of *HeLa* and Conjunctival Cells. J. Exp. Med., **111**: 235, 1960.
127. Chediak, M. M.: Nouvelle Anomalie Leucocytaire de Caractère Constitutionnel et Familial. Rev. Hémat., **7**: 362, 1952.
128. Chernoff, A. J.: Immunological Aspects of the Human Hemoglobins. Conference on Hemoglobin. National Academy of Science, National Research Council, Washington, D.C., **557**: 179, 1958.
129. Chernoff, A. L.: The Alkali Denaturation Procedures, Conference on Hemoglobin. National Academy of Science, National Research Council, Washington, D.C., **557**: 172, 1958.
130. Cherry, W. R., and R. N. Hull: Studies on the Growth of Mammalian Cells in Agitated Fluid Media. Anat. Rec., **124**: 483, 1956.
131. Childs, B., and W. H. Zinkhan: The Genetics of Primaquine Sensitivity of the Erythrocytes, Biochemistry of Human Genetics, Ciba Foundation Symposium on Human Biochemica. Genetics in Relation to the Problem of Gene Action. G. E. W. Wolstenholme, editor, Little, Brown & Co., Boston, Mass., 76, 1959.
132. Chronic Illness in the United States, III. Chronic Illness in a Rural Area. The Hunterdon Study. Harvard University Press, Cambridge, p. 400, 1959.
133. Chronic Illness in the United States, IV. Chronic Illness in an Urban Area. Harvard University Press, Cambridge, p. 624, 1957.
134. Chu, E. H., and N. H. Giles: Human Chromosome Complements in Normal Somatic Cells in Culture. Amer. J. Hum. Genet., **11**: 63, 1959.
135. Chu, E. H., K. K. Sanford, and W. R. Earle: Comparative Chromosomal Studies on Mammalian Cells in Culture. II. Mouse Sarcoma-Producing Cell Strains and Their Derivatives. J. Nat. Cancer Inst., **21**: 729, 1958.
136. Chung, C. S., E. Matsunaga, and N. E. Morton: The *ABO* Polymorphism in Japan (in preparation).

137. Chung, C. S., E. Matsunaga, and N. E. Morton: The *MN* Polymorphism in Japan (in preparation).
138. Chung, C. S., and N. E. Morton: Selection at the *ABO* Locus. Amer. J. Hum. Genet., **13:** 9, 1961.
139. Chung, C. S., and N. E. Morton: Discrimination of Genetic Entities in Muscular Dystrophy. Amer. J. Hum. Genet., **11:** 339, 1959.
140. Chung, C. S., N. E. Morton, and H. A. Peters: Serum Enzymes and Genetic Carriers in Muscular Dystrophy. Amer. J. Hum. Genet., **12:** 52, 1960.
141. Chung, C. S., and N. E. Morton: Discrimination of Genetic Entities in Muscular Dystrophy. Proc. 10th Internat. Congr. Genet., II, 1958.
142. Chung, C. S., O. W. Robison, and N. E. Morton: A Note on Deaf Mutism. Ann. Hum. Genet., **23:** 357, 1959.
143. Clegg, M. D., and W. A. Schroeder: A Chromatographic Study of the Minor Components of Normal Adult Human Hemoglobin Including a Comparison of Hemoglobin from Normal and Phenylketonuric Individuals. J. Amer. Chem. Soc., **81:** 6065, 1959.
144. Coburn, A. G.: The Prevention of Respiratory Tract Bacterial Infections. J. Amer. Med. Assoc., **126:** 88, 1944.
145. Cochran, W. G.: Sampling Techniques. John Wiley and Sons, New York, p. 330, 1953.
146. Cockayne, E. A.: Genetics of Transposition of Viscera. Quart. J. Med., **7:** 479, 1938.
147. Cole, R. D., W. H. Stein, and S. Moore: On the Cysteine Content of Human Hemoglobin. J. Biol. Chem., **233:** 1359, 1958.
148. Conger, A. D., and L. M. Fairchild: A Quick-freeze Method for Making Smear Slides Permanent. Stain Technol., **28:** 281, 1953.
149. Cooke, W. T., A. L. P. Peeney, and C. F. Hawkins: Symptoms, Signs, and Diagnostic Features of Idiopathic Steatorrhoea. Quart. J. Med., **22:** 59, 1953.
150. Coope, R.: Diseases of the Chest, second ed. E. and S. Livingston, Ltd., Edinburgh, p. 442, 1950.
151. Cori, G. T.: Glycogen Structure and Enzyme Deficiencies in Glycogen Storage Disease. Harvey Lect. **48:** 145, 1952.
152. Cotterman, C. W.: Erythrocyte Antigen Mosaicism. J. Cell. Comp. Physiol., **52, Suppl., 1:** 69, 1958.
153. Court-Brown, M., P. A. Jacobs, and R. Doll: Interpretation of Chromosome Counts Made on Bone-Marrow Cells. Lancet, **1:** 160, 1960.
154. Cox, R. P.: Unpublished observation.
155. Craig, L. C., and T. P. King: Some Dialysis Experiments with Polypeptides. J. Amer. Chem. Soc., **77:** 6620, 1955.
156. Crick, F. H. C., J. S. Griffith, and L. E. Orgel: Codes without Commas. Proc. Nat. Acad. Sci. (Wash.), **43:** 416, 1958.
157. Crippled Children's Registry. Annual Report, Vancouver, B.C., 1955.
158. Crooke, A. C., and M. D. Hayward: Mosaicism in Klinefleter's Syndrome. Lancet, **1:** 1198, 1960.
159. Crow, J. F.: Some Possibilities for Measuring Selection Intensities in Man. Hum. Biol., **30:** 1, 1958.
160. Crow, J. F.: A Comparison of Fetal and Infant Death Rates in the Progeny of Radiologists and Pathologists. Amer. J. Roentgenol. Rad. Therap. and Nuclear Med., **73:** 467, 1955.
161. Crow, J. F., and M. Kimura: Some Genetic Problems in Natural Populations. Proc. Third Berkeley Symp. Math. Stat. Prob., **3:** 1, 1955.
162. Crow, J. F., and N. E. Morton: Measurement of Gene Frequency Drift in Small Populations. Evolution, **9:** 202, 1955.
163. Crow, J. F., and N. E. Morton: The Genetic Load Due to Mother-Child Incompatibility. Amer. Nat., **94:** 413, 1960.
164. Crowe, F. W., W. J. Schull, and J. V. Neel: A Clinical Pathological, and Genetic Study of Multiple Neurofibromatosis. C. C. Thomas, Springfield, 9 and 181, 1956.
165. Crumpler, H. R., C. E. Dent, H. Harris, and R. G. Westall: β-Aminoisobutyric Acid (α-methyl-β-Alanine): A new Amino-Acid Obtained from Human Urine. Nature (Lond.), **167:** 307, 1951.
166. Dahlberg, G.: Genetic Investigations in Different Populations. Acta Genet. et Stat. Medica, **3:** 117, 1952.
167. Dahlberg, G.: Mathematical Methods for Population Genetics. Interscience Publishers, New York, N.Y., p. 182, 1948.

168. Dalgaard, O. Z.: Bilateral Polycystic Disease of the Kidneys. Acta Med. Scand., **328** Suppl., p. 255, 1957.

169. Darlington, C. D., and L. F. La Cour: The Handling of Chromosomes, Third Ed. Allen and Unwin, London, p. 180, 1960.

170. David, P. R., and L. S. Snyder: Genetics and Disease. Proc. Second Nat. Cancer Conf., Amer. Cancer Soc., N.Y., 1128, 1954.

171. Davison, A. N., and M. Sandler: Inhibition of 5-Hydroxytryptophan Decarboxylase by Phenylalanine Metabolites. Nature (Lond.), **181**: 186, 1958.

172. Dawe, C. J., and M. Potter: Morphologic and Biologic Progression of a Lymphoid Neoplasm of the Mouse *in vivo* and *in vitro*. Amer. J. Path., **33**: 603, 1957.

173. De Assis, L. M., D. R. Epps, C. Bottura, and I. Ferrari: Chromosomal Constitution and Nuclear Sex of a True Hermaphrodite. Lancet, **2**: 129, 1960.

174. De Carli, L., F. Nuzzo, B. Chiarelli, and E. Poli: Trisomic Condition of a Large Chromosome in a Woman with Mongoloid Traits. Lancet, **1**: 130, 1960.

175. Dean, G.: Porphyria. Brit. Med. J., **2**: 1291, 1953.

176. Debré, R., J. C. Dreyfus, J. Frézal, J. Lamy, P. Maroteaux, F. Schapira, and G. Schapira: Genetics of Hemochromatosis. Ann. Hum. Genet., **23**: 16, 1958.

177. Derrien, Y.: Studies on the Heterogeneity of Adult and Fetal Hemoglobins by Salting-out Alkali Denaturation and Moving Boundary Electrophoresis. National Academy of Science, National Research Council, Washington, D.C., **577**: 183, 1958.

178. Dewey, W. J., and N. E. Morton: Recessive Genes in Low-Grade Mental Deficiency (in preparation).

179. Dherte, P., H. Lehmann, and J. Vandepitte: Hemoglobin P in a Family in the Belgian Congo. Nature (Lond.), **184**: 1133, 1959.

180. DiSant' Agnese, P. A.: Cystic Fibrosis of the Pancreas. Amer. J. Med., **21**: 406, 1956.

181. Dobzhansky, T.: Genetic Loads in Natural Populations. Science, **126**: 191, 1957.

182. Doll, R., and J. Buch: Hereditary Factors in Peptic Ulcer. Ann. of Eugen., **15**: 135, 1950.

183. Donohue, W. L., and H. W. Bain: Chediak-Higashi Syndrome. A Lethal Familial Disease with Anomalous Inclusions in the Leucocytes and Constitutional Stigmata. Pediatrics, **20**: 416, 1957.

184. Dozy, A., and T. H. J. Huisman: The Action of Carboxypeptidase on Different Human Hemoglobins. Biochim. Biophys. Acta, **20**: 400, 1956.

185. Draper, G.: Acute Anterior Poliomyelitis. P. Blakiston's Sons and Co., Philadelphia, 1917.

186. Dubin, I. N.: Chronic Idiopathic Jaundice. Amer. J. Med., **24**: 268, 1958.

187. Dubos, R. J.: In Bacterial and Mycotic Infections of Man. R. J. Dubos, ed., J. P. Lippincott Company, Philadelphia, 785, 1948.

188. Duncan, G. G.: Diabetes in Diseases of Metabolism, Fourth ed. G. G. Duncan, ed., W. B. Saunders Co., Philadelphia, 884, 1959.

189. Dunsford, I., and J. Grant: The Antiglobulin (Coombs) Test in Laboratory Practice. C. C. Thomas, Springfield, p. 120, 1960.

190. Eagle, H.: The Sustained Growth of Human and Animal Cells in a Protein-free Environment. Proc. Nat. Acad. Sci., **46**: 427, 1960.

191. Eagle, H., K. A. Piez, and R. Fleishman: The Utilization of Phenylalanine and Tyrosine for Protein Synthesis by Human Cells in Tissue Culture. J. Biol. Chem., **228**: 847, 1957.

192. Earle, D. P., M. P. Hutt, K. Schmid, and D. Gitlin: Observations on Double Albumin: A Genetically Transmitted Serum Protein Anomaly. J. Clin. Invest., **38**: 1412, 1959.

193. Earle, W. R.: Production of Malignancy *in vitro;* Mouse Fibroblast Cultures and Changes Seen in Living Cells. J. Nat. Cancer Inst., **4**: 165, 1943.

194. Earle, W. R.: Tissue Culture. Laboratory Technique in Biology and Medicine. E. V. Cowdry, ed., Williams and Wilkins, Baltimore, p. 340, 1948.

195. Earle, W. R., J. C. Bryant, E. L. Schilling, and V. J. Evans: Growth of Cell Suspensions in Tissue Culture. Ann. N.Y. Acad. Sci., **63**: 666, 1956.

196. Earle, W. R., E. L. Schilling, J. C. Bryant, and V. J. Evans: The Growth of Pure Strain L Cells in Fluid-suspension Cultures. J. Nat. Cancer Inst., **14**: 1159, 1954.

197. Eaton, J. W., and A. J. Mayer: The Demography of Unique Populations. Hum. Biol. **25**: 206, 1953.

198. Eberlein, W. R., and A. M. Bongiovoni: Pathophysiology of Congenital Adrenal Hyperplasia. Metabolism, **9**: 326, 1960.

199. Edwards, J. H., D. G. Harnden, A. H. Cameron, V. M. Crosse, and O. H. Wolff: A New Trisomic Syndrome. Lancet, **1**: 787, 1960.

200. Efrati, P., and W. Jonas: Chediak's Anomaly of Leucocytes in Malignant Lymphoma Associated with Leukemic Manifestations: Case Report with Necropsy. Blood, 13: 1063, 1958.
201. Eisler, M.: Über ein Gemeinsames Antigen in den Zellen des Menschen und in Shigabazillen. Z. Immunitätsf., 67: 38, 1930.
202. Elkind, M. M.: Cellular Aspects of Tumor Therapy. Radiology, 74: 529, 1960.
203. Elkind, M. M. and H. Sutton: X-ray Damage and Recovery in Mammalian Cells in Culture. Nature (Lond.), 184: 1293, 1959.
204. Elo, J., E. Estola, and N. Malmström: Phytagglutinins Present in Mushrooms. Ann. Med. Exp. Biol. Fenn., 29: 297, 1951.
205. Engelberg, J. A.: A Method of Measuring the Degree of Synchronization of Cell Populations. Exp. Cell Research, 23: 218, 1961.
206. Essen-Möller, E.: Empirische Ähnlichkeits diagnose bei Zwillingen. Hereditas (Lond.), 27: 1, 1941.
207. Evans, D. A. P., K. A. Manley, and V. A. McKusick: Isoniazid Inactivation. A Genetically Determined Phenomenon. Amer. Coll. Phys., 41st Session, 76, 1960.
208. Evans, V. J., J. C. Bryant, W. T. McQuilkin, M. C. Fioramonti, K. K. Sanford, B. B. Westfall, and W. R. Earle: Studies of Nutrient Media for Tissue Cells in vitro. II. An Improved Protein-Free Chemically Defined Medium for Long-term Cultivation of Strain L–929 Cells. Cancer Res., 16: 87, 1956.
209. Evans, V. J., J. C. Bryant, M. C. Fioramonti, W. T. McQuilkin, K. K. Sanford, and W. R. Earle: Studies of Nutrient Media for Tissue Cells in vitro. I. A Protein-free Chemically Defined Media for Cultivation of Strain L Cells. Cancer Res., 16: 77, 1956.
210. Evans, V. J., W. R. Earle, K. K. Sanford, J. E. Shannon, and H. K. Waltz: The Preparation and Handling of Replicate Tissue Cultures for Quantitative Studies. J. Nat. Cancer Inst., 11: 907, 1951.
211. Evans, V. J., H. A. Kerr, W. T. McQuilkin, W. R. Earle, and R. N. Hull: Growth in vitro of a Long-term Strain of Monkey-kidney Cells in Medium NCTC 109 Free of Any Added Protein. Amer. J. Hyg., 70: 297, 1959.
212. Expert Committee on Radiation: Effect of Radiation on Human Heredity: Investigation of Areas of High Natural Radiation. Geneva: World Health Organization Technical Report Series, no. 166, 47, 1959.
213. Fajans, S. S., and J. W. Conn: Approach to Prediction of Diabetes Mellitus by Modification of Glucose Tolerance Test with Cortisone. Diabetes, 3: 96, 1954.
214. Falconer, D. S.: Introduction to Quantitative Genetics. The Ronald Press, New York, p. 365, 1960.
215. Falls, H. F., and J. V. Neel: Genetics of Retinoblastoma. A.M.A. Arch. Ophth., 46: 367, 1951.
216. Farrer, S. M., and C. M. MacLeod: Staphylococcal Infections in a General Hospital. Amer. J. Hyg., 72: 38, 1960.
217. Fearnside, E. G.: Familial Lateral Scierosis with Amyotrophy. Proc. Roy. Soc. Med., London, Sec. Neuro., 5: 143, 1912.
218. Fell, H. B.: Recent Advances in Organ Culture. Sci. Prog., 41: 212, 1958.
219. Fellman, J. H.: Inhibition of Dihydroxphenylalanine Decarboxylase by Aromatic Acids Associated with Phenylpyruvic Oligophrenia. Proc. Soc. Exp. Biol. (N.Y.), 93: 413, 1956.
220. Ferguson-Smith, M. A., A. W. Johnston, and A. N. Weinberg: The Chromosome Complement in True Hermaphroditism. Lancet, 2: 126, 1960.
221. Ferguson-Smith, M. A., B. Lennox, W. S. Mack, and J. S. S. Stewart: Klinefelter's Syndrome, Frequency and Testicular Morphology in Relation to Nuclear Sex. Lancet, 2: 167, 1957.
222. Finch, S. C., and C. A. Finch: Idiopathic Hemochromatosis, an Iron Storage Disease. Medicine, Balt., 34: 381, 1955.
223. Fink, K.: A Substance Occasionally Found as a Major Ninhydrin-reacting Component of Urine. Proc. Soc. Exp. Biol. (N.Y.), 76: 692, 1951.
224. Fioramonti, M. C., V. J. Evans, and W. R. Earle: The Effect of Inoculum Size on Proliferation of NCTC Strain 2071, the Chemically Defined Medium Strain of NCTC Clone 929 (Strain L). J. Nat. Cancer Inst., 21: 579, 1958.
225. Fisher, H. W., T. T. Puck, and G. Sato: Molecular Growth Requirements of Single Mammalian Cells: The Action of Fetuin in Promoting Cell Attachment to Glass. Proc. Nat. Acad. Sci. (Wash.), 44: 4, 1958.
226. Fisher, H. W., T. T. Puck, and G. Sato: III. Quantitative Colonial Growth of Single S3 Cells in a Medium Containing Synthetic Small Molecular Constituents and Two Purified Protein Fractions. J. Exp. Med., 109: 649, 1959.

227. Fisher, H. W., T. T. Puck, and G. Sato: Molecular Growth Requirements of Single Mammalian Cells. J. Exp. Med., **109**: 649, 1959.

228. Fisher, H. W., R. G. Ham, and T. T. Puck: Macromolecular Growth Requirements of Single, Diploid, Mammalian Cells. Fed. Proc., 19, No. 1, Pt. 1, Abstract No. 20, March, 1960.

229. Fisher, H. W., and T. T. Puck: On the Functions of X-irradiated Feeder Cells in Supporting Growth of Single Mammalian Cells. Proc. Nat. Acad. Sci. (Wash.), **42**: 900, 1956.

230. Fisher, R. A.: The Genetical Theory of Natural Selection. The Clarendon Press, Oxford, 1930. Second ed., Dover Pub., New York, p. 291, 1958.

231. Fisher, R. A.: Effect of Methods of Ascertainment Upon Estimation of Frequencies. Ann. Eugen., **6**: 13, 1934.

232. Fisher, R. A.: Statistical Methods and Scientific Inference. Oliver and Boyd, Edinburgh, p. 175, 1956.

233. Fisher, R. A.: Average Excess and Average Effect of a Gene Substitution. Ann. Eugen., **11**: 53, 1941.

234. Fisher, R. A.: Stage of Development as a Factor Influencing the Variance in the Number of Offspring, Frequency of Mutants, and Related Quantities. Ann. Eugen., **9**: 406, 1939.

235. Fogh-Andersen, P.: Inheritance of Harelip and Cleft Palate. Contribution to the Elucidation of the Etiology of the Congenital Clefts of the Face. Copenhagen, Nyt. Nordisk Forlag, A. Busck, p. 266, 1942.

236. Ford, C. E.: Chromosomes. Methodology in Human Genetics. W. J. Burdette, ed., Holden-Day, Inc., San Francisco, 1962.

237. Ford, C. E., and J. L. Hamerton: The Chromosomes of Man. Nature (Lond.), **178**: 1020, 1956.

238. Ford, C. E., and J. L. Hamerton: A Colchicine, Hypotonic Citrate, Squash Sequence for Mammalian Chromosomes. Stain Technol., **31**: 247, 1956.

239. Ford, C. E., J. L. Hamerton, D. W. H. Barnes, and J. F. Loutit: Cytological Identification of Radiation Chimaeras. Nature (Lond.), **177**: 452, 1956.

240. Ford, C. E., P. A. Jacobs, and L. G. Lajtha: Human Somatic Chromosomes. Nature (Lond.), **181**: 1565, 1958.

241. Ford, C. E., K. W. Jones, O. J. Miller, U. Mittwoch, L. S. Penrose, M. Ridler, and A. Shapiro: The Chromosomes of a Patient Showing both Mongolism and the Klinefelter Syndrome. Lancet, **1**: 709, 1959.

242. Ford, C. E., K. W. Jones, P. E. Polani, J. C. de Almedia, and J. H. Briggs: A Sex-chromosome Anomaly in a Case of Gonadal Dysgenesis (Turner's Syndrome). Lancet, **1**: 711, 1959.

243. Ford, C. E., and R. H. Mole: Chromosome Studies in Human Leukemia. Lancet, **2**: 732, 1959 (letter to the editor).

244. Ford, C. E., R. H. Mole, and J. L. Hamerton: Chromosomal Changes in Primary and Transplanted Reticular Neoplasms of the Mouse. J. Cell. Comp. Physiol., **52** Suppl., **1**: 235, 1958.

245. Ford, C. E., P. E. Polani, J. H. Briggs, and P. M. Bishop: A Presumptive Human XXY/XX Mosaic. Nature (Lond.), **183**: 1030, 1959.

246. Ford, D. K.: Personal communication.

247. Ford, E. B.: A Uniform Notation for the Human Blood Groups. Heredity, **9**: 135, 1955.

248. Ford, F. R.: Diseases of the Nervous System in Infancy, Childhood, and Adolescence, First ed. Charles C. Thomas Publisher, Springfield, Ill., 396, 1937.

249. Fortes, M.: A Demographic Field Study in Ashanti. Culture and Fertility. UNESCO, 1954.

250. Foss, G. L., C. B. Perry, and F. J. Y. Wood: Renal Tubular Acidosis. Quart. J. Med., **25**: 185, 1956.

251. Fouquey, C., J. Polonsky, and E. Lederer: Sur la Structure Chimique de L'alcool Ascarylique Isolé de *Parascaris equorum*. (Chemical Structure of Ascarylic Alcohol Isolated from *Parascaris equorum*.) Bull. Soc. Chim. Biol. (Paris), **39**: 101, 1957.

252. Fraccaro, M., C. A. Gemzell, and J. Lindsten: Plasma Level of Growth Hormones and Chromosome Complement in Four Patients with Gonadal Dysgenesis (Turner's Syndrome). Acta Endocr., **34**: 496, 1960.

253. Fraccaro, M., K. Kaijser, and J. Lindsten: Chromosomal Abnormalities in Father and Mongol Child. Lancet, **1**: 724, 1960.

254. Fraccaro, M., K. Kaijser, and J. Lindsten: Chromosome Complement in Gonadal Dysgenesis (Turner's Syndrome), Lancet, **1**: 886, 1959 (letter to the editor).

255. Fraccaro, M., K. Kaijser, and J. Lindsten: Somatic Chromosome Complement in Continuously Cultured Cells in Two Individuals with Gonadal Dysgenesis. Ann. Hum. Genet., **24**: 45, 1960.

256. Fraccaro, M., K. Kaijser, and J. Lindsten: Chromosome Complement in Parents of Patient with Gonadal Dysgenesis (Turner's Syndrome). Lancet, **2**: 1090, 1959.

257. Fraenkel-Conrat, H., J. I. Harris, and A. L. Levy: Recent Developments in Techniques for Terminal Sequence Studies in Peptides and Proteins in Methods of Biochemical Analysis. D. Glick, ed., Interscience Publ. Inc., New York, 359, 1955.

258. Franceschetti, A., and D. Klein: Le Dépistage des Hétérozygotes. Analecta Genetica, **1**: 50, 1954.

259. Fraser, A. S.: Simulation of Genetic Systems by Automatic Digital Computers. I. Introduction. Australian J. Biol. Sci., **10**: 484, 1957. II. Effects of Linkage on Rates of Advance under Selection. Australian J. Biol. Sci., **10**: 492, 1957.

260. Fraser, D.: Hypophosphatasia. Amer. J. Med., **22**: 730, 1957.

261. Fraser, F. C.: Genetic Counseling in Some Common Pediatric Diseases. Pediat. Clin. N. Amer., May: 475, 1958.

262. Fraser, F. C.: Thoughts on the Etiology of Clefts of the Palate and Lip. Acta Genet. Statist. Med., (Basel), **5**: 358, 1955.

263. Fraser, G. R., H. Harris, and E. B. Robson: A New Genetically Determined Plasma Protein in Man. Lancet, **1**: 1023, 1959.

264. Friedberg, C. K.: Diseases of the Heart. Second Ed. W. B. Saunders Co., Philadelphia and London, 864, 1956.

265. Furuhata, T., K. Ichida, and T. Kishi: Heredity and Biochemical Structure of Human Blood; New Theory on Heredity of Blood Groups. Japan Med. World, **7**: 1, 1927.

266. Gardner, E. J.: Mendelian Pattern of Dominant Inheritance for a Syndrome Including Intestinal Polyposis, Osteomas, Fibromas, and Sebaceous Cysts in a Human Family Group. L. Gedda, ed., Novant' Anni Delle Leggi Mendeliane, 321, 1955.

267. Gardner, E. J.: A Genetic and Clinical Study of Intestinal Polyposis, a Predisposing Factor for Carcinoma of the Colon and Rectum. Amer. J. Hum. Genet., **3**: 167, 1951.

268. Gartler, S. M.: An Investigation into the Biochemical Genetics of β-Aminoisobutyric Aciduria. Amer. J. Hum. Genet., **11**: 257, 1959.

269. Gartler, S. M., T. Dobzhansky, and H. K. Berry: Chromatographic Studies on Urinary Excretion Patterns in Monozygotic and Dizygotic Twins. II. Heritability of the Excretion Rates of Certain Substances. Amer. J. Hum. Genet., **7**: 108, 1955.

270. Gates, R. R.: Human Genetics. Macmillan Co., N.Y., p. 310, 1946.

271. Gentry, J. T., E. Parkhurst, and G. V. Bulin, Jr.,: An Epidemiological Study of Congenital Malformations in New York State. Amer. J. Publ. Hlth., **49**: 497, 1959.

272. Gerald, P. S., and P. George: Second Spectroscopically Abnormal Methemoglobin Associated with Hereditary Cyanosis. Science, **129**: 393, 1959.

273. Gerhardt, P. R., I. D. Goldberg, and M. L. Levin; Incidence, Mortality, and Treatment of Breast Cancer in New York State. N.Y. St. J. Med., **55**: 2945, 1955.

274. Gey, G. O.: An Improved Technic for Massive Cell Culture. Amer. J. Cancer, **17**: 752, 1933.

275. Gifford, M. A., W. C. Buss, and R. J. Douds: Data on Coccidioides Fungus Infections. 1901–1936. In Kern County Health Department Annual Report, Bakersfield, California, 39, 1936–1937.

276. Girsh, L. S., and F. E. Karpinski, Jr.: Urinary-tract Malformations. Their Familial Occurrence, with Special Reference to Double Ureter, Double Pelvis, and Double Kidney. New Engl. J. Med., **254**: 854, 1956.

277. Gitlin, D., C. A. Janeway, L. Apt, and J. M. Craig: Agammaglobulinemia in Cellular and Humoral Aspects of the Hypersensitive States. H. S. Lawrence, ed., Hoeber-Harper, New York, 375, 1959.

277a. Gittlesohn, A. M.: Family Limitation Based on Family Composition. Am. J. Human Genet., **12**: 425, 1960.

278. Gohen, J. W.: Genetic Effects in Nonspecific Resistance to Infectious Disease. Bact. Rev., **24**: 192, 1960.

279. Goldman, R., J. L. Reynolds, H. R. Cummings, and S. H. Bassett: Familial Hypoparathyroidism. J. Amer. Med. Ass., **150**: 1104, 1952.

280. Goldstein, M. N.: A Method for More Critical Isolation of Clones Derived from Three Human Cell Strains *in vitro*. Cancer Res., **17**: 357, 1957.

281. Good, R. A., R. A. Bridges, and R. M. Condie: Host-parasite Relationships in Patients with Dysproteinemias. Bact. Rev., **24**: 115, 1960.

282. Goodall, H. B., D. W. W. Hendry, S. D. Lawler, and S. A. Stephen: Data on Linkage in Man: Elliptocytosis and Blood Groups. II. Family 3. Ann. Eugen. Lond., **17**: 272, 1953.

283. Goodall, H. B., D. W. W. Hendry, S. D. Lawler, and S. A. Stephen: Data on Linkage in Man: Elliptocytosis and Blood Groups, III. Family 4. Ann. Eugen., Lond., **18:** 325, 1954.

284. Goodman, H. O., and C. N. Herndon: Genetic Factors in the Etiology of Mental Retardation. Int. Rec. Med., **172:** 61, 1959.

285. Gopal-Ayengar, A. R.: Possible Areas with Sufficiently Different Background-radiation Levels to Permit Detection of Differences in Mutation Rates of Marker Genes. Effect of Radiation on Human Heredity. Geneva: World Health Organization, 115, 1957.

286. Goudie, R. B.: Somatic Segregation of Inagglutinable Erythrocytes. Lancet, **272:** 1333, 1957.

287. Gowen, J. W.: Genetics and Disease Resistance in Genetics in the 20th Century, L. C. Dunn, ed., Macmillan Co., New York, 1951.

288. Gowen, J. W.: Inheritance of Immunity in Animals. Ann. Rev. Microbiol., **2:** 215, 1948.

289. Gowen, J. W.: Genetic Effects in Nonspecific Resistance to Infectious Disease. Bact. Rev., **24:** 192, 1960.

290. Gowen, J. W., and F. L. P. Koch: Spontaneous Ophthalmic Mutation in a Laboratory Mouse. Arch. Path., **28:** 171, 1939.

291. Graff, S., and K. S. McCarty: Sustained Cell Culture. Exp. Cell Res., **13:** 348, 1957.

292. Graham, J. B.: Personal communication.

293. Graham, A. F., and L. Siminovitch: Proliferation of Monkey Kidney Cells in Rotating Cultures. Proc. Soc. Exp. Biol. (N.Y.), **89:** 326, 1955.

294. Greenwood, M.: Epidemics and Crowd Diseases. Williams and Norgate Ltd. (Lond.), 226, 1935.

295. Groen, J.: The Hereditary Mechanism of Gaucher's Disease. Blood, **3:** 1238, 1948.

296. Grubb, R.: Correlation between Lewis Blood Group and Secretor Character in Man. Nature (Lond.), **162:** 933, 1948.

297. Grubb, R.: Some Aspects of Complexity of Human *ABO* Blood Groups. Acta Path. Microbiol. Scand. Suppl., **84:** 1, 1949.

298. Grumbach, M. M., and M. L. Barr: Cytologic Tests of Chromosomal Sex in Relation to Sexual Anomalies in Man. Recent Prog. Hormone Res., **14:** 255, 1958.

299. Gudgent, F., and A. Altschuli: The Nephropathy of Gout. Ann. Intern. Med., **44:** 1182, 1956.

300. Gullbring, B.: Investigation on the Occurrence of Blood Group Antigens in Spermatozoa from Man, and Serological Demonstration of the Segregation of Characters. Acta Med. Scand., **159:** 169, 1957.

301. Gunther, M., and L. S. Penrose: The Genetics of Epiloia. J. Genet. **31:** 413, 1935.

302. Haenszel, W.: Some Problems in the Estimation of Familial Risks of Disease. J. Nat. Cancer Inst., **23:** 487, 1959.

303. Haldane, J. B. S.: Natural Selection in Man. Caryologia, **6:** Suppl. 480, 1954.

304. Haldane, J. B. S.: Natural Selection in Man. Acta Genet. et Stat. Med. (Basel), **6:** 321, 1956.

305. Haldane, J. B. S.: The Rate of Mutation of Human Genes. Proc. Eighth Internat. Congr. Genetics, Hereditas Suppl., 267, 1949.

306. Haldane, J. B. S.: A Mathematical Theory of Natural and Artificial Selection. Proc. Camb. Phil. Soc., **23:** 607, 1927.

307. Haldane, J. B. S.: Disease and Evolution. Suppl., La Ricerca Scientifica, **19:** 68, 1949.

308. Haldane, J. B. S.: The Statics of Evolution. In Evolution as a Process. J. S. Huxley, A. C. Hardy, and E. B. Ford, eds. Allen and Unwin, London, 367, 1954.

309. Haldane, J. B. S.: The Causes of Evolution. Harpers, New York, 234, 1932.

310. Haldane, J. B. S.: The Effect of Variation on Fitness. Amer. Natur., **71:** 337, 1937.

311. Haldane, J. B. S.: A Test for Homogeneity of Records of Familial Abnormalities. Ann. Eugen. (Lond.), **14:** 339, 1947.

312. Haldane, J. B. S.: Parental and Fraternal Correlations for Fitness. Ann. Eugen. (Lond.), **14:** 288, 1949.

313. Haldane, J. B. S.: The Cost of Natural Selection. J. Genet., **55:** 511, 1957.

314. Haldane, J. B. S.: Genetical Evidence for Cytological Abnormality in Man. J. Genet., **26:** 341, 1932.

315. Haldane, J. B. S.: Estimation of the Frequencies of Recessive Conditions in Man. Ann. Eugen. (Lond.), **8:** 255, 1938.

316. Haldane, J. B. S., and C. A. B. Smith: New Estimate of the Linkage between Genes for Color-Blindness and Hemophilia in Man. Ann. Eugen. (Lond.), **14:** 10, 1947.

317. Ham, R. G.: Federation Proceedings, **19,** No. 1, Pt. I, Abstract No. 21, March 1960.

318. Hamburger, J., J. Crosnier, J. Lissac, and J. Naffah: Sur un Syndrome familial de Néphropathie avec Sûrdite. J. Urol., Paris, **62:** 113, 1956.

319. Hamilton, H. L.: Personal communication.
320. Hamilton, M., G. W. Pickering, J. A. F. Roberts, and G. S. C. Sowry: The Etiology of Essential Hypertension. (2) Scores for Arterial Blood Pressures Adjusted for Differences in Age and Sex. Clin. Sci., **13**: 37, 1954.
321. Handa, Y.: Genetic Studies of Pelger-Huet's Nuclear Anomaly in Japanese. Jap. J. Hum. Genet., **4**: 160, 1959.
322. Hansen, M. H., W. N. Hurwitz, and W. G. Madow: Sample Survey Methods and Theory. John Wiley and Sons, New York, p. 970, 1953.
323. Hanson, A.: Action of Phenylalanine Metabolites on Glutamic Acid Decarboxylase and γ-Aminobutyric Acid—α-Ketoglutaric Acid Transaminase in Brain. Acta Chem. Scand., **13**: 1366, 1959.
324. Harnden, D. G.: A Human Skin Culture Technique Used for Cytological Examinations. Brit. J. Exp. Path., **41**: 31, 1960.
325. Harnden, D. G., and C. N. Armstrong: The Chromosomes of a True Hermaphrodite. Brit. Med. J., **7**: 1287, 1959.
326. Harnden, D. G., J. H. Briggs, and J. S. S. Stewart: Nuclear Chromatin of Anencephalic Foetuses. Lancet, **2**: 126, 1959.
327. Harnden, D. G., and J. S. S. Stewart: The Chromosomes in a Case of Pure Gonadal Dysgenesis. Brit. Med. J., **2**: 1285, 1959.
328. Harris, H., U. Mittwoch, E. B. Robson, and F. L. Warren: Pattern of Amino-Acid Excretion in Cystinuria. Ann. Hum. Genet., **19**: 196, 1955.
329. Harris, H. W., R. A. Knight, and M. J. Selin: Studies of Genetic Factors Influencing Isoniazid Blood Levels in Humans, Clin. Res., **7**: 124, 1959.
330. Harris, H., L. S. Penrose, and D. H. Thomas: Cystathioninuria. Ann. Hum. Genet., **23**: 442, 1959.
331. Harris, H., and E. G. Robson: Cystinuria. Amer. J. Med., **22**: 774, 1957.
332. Harris, J. I., F. Sanger, and M. A. Naughton: Species Differences in Insulin. Arch. Biochem. Biophys., **65**: 427, 1956.
333. Hasserodt, U., and J. Vinograd: Dissociation of Human Carbonmonoxyhemoglobin at High pH. Proc. Nat. Acad. Sci. (Wash.), **45**: 12, 1959.
334. Hauge, M., and B. Harvald: Genetics in Intracranial Tumors. Acta. Genet. et Stat. Med. (Basel), **7**: 573, 1958.
335. Hauschka, T. S.: Correlation of Chromosomal and Physiologic Changes in Tumors. J. Cell. Comp. Physiol. Suppl. I., **52**: 197, 1958.
336. Hauser, P. M., and O. D. Duncan: The Study of Populations. An Inventory and Appraisal. The University of Chicago Press, p. 864, 1959.
337. Havinga, E.: Comparison of the Phosphorous Content, Optical Rotation, Separation of Hemes and Globin, and Terminal Amino-Acid Residues of Normal Adult Human Hemoglobin and Sickle-Cell Anemia Hemoglobin. Proc. Nat. Acad. Sci. (Wash.), **39**: 59, 1953.
338. Healy, G. M., D. C. Fisher, and R. C. Parker: Nutrition of Animal Cells, X. Synthetic Medium 858. Proc. Soc. Exp. Biol. (N.Y.), **89**: 71, 1955.
339. Heidelberger, M., and K. Landsteiner: On the Antigenic Properties of Hemoglobin. J. Exp. Med., **38**: 561, 1923.
340. Heitz, E.: Die Nucleal—Quetsch Methode. Ber. D. Bot. Ges., **53**: 870, 1936.
341. Hektoen, L., and K. Schulhof: On Specific Erythroprecipitins (Hemoglobin Percipitins). II. Hemoglobin Percipitins in Identification of Blood. J. Infect. Dis., **33**: 224, 1923.
342. Henningsen, K.: Investigations on Bloodfactor P. Acta Path. Microbiol. Scand., **26**: 639, 1949.
343. Herndon, C. N.: Three North Carolina Surveys. Amer. J. Hum. Genet. **6**: 65, 1954.
344. Herndon, C. N., and R. G. Jennings: Twin-family Study of Susceptibility to Poliomyelitis. Amer. J. Hum. Genet., **3**: 17, 1951.
345. Hertig, A. T.: Pathological Aspects in: The Placenta and Fetal Membranes. C. A. Villee, ed., 109, 1960.
346. Hewitt, H. B., and C. W. Wilson: A Survival Curve for Mammalian Cells Irradiated *in vivo* Nature (Lond.), **183**: 1060, 1959.
347. Higashi, O.: Congenital Gigantism of Peroxidase Granules; The First Case Ever Reported of Qualitative Abnormity of Peroxidase. Tōhoku, J. Exp. Med., **59**: 315, 1954.
348. Hildreth, E. A.: Personal communication.
349. Hill, R., and H. C. Schwartz: A Chemical Abnormality in Hemoglobin *G*. Nature, **184**: 641, 1959.

350. Hill, R. G., and L. C. Craig: Countercurrent Distribution Studies with Adult Human Hemoglobin. J. Amer. Chem. Soc., **81:** 2272, 1959.
351. Hill, R. L., J. R. Kimmel, and E. L. Smith: The Structure of Proteins. Ann. Rev. Biochem., **28:** 97, 1959.
352. Hill, R. L., H. C. Schwartz, and R. T. Swenson: Unpublished observations.
353. Hill, R. L., R. T. Swenson, and H. C. Schwartz: The Nature of the Chemical Abnormality in Hemoglobin. Fed. Proc., **19:** 77, 1960.
354. Hilse, von K., and G. Braunitzer: Hämoglobine. VII. Untersuchungen zur Konstitution der Peptidketten aus Humanhämoglobin A. Z. Naturforsch. **14:** 603, 1959.
355. Hilse, von K., and G. Braunitzer: Hämoglobine. VIII. Zur Chemischen Charakterisierung des Proteins aus Humanhämoglobin A. Z. Naturforsch., **14:** 604, 1959.
356. Hilson, D.: Malformation of Ears as a Sign of Malformation of the Genito-urinary Tract. Brit. Med. J., **22:** 785, 1959.
357. Himes, H. W., and D. Adlersburg: Pathologic Studies in Idiopathic Sprue. The Malabsorption Syndrome. D. Aldersberg, ed., Grune and Stratton Publishers, N.Y., 77, 1960.
358. Hiraizumi, Y., and J. F. Crow: Heterozygous Effects on Viability, Fertility, Rate of Development, and Longevity of *Drosophila* Chromosomes that Are Lethal when Homozygous. Genetics, **45:** 1071, 1960.
359. Hirszfeld, L., and Z. Kostuch: Über das Wesen der Blutgruppe. Klin. Woch., **17:** 1047, 1938.
360. Hitzig, W. H., and R. Gitzelmann: Transplacental Transfer of Leukocyte Agglutinins. Vox Sang., **4:** 445, 1959.
361. Hoagland, M. B.: An Enzymic Mechanism for Amino-Acid Activation in Animal Tissues. Biochim. et Biophys. Acta, **16:** 288, 1955.
362. Hoffenberg, R., and W. P. U. Jackson: Gonadal Dysgenesis in Normal-Looking Females. A Genetic Theory to Explain Variability of the Syndrome. Brit. Med. J., **1:** 1281, 1957.
363. Hollander, W. F., J. W. Gowen, and J. Stadler: A Study of 25 Gynandromorphic Mice of the Bagg Albino Strain. Anat. Rec., **124:** 223, 1956.
364. Holley, E. B.: Syringomyelia, Morvan's Syndrome. J. Pediat., **30:** 96, 1947.
365. Holzel, A., G. M. Komrower, and V. K. Wilson: Amino-Aciduria in Galactosemia. Brit. Med. J., **1:** 194, 1952.
366. Home Nursing Services (Victorian Order of Nurses), Dominion Bureau of Statistics, Ottawa, Canada, 1957.
367. Hommes, F. A., J. Santema-Drinkwaard, and T. H. Huisman: The Sulfhydryl Groups of Four Different Human Hemoglobins. Biochim. Biophys. Acta, **20:** 564, 1956.
368. Hooker, S. B., and L. M. Anderson: Specific Antigenic Properties of Four Groups of Human Erythrocytes. J. Immunol., **6:** 419, 1921.
369. Horowitz, N. H., and M. Fling: The Role of Genes in the Synthesis of Enzymes in Enzymes: Units of Biological Structure and Function. O. H. Gaebler, ed., Academic Press, N.Y., 624, 1956.
370. Hospital Morbidity Study, Report to the Minister of Health. Province of Ontario, Toronto, Ontario, 1954.
371. Hsia, D. Y.: The Laboratory Detection of Heterozygotes. Amer. J. Hum. Genet., **9:** 98, 1957.
372. Hsia, D. Y.: Inborn Errors of Metabolism. Year Book Publishers, Chicago, 1957.
373. Hsia, D. Y., K. W. Driscoll, W. Troll, and W. E. Knox: Detection by Phenylalanine Tolerance Tests of Heterozygous Carriers of Phenylketonuria. Nature (Lond.), **178:** 1239, 1956.
374. Hsia, D. Y.: I. Huang, and S. G. Driscoll: The Heterozygous Carrier in Galactosemia. Nature (Lond.), **182:** 1289, 1958.
375. Hsu, T. C.: Mammalian Chromosomes *in vitro*. 1. The Karyotype of Man. J. Hered., **43:** 167, 1952.
376. Hughes, A.: Some Effects of Abnormal Tonicity on Dividing Cells in Chick Tissue Cultures. Quart. S. Micr. Sci., **93:** 207, 1952.
377. Hughes, H. B., L. H. Schmidt, and J. P. Biehl: The Metabolism of Isoniazid: Its Implications and Therapeutic Use. Fourteenth Conf. on Chemotherapy of TB. Veterans Adm. and Armed Forces, 217, 1955.
378. Huisman, T. H. J., and A. Dozy: The Action of Carboxypeptidase on Different Human Hemoglobins. Biochim. Biophys. Acta, **20:** 400, 1956.
379. Huisman, T. H. J., and J. Drinkwaard: The *N*-terminal Residues of Five Different Human Hemoglobins. Biochim. Biophys. Acta, **18:** 588, 1955.
380. Huisman, T. H. J., J. H. P. Jonxis, and P. C. Van Der Schaaf: Amino-acid Composition of Four Different Kinds of Human Hemoglobin. Nature (Lond.), **175:** 902, 1955.

381. Huisman, T. H. J., E. A. Martis, and A. Dozy: Chromatography of Hemoglobin Types on Carboxymethylcellulose. J. Lab. Clin. Med., **52**: 312, 1958.

382. Huguley, C. M., Jr., J. A. Bain, S. L. Rivers, and R. B. Scoggins: Refractory Megaloblastic Anemia Associated with Excretion of Orotic Acid. Blood, **14**: 615, 1959.

383. Hungerford, D. A., A. J. Donnelly, P. C. Nowell, and S. Beck: The Chromosome Constitution of a Human Phenotypic Intersex. Amer. J. Hum. Genet., **11**: 215, 1959.

384. Hungerford, D. A., and J. J. Freed: Content of the Nuclei and Chromosome Number in Sublines of the Ehrlich Ascites Carcinoma. Cancer Res., **17**: 177, 1957.

385. Hunt, J. A.: Identity of the α-Chains of Adult and Fetal Human Hemoglobins. Nature (Lond.), **183**: 1373, 1959.

386. Hunt, J. A., and V. M. Ingram: Allelomorphism and the Chemical Differences of the Human Hemoglobins A, S, and C. Nature (Lond.), **181**: 1062, 1958.

387. Hunt, J. A., and V. M. Ingram: The Genetical Control of Protein Structure: The Abnormal Human Hemoglobins. Ciba Foundation Symposium, Biochemistry of Human Genetics. Little, Brown and Company, Boston, Massachusetts, 114, 1959

388. Hunt, J. A., and V. M. Ingram: Human Hemoglobin E: The Chemical Effect of Gene Mutation. Abnormal Human Hemoglobins. Nature, **184**: 870, 1959.

389. Hunt, J. A., and V. M. Ingram: Abnormal Human Hemoglobins. II. The Chymotryptic Digestion of the Trypsin-Resistant Core of Hemoglobins A and S. Biochim. Biophys. Acta, **28**: 546, 1958.

390. Hunt, J. A., and H. Lehmann: Hemoglobin "Bart's": A Fetal Hemoglobin without α-Chains. Abnormal Human Hemoglobins. Nature, **184**: 872, 1959.

391. Huntley, C. C., and S. C. Dees: Eczema Associated with Thrombocytopenic and Purpura and Purulent Otitis Media. Pediatrics, **19**: 351, 1957.

392. Huron, R., and J. Ruffie: Les Méthodes en Génétique Générale et en Génétique Humaine. Masson, Paris, 1959.

393. Huth, E. J., G. D. Webster, and J. R. Elkinton: The Renal Excretion of Hydrogen Ion in Renal Tubular Acidosis, Amer. Med., **29**: 586, 1960.

394. Illness in the Civil Service. Dominion Bureau of Statistics, Ottawa Canada, Catalogue No. 82–203, 1958.

395. Ingalls, T. H., F. R. Avis, F. J. Curley, and H. M. Temin: Genetic Determinants of Hypoxia-induced Congenital Anomalies. J. Hered., **44**: 185, 1953.

396. Ingenito, E. F., J. M. Craig, J. Labesse, M. Cautier, and D. D. Rutstein: Cells of Human Heart and Aorta Grown in Tissue Culture. A. M. A. Arch. of Path., **65**: 355, 1958.

397. Ingram, V. M.: How Do Genes Act? Scient. Amer., **198**: 68, 1958.

398. Ingram, V. M.: Abnormal Human Hemoglobins. I. The Comparison of Normal Human and Sickle-Cell Hemoglobins by "Fingerprinting." Biochim. Biophys. Acta, **28**: 539, 1958.

399. Ingram, V. M.: Abnormal Human Hemoglobins. III. The Chemical Difference between Normal and Sickle-cell Hemoglobins. Biochim. Biophys. Acta, **36**: 402, 1959.

400. Ingram, V. M.: Specific Chemical Difference between the Globins of Normal Human and Sickle-cell Anemia Hemoglobin Nature, (Lond.), **178**: 792, 1956.

401. Ingram, V. M.: Biochem. of Human Genetics. Ciba Foundation Symposium. G. E. W. Wolstenholme and C. M. O'Connor, eds., Little, Brown & Co., Churchill, London, 128, 1959.

402. Ingram, V. M.: Gene Mutations in Human Hemoglobin: The Chemical Difference between Normal and Sickle-cell Hemoglobin. Nature (Lond.), **180**: 326, 1957.

403. Ingram, V. M.: Constituents of Human Hemoglobin. Nature, **183**: 1795, 1959.

404. Ingram, V. M.: Separation of the Peptide Chains of Human Globin. Nature, **182**: 1795, 1959.

405. Ingram, V. M., and A. O. Stretton: Genetic Basis of the Thalassemia Diseases. Nature (Lond.), **184**: 1903, 1959.

406. Itano, H.: Electrophoretic Analysis of the Abnormal Human Hemoglobins. Conference on Hemoglobin. Natl. Acad. of Sci. Natl. Res. Coun., Washington, D.C., **577**: 144, 1958.

407. Itano, H.: Genetics. Transactions of the First Conference. Josiah Macy Foundation, New York, 1960.

408. Itano, H. A.: In Abnormal Hemoglobins. J. H. P. Jonxis and J. F. Delafresnaye, eds., C. C. Thomas, Springfield, Ill., 427, 1959.

409. Itano, H. A.: The Human Hemoglobins: Their Properties and Genetic Control. Advanc. Protein Chem., **12**: 215, 1957.

410. Itano, H. A., and E. Robinson: Genetic Control of the α and β-chains of Hemoglobin. Fed. Proc., **19**: 193, 1960.

411. Itano, H. A., and E. Robinson: Properties and Inheritance of Hemoglobin by Asymmetric Recombination. Nature (Lond.), **184**: 1468, 1959.

412. Itano, H. A., and E. Robinson: Formation of Normal and Doubly Abnormal Hemoglobins by Recombination of Hemoglobin I with S and C. Nature (Lond.), **183**: 1799, 1959.

413. Itano, H. A., and S. J. Singer: On Dissociation and Recombination of Human Adult Hemoglobins A, S, and C. Proc. Soc. Nat. Acad. Sci. (Wash.), **44**: 522, 1958.

414. Itano, H., S. J. Singer, and E. Robinson: Chemical and Genetical Units of the Hemoglobin Molecule. In Ciba Foundation Symposium on Human Biochem. of Human Genet. G. E. W. Wolstenholme and C. M. O'Connor, eds., Little, Brown & Co., Boston, Mass., and Churchill, London, 96, 1959 and 1960.

415. Jacobs, P. A., A. G. Baikie, W. M. Court-Brown, D. G. Harnden, T. N. MacGregor, and N. Maclean: Use of the Term Superfemale. Lancet, **2**: 1145, 1959.

416. Jacobs, P. A., A. G. Baikie, W. M. Court-Brown, T. N. MacGregor, N. Maclean, and D. G. Harnden: Evidence for the Existence of the Human Superfemale. Lancet, **2**: 423, 1959.

417. Jacobs, P. A., A. G. Baikie, W. M. Court-Brown, H. Forrest, J. R. Roy, J. S. S. Stewart, and B. Lennox: Chromosomal Sex in the Syndrome of Testicular Feminization. Lancet, **2**: 591, 1959.

418. Jacobs, P. A., A. G. Baikie, W. M. Court-Brown, and J. A. Strong: The Somatic Chromosomes in Mongolism. Lancet, **1**: 710, 1959.

419. Jacobs, P. A., D. G. Harnden, W. M. Court-Brown, J. Goldstein, H. G. Close, T. N. MacGregor, N. Maclean and J. A. Strong: Abnormalities Involving the X-Chromosome in Women. Lancet, **1**: 1213, 1960.

420. Jacobs, P. A., and A. J. Keay: Somatic Chromosomes in a Child with Bonnevie-Ullrich Syndrome. Lancet, **2**: 732, 1959.

421. Jacobs, P. A., and J. A. Strong: A Case of Human Intersexuality Having a Possible *XXY* Sex-determining Mechanism. Nature (Lond.), **183**: 302, 1959.

422. Jacobsen, O.: Heredity in Breast Cancer. A Genetic and Clinical Study of Two Hundred Probands. Nyt Nordisk Forlag, Copenhagen, p. 306, 1946.

423. Jacquot, A. Y., L. G. Aubel, and R. Wurmser: Physico-chemical Study of Human Iso-hemagglutination. Ann. Eugen., **18**: 183, 1954.

424. Janeway, C. A., J. Craig, M. Davidson, W. Downey, D. Gitlin, and J. C. Sullivan: Hyper-gammaglobulinemia Associated with Severe Recurrent and Chronic Nonspecific Infection. A. M. A. Amer. J. Dis. Child., **88**: 388, 1954.

425. Jenne, J. W.: Studies of Human Patterns of Isoniazid Metabolism Using as Intravenous Fall-off Technique with a Chemical Method. Amer. Rev. of Respiratory Diseases. **81**: 1, 1960.

426. Jepson, J. B., W. Lovenberg, P. Zaltzman, J. A. Oates, A. Sjoerdsma, and S. Udenfriend: Amine Metabolism, Studied in Normal and Phenylketonuric Humans by Monoamine Oxidase Inhibition. Biochem. J., **74**: 5, 1960.

427. Jones, R. T., and W. A. Schroeder: unpublished work, and Jones, R. T.: Ph.D. Thesis, California Institute of Technology, Pasadena, 1961.

428. Jones, R. T., W. A. Schroeder, J. E. Balog, and J. R. Vinograd: Gross Structure of Hemoglobin H. J. Amer. Chem. Soc., **81**: 3161, 1959.

429. Jones, R. T., W. A. Schroeder, and J. R. Vinograd: Identity of the α-Chains of Hemoglobins A and F. J. Amer. Chem. Soc., **81**: 4749, 1959.

430. Jost, A.: Recherches sur le Controle Hormonal de C'Organogénese Sexuelle du Lapin et Remarques sur Certaines Malformations de C'Appareil Génital Humain. Gynecol. et Obstet., **49**: 44, 1960.

431. Kabat, E. A.: Blood Group Substances, their Chemistry and Immunochemistry. Acad. Press, New York, p. 330, 1956.

432. Kalckar, H. M., E. P. Anderson, and K. J. Isselbacher: Galactosemia, A Congenital Defect in a Nucleotide Transferase: A Preliminary Report. Proc. Nat. Acad. Sci. (Wash.), **42**: 49, 1956.

433. Kallmann, F. J., and D. Reisner: Twin Studies on Genetic Variations in Resistance to Tuberculosis. J. Hered., **34**: 269, 1943.

434. Kalow, W.: Cholinesterase Types. Biochemistry of Human Genetics. Ciba Foundation Symposium, C. E. W. Wolstenholme and C. M. O'Connor, eds., Little, Brown & Co., Boston, Mass., and Churchill, London, 39, 1959.

435. Kalow, W., and N. Staron: On Distribution and Inheritance of Atypical Forms of Human Serum Cholinesterase, as Indicated by Dibucaine Numbers. Canad. J. Biochem. Physiol., **35**: 1305, 1957.

436. Kaplan, I. I.: The Treatment of Female Sterility with X-rays to the Ovaries and the Pituitary with Special Reference to Congenital Anomalies of the Offspring. Canad. Med. Assoc. J., **76**: 43, 1957.
437. Kass, E. H.: Asymptomatic Infections of the Urinary Tract. Trans. Assoc. Amer. Phycns., **69**: 56, 1956.
438. Kato, R., and G. Yerganian: Further Observations on the Sex Chromosomes of the Chinese Hamster. Records Genet. Soc. Amer., **28**: 79, 1959.
439. Katz, A. M., and A. I. Chernoff: Structural Similarities between Hemoglobins A and F. Science, **130**: 1574, 1959.
440. Katz, A. M., W. J. Dreyer, and C. B. Anfinsen: Peptide Separation by Two-dimensional Chromatography and Electrophoresis. J. Biol. Chem., **234**: 2897, 1959.
441. Kempthorne, O.: An Introduction to Genetic Statistics. John Wiley and Sons, New York, p. 545, 1957.
442. Kendrew, J. C., R. E. Dickerson, B. E. Strandberg, R. G. Hart, D. R. Davies, D. C. Phillips, and V. C. Shore: Structure of Myoglobin. A Three-dimensional Fourier Synthesis at 2Å Resolution. Nature, **185**: 422, 1960.
443. Kennard, J. H.: Cholelithiases in Identical Twins. New Engl. J. Med., **252**: 1131, 1955.
444. Kety, S. S.: Biochemical Theories of Schizophrenia. Science, **129**: 1528, 1959.
445. Keyfitz, N.: A Factorial Arrangement of Comparisons of Family Size. Amer. J. Soc., **58**: 470, 1953.
446. Kimura, M.: A Model of a Genetic System which Leads to Closer Linkage by Natural Selection. Evolution, **10**: 278, 1956.
447. Kimura, M.: Optimum Mutation Rate and Degree of Dominance as Determined by the Principle for Minimum Genetic Load. J. Genet., 1961 (in press).
448. Kimura, M.: Rules for Testing Stability of a Selective Polymorphism. Proc. Nat. Acad. Sci. (Wash.), **42**: 336, 1956.
449. Kimura, M.: On the Change of Population Fitness by Natural Selection. Heredity, **12**: 145, 1958.
450. Kinosita, R., S. Ohno, W. D. Kaplan, and J. P. Ward: Meioticlike Divisions in Normal Myelocytes. Exp. Cell Res., **6**: 557, 1954.
451. Kirkman, H. N.: Characterization of Partially Purified Glucose-6-Phosphate Dehydrogenase from Normal and Primaquine-sensitive Erythrocytes. Fed. Proc., **18**: 261, 1959.
452. Kirkman, H. N., and E. Bynum: Enzymic Evidence of a Galactosemic Trait in Parents of Galactosemic Children. Ann. Hum. Genet., **23**: 117, 1959.
453. Kligman, A. M.: Personal communication.
454. Kligman, A. M., and W. Epstein: Personal communication.
455. Knight, R. A., M. J. Selin, and H. W. Harris: Comparison of Isoniazid Concentration in the Blood of People of Japanese and European Descent. Therapeutic and Genetic Implications, Amer. Rev. Tuberculosis and Pulm. Dis., **78**: 944, 1958.
456. Knox, W. E.: Sir Archibald Garrod's Inborn Errors of Metabolism. II. Alkaptonuria. Amer. J. Hum. Genet., **10**: 95, 1958.
457. Knox, W. E., and E. C. Messinger: The Detection in the Heterozygote of the Metabolic Effect of the Recessive Gene for Phenylketonuria. Amer. J. Hum. Genet., **10**: 53, 1958.
458. Kodani, M.: The Supernumerary Chromosome of Man. Amer. J. Hum. Genet., **10**: 125, 1958.
459. Kodani, M.: Three Diploid Chromosome Numbers of Man. Proc. Nat. Acad. Sci. (Wash.), **43**: 285, 1957.
460. Kojima, K.: Stable Equilibria for the Optimum Model. Proc. Nat. Acad. Sci. (Wash.), **45**: 989, 1959.
461. Konigsberg, W. H., and R. J. Hill: The Separation of Tryptic Peptides from the Alpha Chain of Human Hemoglobin *A* by Countercurrent Distribution. Fed. Proc., **19**: 342, 1960.
462. Kratchman, J., and D. Grahn: Relationships between the Geologic Environment and Mortality from Congenital Malformation. TID–8204. United States Atomic Energy Commission, Wash., D.C., 1959.
463. Krooth, R. S.: Use of Fertilities of Affected Individuals and Their Unaffected Sibs in the Estimation of Fitness. Amer. J. Hum. Genet., **7**: 325, 1955.
464. Krooth, R. S.: On the Estimation of the Frequency of Genetic Carriers. Amer. J. Hum. Genet., **9**: 170, 1957.
465. Krüpe, M.: Blutgruppenspezifische Pflanzliche Eiweiszkörper (Phytagglutinine). Ferdinand Enke, Stuttgart, 1956.

466. Krüpe, M.: Weitere Beobachtungen über die Reaktionsweise des Hämagglutinierenden Extraktes aus Samen von Sophora Japonica (Schnurbaum). Z. Hyg. Infekt.-Kr. 167, 1953.

467. Kuchler, R. J., and D. J. Merchant: Propagation of Strain L (Earle) Cells in Agitated Fluid Suspension Cultures. Proc. Soc. Exper. Biol. and Med (N.Y.), 92: 803, 1956.

468. Kuhn, R.: Aminozucker. Angew. Chem., 69: 23, 1957.

469. Kunkel, H. G.: Zone Electrophoresis and the Minor Hemoglobin Components of Normal Human Blood. Conference on Hemoglobin. Nat. Acad. of Sci., Nat. Res. Coun., Wash., D.C., 577: 157, 1958.

470. Kunkel, H. G., and G. Wallenius: New Hemoglobin in Normal Adult Blood. Science, 122: 288, 1955.

471. Kurland, L. T.: Epidemiologic Investigations of Amyotrophic Lateral Sclerosis. III. A Genetic Interpretation of Incidence and Geographic Distribution. Proc. Staff Meet. Mayo Clin., 32: 449, 1957.

472. La Du, B. N., V. C. Zannoni, L. Laster, and J. E. Seegmuller: The Nature of the Defect in Tyrosine Metabolism in Alkaptonuria. J. Biol. Chem., 230: 251, 1958.

473. Lajtha, L. G.: Culture of Human Bone Marrow *in vitro:* The Reversibility between Normoblastic and Megaloblastic Series of Cells. J. Clin. Path., 5: 67, 1952.

474. Lajtha, L. G.: Bone Marrow Culture. Meth. Med. Res., 8: 12, 1960.

475. Lalezari, P., M. Nussbaum, S. Gelman, and T. H. Spaet: Neonatal Neutropenia Due to Maternal Isoimmunization. Blood, 15: 236, 1960.

476. Landsteiner, K.: Über Unterschiede des Fötalen und Mütterlichen Blutserums und über eine Agglutinations und Fällungshemmende Wirkung des Normalserums. München. Med. Wchnschr., 49: 473, 1902.

477. Landsteiner, K.: The Specificity of Serological Reactions, Second ed. Harvard University Press, p. 310, 1945.

478. Landsteiner, K., and P. Levine: Further Observations on Individual Differences of Human Blood. Proc. Soc. Exp. Biol. and Med. (N.Y.), 24: 941, 1927.

479. Landsteiner, K., L. G. Longsworth, and J. Van Der Scheer: Electrophoresis Experiments with Egg Albumins and Hemoglobins. Science, 88: 83, 1938.

480. Landsteiner, K., and A. S. Wiener: An Agglutinable Factor in Human Blood Recognized by Immune Sera for Rhesus. Proc. Soc. Exp. Biol. and Med. (N.Y.), 43: 223, 1940.

481. Lawler, S. D., and J. H. Renwick: Blood Groups and Genetic Linkage. Brit. Med. Bull., 15: 145, 1959.

482. Lawler, S. D., and M. Sandler: Data on Linkage in Man: Elliptocytosis and Blood Groups: Families 5, 6, and 7. Ann. Eug. (Lond.), 18: 328, 1954.

483. Lawrence, R. D.: Three Types of Human Diabetes. Ann. Intern. Med., 43: 1199, 1955.

484. Lederberg, J.: Prospects for a Genetics of Somatic and Tumor Cells. Ann. N.Y. Acad. Sci., 63: 662, 1956.

485. Ledley, R. S.: Digital Electronic Computers in Biomedical Science. Science, 130: 1225, 1959.

486. Lee, H. H., and T. T. Puck: Abstract No. 150. Single Cell Techniques in the Study of Ultraviolet Irradiation of Mammalian Cells. Radiat. Res., 9: 142, 1958.

487. Lehmann, U., and H. Forssman: Chromosome Complement in a Mongoloid Mother, her Child and the Child's Father. Lancet, 1: 498, 1960.

488. Lejeune, J., M. Gautier, and R. Turpin: Études des Chromosomes Somatiques de Neuf Enfants Mongoliens. C. R. Acad. Sci. (Paris), 248: 1721, 1959.

489. Lejeune, J., M. Gautier, and R. Turpin: Les Chromosomes Humains en Culture de Tissus. C. R. Acad. Sci. (Paris), 248: 602, 1959.

490. Lejeune, J., R. Turpin, and M. Gautier: Le Mongolisme, Premier Exemple d'Aberration Autosomique Humaine. Ann. Génét., 2: 49, 1959.

491. Lennox, B.: Use of the Term Superfemale. Lancet, 1: 55, 1960.

492. Leong, G. F., R. L. Pessotti, and R. W. Brauer: Liver Cell Proliferation in X-Irradiated Rats after Single and Repetitive Partial Hepatectomy. Radiat. Res., 12: 451, 1960.

493. Levan, A., and J. J. Biesele: Role of Chromosomes in Cancerogenesis, as Studied in Serial Tissue Culture of Mammalian Cells. Ann. Acad. Sci. (N.Y.), 71: 1022, 1958.

494. Levan, A., and T. C. Hsu: The Human Idiogram. Hereditas, 45: 665, 1959.

495. Levine, R., L. Burnham, N. J. Englewood, E. M. Katzin, and P. Vogel: The Role of Isoimmunization in the Pathogenesis of Erythroblastosis Fetalis. Amer. J. Obstet. Gynec., 42: 925, 1941.

496. Levine, P., M. J. Celano, S. Lang, and V. Berliner: On Anti-*M* in Horse Sera. Vox Sang. (Basel), 2: 433, 1957.

497. Levine, P., R. Ottensooser, M. F. Celano, and W. Pollitzer: On Reactions of Plant Anti-N with Red Cells of Chimpanzees and Other Animals. Amer. J. Phys. Anthropol., **13:** 29, 1955.

498. Levine, P., and R. E. Stetson: An Unusual Case of Intra-group Agglutination. J. Amer. Med. Assoc., **113:** 126, 1939.

499. Levy, A. L., and D. Chung: Two-dimensional Chromatography of Amino Acids on Buffered Papers. Analyt. Chem., **25:** 396, 1953.

500. Lewontin, R. C.: A General Method for Investigating the Equilibrium of Gene Frequency in a Population. Genetics, **43:** 419, 1958.

501. Lewontin, R. C., and L. C. Dunn: The Evolutionary Dynamics of a Polymorphism in the House Mouse. Genetics, **45:** 705, 1960.

502. Li, C. C.: Population Genetics. Univ. of Chicago Press, 1955.

503. Li, C. H.: Methods for Study of Peptide Structure. Methods in Medical Research. R. W. Gerard, ed.-in-chief. The Year Book Publishers, Inc., Chicago, **3:** 219, 1950.

504. Li, C. H.: Species Variation and Structural Aspects in Some Pituitary Hormones. Symposium on Protein Structure. A. Neuberger, ed., J. Wiley and Sons, Inc., N.Y., 1958.

505. Lieberman, I., and P. Ove: Estimation of Mutation Rates with Mammalian Cells in Culture. Proc. Nat. Acad. Sci. (Wash.), **45:** 872, 1959.

506. Lieberman, I., and P. Ove: Isolation and Study of Mutants from Mammalian Cells in Culture. Proc. Nat. Acad. Sci. (Wash.), **45:** 867, 1959.

507. Liebold, B., K. Hilse, K. Simon, and G. Braunitzer: Die Isolierung Einiger Peptide nach Einwirkung von Trypsin auf das Protein des Drist. Hum. Hämoglobins. A. Z. Physiol. Chem., **315:** 278, 1959.

508. Lilienfeld, A. M.: Diagnostic and Therapeutic X-Radiation in an Urban Population. Publ. Hlth. Rep., **74:** 29, 1959.

509. Lilienfeld, A. M.: Epidemiologic Methods and Inferences. H. E. Hilleboe, and G. W. Larimore, eds., Preventive Medicine, chap. 40. W. B. Saunders Co., Philadelphia, 1959.

510. Lima-De-Faria, A.: Incorporation of Tritiated Thymidine into Meiotic Chromosomes. **130:** 503, 1959.

511. Littell, A. S.: Personal communication.

512. Littler, T. R., and R. G. Ellis: Gallstones: A Clinical Study. Brit. Med. J., **1:** 842, 1952.

513. Lotka, A. J.: Elements of Physical Biology. Williams and Wilkins, Baltimore, 1925.

514. Lowe, C. U., M. Terrey, and E. A. MacLachian: Organic-Aciduria, Decreased Renal Ammonia Production. Hydrophthalmos and Mental Retardation. A Clinical Entity. A. M. A., Amer. J. Dis. Child., **83:** 164, 1952.

515. Luria, S. E., and M. Delbrück: Mutations of Bacteria from Virus Sensitivity to Virus Resistance. Genetics, **28:** 491, 1943.

516. Macht, S. H., and P. S. Lawrence: National Survey of Congenital Malformations Resulting from Exposure to Roentgen Radiation. Amer. J. Roent., **73:** 442, 1955.

517. Macklin, M. T.: Inheritance of Retinoblastoma in Ohio. A. M. A. Arch. Ophthal., **62:** 842, 1959.

518. Macklin, M. T.: A Study of Retinoblastoma in Ohio. Amer. J. Hum. Genet., **12:** 1, 1960.

519. Macklin, M. T.: Methods of Selection of Probands and Controls. Amer. J. Hum. Genet., **6:** 86, 1954.

520. MacLeod, C. M.: Bacterial and Mycotic Infections of Man. R. J. Dubos, ed., Lippincott Co., Philadelphia, 83, 1948.

521. MacLeod, C. M., and M. R. Krauss: Relations of Virulence of Pneumococcal Strains for Mice to the Quantity of Capsular Polysaccharide. J. Exp. Med., **92:** 1, 1950.

522. Mäkelä, O.: Studies in Hemagglutinins of Leguminosae Seeds. Ann. Med. Exp. Fenn., **11** Suppl., 1, 1957.

523. Mäkelä, O., P. Mäkelä and R. Lehtovaara: Sugar Specificity of Plant Hemagglutinins. Ann. Med. Exp. Fenn., **37:** 328, 1959.

524. Mäkelä, O., and P. Mäkelä: Some New Blood Group Specific Phytagglutinin; a Preliminary Report. Ann. Med. Exp. Fenn., **34:** 402, 1956.

525. Makino, S., and I. Nishimura: Water-pretreatment Squash Technic; a New and Simple Practical Method for the Chromosome Study of Animals. Stain Technol., **27:** 1, 1951.

526. Makino, S., and M. Sasaki: A Study of Somatic Chromosomes in a Japanese Population. Amer. J. Hum. Genet., **13:** 47, 1961.

527. Makino, S., and M. Sasaki: Chromosome Constitution in Normal Human Subjects and in One Case of True Hermaphroditism. Proc. Japan Acad., **36:** 156, 1960.

528. Manning, G. W.: Cardiac Manifestations in Friedreich's Ataxia. Amer. Heart J., **39:** 799, 1950.

529. Martin, L.: The Hereditary and Familial Aspects of Exophthalmic Nodular Goitre. Quart. J. Med. N. S., **14**: 207, 1945.
530. Masri, M. S., and K. Singer: Studies on Abnormal Hemoglobins, XII. Terminal and Free Amino Groups of Various Types of Human Hemoglobin. Arch. Biochem. Biophys., **58**: 414, 1955.
531. Mather, K.: Statistical Analysis in Biology. Interscience, N.Y., p. 267, 1947.
532. Matsuda, G., W. A. Schroeder, R. T. Jones, and N. Weliky: Is There an Embryonic or Primative Human Hemoglobin? Blood, **16**: 984, 1960.
533. Matsunaga, E., and H. Ogyu: Genetic Study of Retinoblastomia in a Japanese Population. Jap. J. Hum. Genet., **4**: 156, 1959.
534. Meehl, P. E.: Clinical Versus Statistical Prediction. University of Minnesota Press, Minneapolis, pp. 149, 1954.
535. Menolasino, N. J., I. Davidsohn, and D. E. Lynch: A Simplified Method for the Preparation of Anti-*M* and Anti-*N* Typing Sera. J. Lab. Clin. Med., **44**: 495, 1954.
536. Merch, T.: Chondrodystrophic Dwarfs in Denmark. E. Munksgaard, Copenhagen, **3**: 200, 1941.
537. Merchant, D. C., R. H. Kahn, and W. H. Murphy: Handbook of Cell and Organ Culture. Burgess Publishing Co., Minneapolis, p. 188, 1960.
538. Meyers, S. N., and C. M. Jensen: Significance of Familial Factors in the Development of Tuberculin Allergy. Amer. J. of Hum. Genet., **3**: 325, 1951.
539. Miller, M. J., J. V. Neel, and F. B. Livingstone: Distribution of Parasites in the Red Cells of Sickle-cell Trait Carriers Infected with Plasmodium Falciparum. Trans. Roy. Soc. Trop. Med. Hyg. **50**: 294, 1956.
540. Mitchell, R. S., and J. C. Bell: Clinical Implications of Isoniazid Blood Levels in Pulmonary Tuberculosis. New Engl. J. Med., **257**: 1066, 1957.
541. Mitoma, C., R. M. Auld, and S. Udenfriend: On the Nature of Enzymatic Defect in Phenylpyruvic Oligophrenia. Proc. Soc. Exper. Biol. (N.Y.), **94**: 634, 1957.
542. Moen, J. K.: The Development of Pure Cultures of Fibroblasts from Single Mononuclear Cells. J. Exper. Med., **61**: 247, 1935.
543. Moldawer, M. P., G. L. Nardi, and J. W. Raker: Concomitance of Multiple Adenomas of the Parathyroids and Pancreatic Islets with Tumor of the Pituitary: A Syndrome of Familial Incidence. Amer. J. Med. Sci., **228**: 190, 1954.
544. Mollenbach, C. J.: Congenital Defects in the Internal Membrane of the Eye, Clinical and Genetic Aspects. Medfadte Defekter i ø Jeto Indre Hinder Klinik og Arvelighedsforhold. Copenhagen Nyt Nordisk Forlag. A. Busck, p. 164, 1947.
545. Mommaerts, W. F. H., M. B. Illingworth, C. M. Pearson, R. J. Guillory, and K. Seraydarian: A Functional Disorder of Muscle Associated with the Absence of Phosphorylase. Proc. Nat. Acad. Sci. (Wash.), **45**: 791, 1959.
546. Monod, J.: Remarks on the Mechanism of Enzyme Induction in Enzymes in Units of Biological Structure and Function. O. H. Gaebler, ed., Academic Press, New York, 7, 1956.
547. Mood, A. M.: Introduction to the Theory of Statistics, First ed. McGraw-Hill, N.Y., p. 433, 1950.
548. Moore, A. E., C. M. Southam, and S. Sternberg: Neoplastic Changes Developing in Epithelial Cell Lines Derived from Normal Persons. Science, **124**: 127, 1956.
549. Moore, K. J.: Sex Reversal in Newborn Babies. Lancet, **1**: 217, 1959.
550. Moore, S., D. H. Spackman, and W. H. Stein: Chromatography of Amino Acids on Sulfonated Polystyrene Resins: An Improved System. Analyt. Chem., **30**: 1185, 1958.
551. Moore, S., and W. H. Stein: Column Chromatography of Peptides and Proteins. Adv. Protein, Chem., **11**: 191, 1956.
552. Moore, S., and W. H. Stein: Procedures for the Chromatographic Determination of Amino Acids on Four Per Cent Cross-linked Sulfonated Polystyrene Resins. J. Biol. Chem., **211**: 893, 1954.
553. Moorhead, P. S., P. C. Nowell, W. J. Mellman, D. M. Batipps, and D. A. Hungerford: Chromosome Preparations of Leucocytes Cultured from Human Peripheral Blood. Exp. Cell Res., **20**: 613, 1960.
554. Morch, T.: Chondrodystrophic Dwarfs in Denmark. Opera ex Domo Biologiae Hereditariae Universitatis Hafniensis. Copenhagen: E. Munksgaard, **3**: 200, 1941.
555. Morgan, J. F., M. E. Campbell, and H. J. Morton: The Nutrition of Animal Tissues Cultivated *in vitro*. I. A Survey of Natural Materials as Supplements to Synthetic Medium 199. J. Nat. Cancer Inst., **16**: 557, 1955.

556. Morgan, J. F., H. J. Morton, and R. C. Parker: Nutrition of Animal Cells in Tissue Cuture. I. Initial Studies on a Synthetic Medium. Proc. Soc. Exper. Biol. (N.Y.), **73:** 1, 1950.

557. Morgan, J. F.: Tissue Culture Nutrition. Bact. Rev., **22:** 20, 1958.

558. Morgan, T. H., C. B. Bridges, and A. H. Sturtevant: The Genetics of Drosophila. Bibl. Genet., **2:** 1, 1925.

559. Morgan, W. T. J., and H. K. King: Studies in Immunochemistry. Isolated from Hog Gastric Mucin of Polysaccharide-amino Acid Complex Possessing Blood Group *A* Specificity. Biochem. J., **37:** 640, 1943.

560. Morgan, W. T. J., and W. M. Watkins: The Product of the Human Blood Group *A* and *B* Genes in Individuals Belonging to Group *AB*. Nature (Lond.), **177:** 521, 1956.

561. Morgan, W. T. J., and W. M. Watkins: The Inhibition of the Hemogglutinins in Plant Seeds by Human Blood Group Substances and Simple Sugars. Brit. J. Exp. Path., **34:** 94, 1953.

562. Morgan, W. T. J., and W. M. Watkins: Some Aspects of the Biochemistry of the Human Blood Group Substances. Brit. Med. Bull., **15:** 109, 1959.

563. Morkovin, D., P. I. Marcus, T. T. Puck, and S. J. Cieciura: Action of X Rays on Mammalian Cells. II. Survival Curves of Cells from Normal Human Tissues. J. of Exp. Med., **106:** 485, 1957.

564. Morkovin, D., and T. T. Puck: Single Cell Techniques in the Study of Radioprotective Action. Radiat. Res., **9:** 155, 1958.

565. Morrison, H.: The Familial Incidence of Exophthalmic Goiter. New Engl. J. Med., **199:** 85, 1928.

566. Morrison, M., and J. L. Cook: Chromatographic Fractionation of Normal Adult Oxyhemoglobin. Science, **122:** 920, 1955.

567. Morton, N. E.: The Detection and Estimation of Linkage between the Genes for Elliptocytosis and the Rh Blood Type. Amer. J. Hum. Genet., **8:** 80, 1956.

568. Morton, N. E.: The Mutational Load Due to Detrimental Genes in Man. Amer. J. Hum. Genet., **12:** 348, 1960.

569. Morton, N. E.: Methods of Study in Human Genetics. Genetics and Cancer, University of Texas Press, Austin, 391, 1959.

570. Morton, N. E.: Genetic Tests Under Incomplete Ascertainment. Amer. J. Hum. Genet., **11:** 1, 1959.

571. Morton, N. E.: Further Scoring Types in Sequential Linkage Tests, with a Critical Review of Autosomal and Partial Sex Linkage in Man. Amer. J. Hum. Genet., **9:** 55, 1957.

572. Morton, N. E.: Sequential Tests for the Detection of Linkage. Amer. J. Hum. Genet., **7:** 277, 1955.

573. Morton, N. E., and C. S. Chung: Formal Genetics of Muscular Dystrophy. Amer. J. Hum. Genet., **11:** 360, 1959.

574. Morton, N. E., and C. S. Chung: Are the *MN* Blood Groups Maintained by Selection. Amer. J. Hum. Genet., **11:** 237, 1959.

575. Morton, N. E., J. F. Crow, and H. J. Muller: An Estimate of the Mutational Damage in Man from Data on Consanguineous Marriages. Proc. Nat. Acad. Sci. (Wash.), **42:** 855, 1956.

576. Mosbech, J., and A. Videbaek: On the Etiology of Esophogeal Carcinoma. J. Nat. Cancer Inst., **15:** 1665, 1955.

577. Motulsky, A. G.: Metabolic Polymorphisms and the Role of Infectious Diseases in Human Evolution. Hum. Biol., **32:** 28, 1960.

578. Motulsky, A. G.: Discussion in Methodology in Human Genetics. W. J. Burdette, ed., Holden-Day, Inc., San Francisco, 1962.

579. Motulsky, A. G.: Population Genetics of Glucose-6-Phosphate Dehydrogenase Deficiency of the Red Cell. Proc. Conf. on Genetic Polymorphism and Geographic Variations in Disease. N. I. H., 258, 1960.

580. Mourant, A. E.: New Human Blood Group Antigen of Frequent Occurrence. Nature (Lond.), **158:** 237, 1946.

581. Murayama, M.: The Chemical Difference between Normal Human Hemoglobin and Hemoglobin I. Fed. Proc., **19:** 78, 1960.

582. Murayama, M., and V. M. Ingram: Comparison of Normal Adult Human Hemoglobin with Hemoglobin *I* by Fingerprinting. Nature (Lond.), **183:** 1798, 1959.

583. Murphy, D. P.: Congenital Malformations, Second ed. Lippincott, Philadelphia, p. 127, 1947.

584. Murray, M. R., and G. Kopech: A Bibliography of the Research in Tissue Culture.   Acad. Press Inc., New York, p. 1741, 1953.

585. McCarty, M.: The Immune Response in Rheumatic Fever.   Rheumatic Fever, A Symposium. L. Thomas, ed., Univ. of Minn. Press, Minneapolis, 136, 1952.

586. McCracken, D. D.: Digital Computer Programming.   John Wiley and Sons, Inc., New York, N.Y., p. 253, 1957.

587. McCullock, E. A. and J. E. Till: (to be published).

588. McElroy, W. D., and B. Glass, eds.: A Symposium on the Chemical Basis of Heredity.   The Johns Hopkins Press, Baltimore, p. 848, 1957.

589. McKusick, V. A.: Heritable Disorders of Connective Tissue, Second ed.   C. V. Mosby Co., St. Louis, 42, 1960.

590. McLimans, W. F., E. V. Davis, F. L. Glover, and G. W. Rake: The Submerged Culture of Mammalian Cells: The Spinner Culture.   J. Immunol., 79: 428, 1957.

591. McQuilkin, W. T., W. R. Earle, and V. J. Evans: Cinemicrographic Comparison of NCTC Clone 929 (Strain L) Mouse Cells Grown in Protein Medium and in Chemically Defined Protein-Free Medium.   Excerpta Medica, 1961.

592. McQuilkin, W. T., V. J. Evans, and W. R. Earle: The Adaptation of Additional Lines of NCTC Clone 929 (Strain L) Cells to Chemically Defined Protein-Free Medium NCTC 109.   J. Nat. Cancer Inst., 19: 885, 1957.

593. Nachtsheim, H.:   Chromosomenaberrationen beim Sauger und Ihre Bedeutung für die Entstehung von Missbildungen.   Naturwiss., 46: 637, 1959.

594. Nachtsheim, H.: Die Mutationsrate Menschlicher Gene.   Naturwiss., 17: 385, 1954.

595. Nachtsheim, H.: Atomic Energie und Erbgut.   Münch. Med. Wschr., 99: 1283, 1957.

596. Neel, J. V.: Studies on the Interaction of Mutations Affecting the Chetae of *Drosophila melanogaster*.   I. The Interaction of Hairy, Polychetoid, and Hairy Wing.   Genetics, 26: 52, 1941.

597. Neel, J. V.: Studies on the Interaction of Mutations Affecting the Chaetae of *Drosophila melanogaster*.   II. The Relation of Character Expression to Size in Flies Homozygous for Polychetoid, Hairy, Hairy Wing, and the Combinations of these Factors.   Genetics, 28: 49, 1943.

598. Neel, J. V.: Some Problems in the Estimation of Spontaneous Mutation Rates in Animals and Man.   Effect of Radiation on Human Heredity.   World Health Organization, Geneva, 139, 1957.

599. Neel, J. V.: Genetic Aspects of Abnormal Hemoglobins: Conference on Hemoglobin.   Nat. Acad. Sci., Nat. Res. Coun., Wash., D.C., 557: 253, 1958.

600. Neel, J. V.: The Study of Human Mutation Rates.   Amer. Nat., 86: 129, 1952.

601. Neel, J. V.: On Some Pitfalls in Developing an Adequate Genetic Hypothesis.   Amer. J. Hum. Genet., 7: 1, 1955.

602. Neel, J. V.: The Study of Natural Selection in Primitive and Civilized Human Populations. Hum. Biol., 30: 43, 1958.

603. Neel, J. V.: In Biochemistry of Human Genetics.   Ciba Foundation Symposium.   G. E. W. Wolstenholme and C. M. O'Connor, eds., Little, Brown Co., Boston, Mass., and Churchill, London, 131, 1959.

604. Neel, J. V.: Detection of the Genetic Carriers of Hereditary Disease.   Amer. J. Hum. Genet., 1: 19, 1949.

605. Neel, J. V.: The Detection of the Genetic Carriers of Inherited Disease in Clinical Genetics. A. Sorsby, ed., C. V. Mosby, St. Louis, 27, 1953.

606. Neel, J. V.: A Study of Major Congenital Defects in Japanese Infants.   Amer. J. Hum. Genet., 10: 398, 1958.

607. Neel, J. V.: Human Hemoglobin Types.   New Engl. J. of Med., 256: 161, 1957.

608. Neel, J. V.: The Inheritance of Sickle-cell Anemia.   Science, 110: 64, 1949.

609. Neel, J. V., and H. F. Falls: The Rate of Mutation of the Gene Responsible for Retinoblastoma in Man.   Science, 114: 419, 1951.

610. Neel, J. V., and W. J. Schull: The Effect of Exposure to the Atomic Bombs on Pregnancy Termination in Hiroshima and Nagasaki.   Nat. Acad. of Sci.-Nat. Res. Coun. Publ., p. 702, 1956.

611. Neel, J. V., and W. J. Schull: Human Heredity.   University of Chicago Press, p. 361, 1954.

612. Neel, J. V., W. J. Schull, and K. Takeshima: A Note on Achondroplasia in Japan.   Jap. J. Hum. Genet., 4: 165, 1959.

613. Neel, J. V., W. J. Schull, and H. S. Shapiro: Absence of Linkage between the Genes for Sickling Phenomenon, the MN Blood Types, and the S-agglutinogen. Amer. J. Hum. Genet., 4: 204, 1952.

614. Neilson, W. E., ed.: Webster's New International Dictionary of the English Language. G. & C. Merriam Co., Springfield, p. 3445, 1957.

615. Nelson, W. O.: Sex Differences in Human Nuclei with Particular Reference to the Klinefelter Syndrome, Gonadal Agenesis, and Other Types of Hermaphroditism. Acta Endocr., 23: 227, 1956.

616. Newcombe, H. B.: Feasibility of Estimating Consequences of an Increased Mutation Rate. Symposium on Molecular Genetics and Human Disease. L. I. Gardner, ed., C. C. Thomas, Springfield, p. 297, 1961.

617. Newcombe, H. B.: Genetic Effects of Ionizing Radiation. Canad. J. Biochem., 38: 330, 1960.

618. Newcombe, H. B.: Population Records in Methodology in Human Genetics. W. J. Burdette, ed., Holden-Day, Inc., San Francisco, 1962.

619. Newcombe, H. B., A. P. James, and S. J. Axford: Family Linkage of Vital and Health Records. AECL 470, Atomic Energy of Canada Limited, Chalk River, Ontario, 1957.

620. Newcombe, H. B., J. M. Kennedy, S. J. Axford, and A. P. James: Automatic Linkage of Vital Records. Science, 130: 954, 1959.

621. Newcombe, H. B., and P. O. W. Rhynas: in preparation.

622. Newton, E. J.: Hematogenous Brain Abscess in Cyanotic Congenital Heart Disease. Quart. J. Med., 25: 201, 1956.

623. Nilsson, I. M., S. Bergman, J. Reitalu, and J. Waldenstrom: Hemophilia *A* in a Girl with Male Sex Chromatin Pattern. Lancet, 2: 264, 1959.

624. Nishimura, E. T., H. B. Hamilton, T. Y. Kobara, S. Takahara, Y. Ogura, and K. Doi: Carrier State in Human Acatalasemia. Science, 130: 333, 1959.

625. Norton, H. W.: Estimation of Linkage in Rucker's Pedigree of Nystagmus and Colorblindness. Amer. J. Hum. Genet., 1: 55, 1949.

626. Novitski, E., and E. R. Dempster: An Analysis of Data from Laboratory Populations of *Drosophila melanogaster*. Genetics, 43: 470, 1958.

627. Novitski, E., and A. W. Kimball: Birth Order, Parental Ages, and Sex of Offspring. Amer. J. Hum. Genet., 10: 268, 1958.

628. Novitski, E., and L. Sandler: Relationship between Parental Age, Birth Order, and Secondary Sex Ratio in Humans. Ann. Human Genet., 21: 123, 1956.

629. Nowakowski, H., W. Lenz, S. Bergman, and J. Reitalu: Chromosomenbefunde beim Echten Klinefelter-Syndrom. Acta Endocrinologica, 34: 483, 1960.

630. Nowell, P. C., and D. A. Hungerford: Chromosome Studies on Normal and Leukemic Human Leucocytes. J. Nat. Cancer Inst., 25: 85, 1960.

631. Ohno, S., and T. S. Hauschka: Allocycly of the X-Chromosome in Tumors and Normal Tissues. Cancer Res., 20: 541, 1960.

632. Ohno, S., W. D. Kaplan, and R. Kinosita: Formation of the Sex Chromatin by a Single X-Chromosome in Liver Cells of *Rattus norvegicus*. Exp. Cell Res., 18: 415, 1959.

633. Olsen, A. M.: Bronchiectasis and Dextrocardia. Amer. Rev. Tuberc., 47: 435, 1943.

634. Oncley, J. L. ed.: Biophysical Science. A Study Program. Revs. Modern Phys., 31: No. 1, 1959; No. 2, 1959.

635. Ottensooser, F., and K. Silberschmidt: Hemagglutinin Anti-*N* in Plant Seeds. Nature (Lond.), 172: 914, 1953.

636. Owens, O. V. H., G. O. Gey, and M. K. Gey: A New Method for the Cultivation of Mammalian Cells Suspended in Agitated Fluid Medium (Abstract). Proc. Amer. Assn. Cancer Res., 13: 41, 1953.

637. Panel Discussion on Tissue Culture Genetics. Biochemistry of Human Genetics, Ciba Foundation Symposium. G. E. W. Wolstenholme and C. M. O'Connor, eds., Little, Brown & Co., Boston, Mass., and Churchill, London, 278, 1959.

638. Pardee, A. B., F. Jacob, and J. Monod: The Genetic Control and Cytoplasmic Expression of Inducibility in the Synthesis of β-Galactosidase by *E. coli*. J. Mol. Biol., 1: 165, 1959.

639. Pare, C. M., M. Sandler, and R. S. Stacey: 5-Hydroxytryptamine Deficiency in Phenylketonuria. Lancet, 272: 551, 1957.

640. Parker, R. C.: Methods of Tissue Culture, Second ed. Hoeber, New York, pp. 294, 1950.

641. Paschkis, K. E., A. E. Rakoff, and A. Cantarow: Clinical Endocrinology, second ed. Hoeber and Harper Publishers, New York, 305, 1958.

642. Patau, K.: The Identification of Individual Chromosomes, Especially in Man. Amer. J. Hum. Genet., **12:** 250, 1960.

643. Patau, K., and H. Nachtsheim: Mutations und Selektionsdruck beim Pelgar-Gen des Menschen. Zeit. Naturforsch., **1:** 345, 1946.

644. Patau, K., D. W. Smith, E. Therman, S. L. Inhorn, and H. P. Wagner: Multiple Congenital Anomalies Caused by an Extra Autosome. Lancet, **1:** 790, 1960.

645. Paul, J. R.: Cell and Tissue Culture. Second ed., Williams & Wilkins Co., Baltimore, p. 312, 1960.

646. Pauling, L., H. Itano, S. J. Singer, and I. C. Wells: Sickle-cell Anemia, A Molecular Disease. Science, **110:** 543, 1949.

647. Pennell, R. B.: Fractionation and Isolation of Purified Components by Precipitation Methods. The Plasma Proteins. F. W. Putnam, ed., Acad. Press, New York, 1960.

648. Penrose, L. S.: Genetics of Anencephaly. J. Mental Def. Res., **1:** 4, 1957.

649. Penrose, L. S.: A Clinical and Genetical Study of 1280 Cases of Mental Defect. Gt. Brit. Med. Res. Coun., Sp. Rep. Ser., **229:** 1, 1938.

650. Penrose, L. S.: Parental Age and Mutation. Lancet, **2:** 312, 1955.

651. Penrose, L. S.: Mutation in Man. Acta Genet. (Basel), **6:** 169, 1956.

652. Penrose, L. S.: Autosomal Mutation and Modification in Man with Special Reference to Mental Defect. Ann. Eugen., **7:** 1, 1936.

653. Penrose, L. S., S. M. Smith, and D. A. Sprott: On the Stability of Allelic Systems with Special Reference to Hemoglobins A, S, and C. Ann. Hum. Genet., **21:** 90, 1956.

654. Penrose, L. S., and C. Stern: Reconsideration of the Lambert Pedigree (Ichthyosis hystrix gravior). Ann. Hum. Genet., **22:** 258, 1959.

655. Perkoff, G. T., F. E. Stephens, D. A. Dolowitz, and F. H. Tyler: A Clinical Study of Hereditary Interstitial Pyelonephritis. A. M. A. Arch. of Int. Med., **88:** 191, 1951.

656. Perla, D., and J. Marmorston: Natural Resistance and Clinical Medicine. Little, Brown & Co., Boston, p. 534, 1941.

657. Perry, V. P., V. J. Evans, W. R. Earle, G. W. Hyatt, and W. C. Bedell: Long-term Tissue Culture of Human Skin. Amer. J. Hyg., **63:** 52, 1956.

658. Perry, V. P., K. K. Sanford, V. J. Evans, G. W. Hyatt, and W. R. Earle: Establishment of Clones of Epithelial Cells from Human Skin. J. Nat. Cancer Inst., **18:** 709, 1957.

659. Perutz, M. F., and J. M. Mitchison: State of Hemoglobin in Sickle-Cell Anemia. Nature (Lond.), **166:** 677, 1950.

660. Perutz, M. F., M. G. Rossman, A. F. Cullis, H. Muirhead, G. Will, and A. C. T. North: Structure of Hemoglobin. A Three-dimensional Fourier Synthesis at 5.5 Å Resolution Obtained by X-ray Analysis. Nature, **185:** 416, 1960.

661. Peterman, M. G.: Chronic Pyelonephritis with Renal Acidosis. A. M. A. Amer. J. Dis. Child., **69:** 291, 1945.

662. Pfandler, U.: La Manifestation Hétérozygote et Homozygote de Certains Troubles du Métabolisme; Porphyrie Chronique, Cystinose, Maladie de Niemann-Pick. J. Génét. Hum., **5:** 248, 1956.

663. Philip, U., and A. Sorsby: The Mutation Rate of Retinoblastoma. Unpublished manuscript presented before the Genetical Society of Great Britain, 1944.

664. Physicians' Notice of Birth Statistics. Health Branch, Department of Health and Welfare. Victoria, B.C., 1953.

665. Pickles, M. M.: Effect of Cholera Filtrate on Red Cells as Demonstrated by Incomplete Rh Antibodies. Nature (Lond.), **158:** 880, 1946.

666. Pillsbury, D. M., W. B. Shelley, and A. M. Kligman: Dermatology. W. B. Saunders Co., Philadelphia, 1208, 1956.

667. Polani, P. E., J. H. Briggs, C. E. Ford, C. M. Clarke, and J. M. Berg: A Mongol Girl with 46 Chromosomes. Lancet, **1:** 721, 1960.

668. Pomerat, C. M., assoc. ed.: Tissue Culture Methods. Methods in Medical Research. M. B. Visscher, ed., Year Book Publishers, Chicago, **4:** 198, 1951.

669. Porath, J.: Gel Filtration of Proteins, Peptides and Amino Acids. Biochem. Biophys. Acta, **39:** 193, 1960.

670. Porath, J., and P. Flodin: Gel Filtration: A Method for Desalting and Group Separation. Nature (Lond.), **183:** 1657, 1959.

671. Porter, R. R.: Globulin and Antibodies. The Plasma Proteins. F. W. Putnam, ed., Acad. Press. New York, **1:** 241, 1960.

672. Porter, R. R., and F. Sanger: The Free Amino Groups of Hemoglobins. Biochem. J., **42:** 287, 1948.

673. Potter, E. L.: Pathology of the Fetus and Newborn. Year Book Publishers, Chicago, Ill. 363, 1952.

674. Powys, A. O.: Data for the Problem of Evolution in Man. Biometrika, **4**: 233, 1905.

675. Prader, A., J. Schneider, J. M. Frances, and W. Zublin: Frequency of the True (Chromatin-positive) Klinefelter's Syndrome. Lancet, **1**: 968, 1958.

676. Price, B.: Primary Biases in Twin Studies: A Review of Prenatal and Natal Difference-producing Factors in Monozygotic Pairs. Amer. J. Hum. Genet., **2**: 293, 1950.

677. Price-Evans, D. A.: Liverpool Medical Institution: The Role of Inheritance in Common Diseases. Lancet, **2**: 526, 1959.

678. Prins, H. K., and T. H. Huisman: Some Observations about the Heterogeneity of Hemo-globin in Aluminum Oxide Chromatography. Biochim. Biophys. Acta, **20**: 570, 1956.

679. Prins, H. K., and T. H. Huisman: Chromotagraphic Behavior of Hemoglobin E. Nature, (Lond.) **177**: 840, 1956.

680. Puck, T. T.: *In vitro* Studies on the Radiation Biology of Mammalian Cells, Progress in Bio-physics and Biophysical Chemistry. Pergamon Press, London, **10**: 237, 1950.

681. Puck, T. T.: The Action of Radiation on Mammalian Cells. Amer. Nat., **94**: 95, 1960.

682. Puck, T. T.: Action of Radiation on Mammalian Cells. III. Relationship between Repro-ductive Death and Induction of Chromosome Anomalies by X-Irradiation of Euploid Human Cells *in vitro*. Proc. Nat. Acad. Sci. (Wash.), **44**: 772, 1958.

683. Puck, T. T.: Lethal Effects of Radiation on Cell Reproduction. Radiat. Res., **11**: 462, 1959.

684. Puck, T. T.: Quantitative Studies on Mammalian Cells *in vitro*. Rev. Modern Physics, **31**: 433, 1959.

685. Puck, T. T.: Growth and Genetics of Somatic Mammalian Cells *in vitro*. J. Cell. Comp. Physiol., **52** Suppl., 1: 287, 1958.

686. Puck, T. T., S. J. Cieciura, and A. Robinson: Genetics of Somatic Mammalian Cells. III. Long-Term Cultivation of Euploid Cells from Human and Animal Subjects. J. Exp. Med., **108**: 945, 1958.

687. Puck, T. T., S. J. Cieciura, and H. W. Fisher: Clonal Growth *in vitro* of Human Cells with Fibroblastic Morphology. Comparison of Growth and Genetic Characteristics of Single Epithelioid and Fibroblast-like Cells from a Variety of Human Organs. J. Exp. Med., **106**: 145, 1957.

688. Puck, T. T., P. I. Marcus, and S. J. Cieciura: Clonal Growth of Mammalian Cells *in vitro;* Growth Characteristics of Colonies from Single HeLa Cells with and without a Feeder Layer. J. Exp. Med., **103**: 273, 1956.

689. Puck, T. T., and P. I. Marcus: Action of X Rays on Mammalian Cells. J. Exp. Med., **103**: 653, 1956.

690. Puck, T. T., and P. I. Marcus: A Rapid Method for Viable Cell Titration and Clone Pro-duction with HeLa Cells in Tissue Culture: The Use of X-irradiated Cells to Supply Conditioning Factors. Proc. Nat. Acad. Sci. (Wash.) **41**: 432, 1955.

691. Puck, T. T., D. Morkovin, P. I. Marcus, and S. J. Cieciura: Action of X Rays on Mam-malian Cells. II. Survival Curves of Cells from Normal Human Tissues. J. Exp. Med., **106**: 485, 1957.

692. Puck, T. T., A. Robinson, and J. DeMetry: unpublished.

693. Puck, T. T., A. Robinson, and J. H. Tjio: Familial Primary Amenorrhea due to Testicular Feminization: A Human Gene Affecting Sex Differentiation. Proc. Soc. Exp. Biol. Med., (N.Y.), **103**: 192, 1960.

694. Querido, A., J. B. Stanbury, A. A. H. Kassenaar, and J. W. A. Meijer: The Metabolism of Iodotyrosines. III. Di-iodotyrosine Deshalogenating Activity of Human Thyroid Tissue. J. Clin. Endocr., **16**: 1096, 1956.

695. Rabson, A. S.: Personal communication.

696. Race, R. R.: Incomplete Antibody in Human Serum. Nature (Lond.) **153**: 771, 1944.

697. Race, R. R.: Genes Modifying the Expression of the *A1 A2 BO* Genes. Vox Sang. (Basel), **2**: 2, 1957.

698. Race, R. R.: On Inheritance and Linkage Relations of Acholuric Jaundice. Ann. Eugen., **11**: 365, 1942.

699. Race, R. R., and R. Sanger: Blood Groups in Man, First ed. C. C. Thomas, Springfield, Ill., p. 290, 1950.

700. Race, R. R., and R. Sanger: Blood Groups in Man, Third ed. C. C. Thomas, Springfield, Ill., p. 377, 1958.

701. Race, R. R., and R. Sanger: The Inheritance of Blood Groups. Brit. Med. Bull., **15**: 99, 1959.

702. Ranney, H. M., D. L. Larson, and G. H. McCormack: Some Clinical, Biochemical, and Genetic Observations on Hemoglobin *C*. J. Clin. Investigation, **32**: 1277, 1953.

703. Rantz, L. A., W. P. Creger, and S. H. Choy: Immunologic Hyperreactivity in the Pathogenesis of Rheumatic Fever and Other Diseases. Amer. J. Med., **12**: 115, 1952.

704. Rantz, L. A., M. Maroney, and J. M. DiCaprio: Infection and Reinfection by Hemolytic Streptococci in Early Childhood. Rheumatic Fever, A Symposium. L. Thomas, ed., Univ. of Minn. Press, Minneapolis, 90, 1952.

705. Rao, C. R.: Advanced Statistical Methods in Biometric Research. Wiley and Sons, N.Y., p. 390, 1952.

706. Raper, A. B.: Sickling in Relation to Morbidity from Malaria and Other Diseases. Brit. Med. J., **1**: 965, 1956.

707. Rapoport, M.: Cystinuria. Textbook of Pediatrics, Mitchell and Nelson, Fifth ed., W. Nelson, ed., W. B. Saunders Co., Philadelphia, 288, 1950.

708. Read, F. E. M., A. Ciocco, and H. B. Taussig: The Frequency of Rheumatic Manifestations among Siblings, Parents, Uncles, Aunts, and Grandparents of Rheumatic and Control Patients. Amer. J. Hyg., **27**: 719, 1938.

709. Reed, S. C.: personal communication.

710. Reed, T. E.: The Definition of Relative Fitness of Individuals with Specific Genetic Traits. Amer. J. Hum. Genet., **11**: 137, 1959.

711. Reed, T. E., and J. V. Neel: Huntington's Chorea in Michigan. II. Selection and Mutation. Amer. J. Hum. Genet., **11**: 107, 1959.

712. Registry for Handicapped Children. Annual Report (Special Report No. 37), Health Branch, Department of Health Services and Hospital Insurance, Vancouver, B.C., 1958.

713. Reichert, E. T., and A. P. Brown: The Crystallography of Hemoglobin. Carnegie Inst. of Wash., D.C., 116, 1909.

714. Renkonen, K. O.: Studies on Hemagglutinins Present in Seeds of Some Representatives of Family of Leguminoseae. Ann. Med. Exp. Biol. Fenn., **26**: 66, 1948.

715. Reyersbach, G. C., and A. M. Butler: Congenital Hereditary Hematuria. New Engl. J. Med., **251**: 377, 1954.

716. Rhinesmith, H. S., W. A. Schroeder, and L. Pauling: A Quantitative Study of the Hydrolysis of Human Dinitrophenyl (DNP) Globin: The Number and Kind of Polypeptide Chains in Normal Adult Human Hemoglobin. J. Amer. Chem. Soc., **79**: 4682, 1957.

717. Rhinesmith, H. S., W. A. Schroeder, and L. Pauling: The N-terminal Amino-acid Residues of Normal Adult Human Hemoglobin: A Quantitative Study of Certain Aspects of Sanger's DNP-method. J. Amer. Chem. Soc., **79**: 609, 1957.

718. Rhinesmith, H. S., W. A. Schroeder, and N. Martin: The N-terminal Sequence of the β-chains of Normal Adult Human Hemoglobin. J. Amer. Chem. Soc., **80**: 3358, 1958.

719. Rife, D. C.: Genetic Studies on Monozygotic Twins; Diagnostic Formula. J. Hered., **24**: 339, 1933.

720. Rinaldini, L. M.: An Improved Method for the Isolation and Quantitative Cultivation of Embryonic Cells. Exp. Cell Res., **16**: 477, 1959.

721. Rinaldo, J. A., and J. I. Baltz: Familial Cholelithiasis with Special Reference to Its Relation to Familial Pancreatitis. Amer. J. Med., **23**: 880, 1957.

722. Ritter, A.: Vererbung von Ureter— und Nierenbeckenanomalien und ihre Klinische Bedeutung. Helv. Med. Acta, **2**: 169, 1935.

723. Roberts, E., and S. Frankel: Glutamic Acid Decarboxylase in Brain. J. Biol. Chem., **188**: 789, 1951.

724. Roberts, J. A. F.: Associations between Blood Groups and Disease. Acta Genet. (Basel), Statist. Med., **6**: 549, 1956–1957.

725. Roberts, J. A. F.: An Introduction to Medical Genetics. Oxford University Press, London, p. 263, 1959.

726. Roberts, J. A. F.: The Use of Regressions Involving Variances of Dependent Variates for Calculating Age-Corrected Scores (Abstract). Biometrics, **9**: 267, 1953.

727. Roberts, J. A. F., and M. A. Mellone: On the Adjustment of Terman-Merrill I.Q.'s to Secure Comparability at Different Ages. Brit. J. of Psych., **5**: 65, 1952.

728. Roberts, J. A. F., R. M. Norman, and R. Griffiths: Studies on a Child Population. I. Definition of the Sample, Method of Ascertainment, and Analysis of the Results of Group Intelligence Test. Ann. Eugen., **6**: 319, 1935.

729. Robinow, M.: Statistical Diagnosis of Zygosity in Multiple Human Births. Hum. Biol., **15**: 221, 1943.

730. Rood, J. J., A. van Leeuwen, and J. G. Eernisse: Leucocyte Antibodies in Sera of Pregnant Women. Vox Sang., **4:** 427, 1959.

731. Ross, L. J.: Congenital Cardiovascular Anomalies in Twins. Circulation, **20:** 327, 1959.

732. Rossi-Fanelli, A., D. Cavallini, and C. de Marco: Amino-Acid Composition of Human Crystallized Myoglobin and Hemoglobin. Biochim. Biophys. Acta, **17:** 377, 1955.

733. Rossi-Fanelli, A., C. de Marco, A. S. Benerecetti, and L. Guacci: Amino-Acid Composition of Hemoglobin A_2. Biochim. Biophys. Acta, **38:** 380, 1960.

734. Rothfels, K. H., and L. Siminovitch: An Air-drying Technique for Flattening Chromosomes in Mammalian Cells Grown *in vitro*. Stain Technol., **33:** 73, 1958.

735. Rothfels, K. H., and L. Siminovitch: The Chromosome Complement of the Rhesus Monkey (*Macaca mulatta*) Determined in Kidney Cells Cultivated *in vitro*. Chromosoma, **9:** 163, 1958.

736. Rowland, L. P.: Muscular Dystrophies, Polymyositis, and Other Myopathies. J. Chron. Dis., **8:** 510, 1958.

737. Rubin, E. H.: Diseases of the Chest. Saunders Publishing Co., Philadelphia, p. 431, 1947.

738. Rucknagel, D. L., and J. V. Neel: The Hemoglobinopathies. Advances in Medical Genetics. A. Steinberg and D. Adlersberg, eds. Grune and Stratton, New York, 1961.

739. Russell, W. L., L. B. Russell, and E. M. Kelly: Radiation Dose Rate and Mutation Frequency. Science, **128:** 1546, 1958.

740. Russell, W. L., and L. B. Russell: Radiation-inducted Genetic Damage in Mice. Second U. N. Int. Conf. on Peaceful Uses of Atomic Energy. (A/Conf. 15/P/897/U.S.A.), 1958.

741. Russell, W. L., and L. B. Russell: In Radiation Biology. A. Hollaender, ed., McGraw-Hill Book Co., Inc., New York, Toronto, London, 87, 1954.

742. Russell, W. L., and L. B. Russell: The Genetic and Phenotypic Characteristics of Radiation-induced Mutations in Mice. Radiat. Res., 1 Suppl., 296, 1959.

743. Russell, W. L., L. B. Russell, and M. B. Cupp: Dependence of Mutation Frequency on Radiation Dose Rate in Female Mice. Proc. Nat. Acad. Sci. (Wash.), **45:** 18, 1959.

744. Russell, W. R.: Poliomyelitis, Second ed. Edward Arnold Ltd., 46, 1956.

745. Sandberg, A. A., G. Koepf, L. H. Crosswhite, and T. S. Hauschka: The Chromosome Constitution of Human Marrow in Various Developmental and Blood Disorders. Amer. J. Hum. Genet., **12:** 231, 1960.

746. Sandler, L., and E. Novitski: Meiotic Drive as an Evolutionary Force. Amer. Natur., **91:** 105, 1957.

747. Sands, J. H., P. O. Palmer, R. L. Mayock, and W. P. Creger: Evidence for Serologic Hyperreactivity in Sarcoidosis. Amer. J. Med., **19:** 401, 1955.

748. Sanford, K. K., W. R. Earle, and G. D. Likely: Growth *in vitro* of Single Isolated Tissue Cells, J. Nat. Cancer Inst., **9:** 229, 1948.

749. Sanford, K. K., W. R. Earle, V. J. Evans, H. K. Waltz, and J. E. Shannon: The Measurement of Proliferation in Tissue Cultures by Enumeration of Cell Nuclei. J. Nat. Cancer Inst., **11:** 773, 1951.

750. Sanford, K. K.: Clonal Studies on Normal Cells and on Their Neoplastic Transformation *in vitro*. Cancer Res., **18:** 747, 1957.

751. Sanford, K. K., A. B. Covalesky, L. T. Dupree, and W. R. Earle: Cloning of Mammalian Cells by a Simplified Capillary Technique. Exp. Cell Research, **23:** 261, 1961.

752. Sanford, K. K., G. D. Likely, and W. R. Earle: The Development of Variations in Transplantability and Morphology within a Clone of Mouse Fibroblasts Transformed to Sarcoma-producing Cells *in vitro*. J. Nat. Cancer Inst., **15:** 215, 1954.

753. Sanford, K. K., R. M. Merwin, G. L. Hobbs, M. C. Fioramonti, and W. R. Earle: Studies on the Difference in Sarcoma-producing Capacity of Two Lines of Mouse Cells Derived *in vitro* for One Cell. J. Nat. Cancer Inst., **20:** 121, 1958.

754. Sanford, K. K., R. M. Merwin, G. L. Hobbs, J. M. Young, and W. R. Earle: Clonal Analysis of Variant Cell Lines Transformed to Malignant Cells in Tissue Culture. J. Nat. Cancer Inst., **23:** 1035, 1959.

755. Sanford, K. K., B. B. Westfall, E. H. Y. Chu, A. B. Covalesky, L. T. Dupree, G. L. Hobbs, and W. R. Earle: Alterations *in vitro* within a Clone of Mouse Tumor Cells. (In preparation.)

756. Sanford, K. K., B. B. Westfall, M. C. Fioramonti, W. T. McQuilkin, J. C. Bryant, E. V. Peppers, V. J. Evans, and W. R. Earle: The Effect of Serum Fractions on the Proliferation of Strain L Mouse Cells *in vitro*. J. Nat. Cancer Inst., **16:** 78, 1955.

757. Sanger, F.: The Arrangement of Amino Acids in Proteins. Adv. Protein Chem., **7:** 1, 1952.

758. Sanger, R.: An Association between the *P* and *Jay* Systems of Blood Groups. Nature (Lond.), **176:** 1163, 1955.

759. Sanger, R., and R. R. Race: Subdivisions of *MN* Blood Groups in Man. Nature (Lond.), **160:** 505, 1947.

759a. Sasaki, M.: Observations on the Modification in Size and Shape of Chromosomes Due to Technical Procedure. Chromosoma, **11:** 514, 1961.

760. Sato, G., H. W. Fisher, and T. T. Puck: Molecular Growth Requirements of Single Mammalian Cells. Science, **126:** 961, 1957.

761. Schapira, G., and J. C. Dreyfus: Groups N. Terminaux de l'Hémoglobine de la Maladie de Cooley. Académie des Sciences, Paris. Société de Biologie Comptes Rendus, **148:** 895, 1954.

762. Scheinberg, F. H., R. S. Harris, and J. L. Spitzer: Differential Titration by Means of Paper Electrophoresis and the Structure of the Human Hemoglobins. Proc. Nat. Acad. Sci. (Wash.), **40:** 777, 1954.

763. Scheinberg, S. L., and R. P. Reckel: Induced Somatic Mutations Affecting Erythrocyte Antigens. Rec. Gen. Soc. Amer., **28:** 93, 1959.

764. Schenck, D. M., and M. Moskowitz: Method for Isolating Single Cells and Preparation of Clones from Human Bone Marrow Cultures. Proc. Soc. Exp. Biol. and Med., **99:** 30, 1958.

765. Schenk, S. G.: Congenital Cystic Disease of Lungs; Clinical Copathological Study. Amer. J. Roentgenol., **35:** 604, 1936.

766. Scherer, W. F., ed.: Introduction to Cell and Tissue Culture. Burgess, Minneapolis, 1955.

767. Schiff, F.: Die Diagnose des Serologischen Ausscheidungstypus in der Blutgruppe O Mittels Heterogenetischen Immuserums. Z. Immunitätsf., **82:** 302, 1934.

768. Schiff, F., and L. Adelsberger: Über Blutgruppenspezifische Antikörper und Antigene. I. Z. Immunitätsf., **40:** 355, 1924.

769. Schiff, F., and W. C. Boyd: Blood Grouping Technic. Interscience Publishers, New York, 1942.

770. Schiff, F., and H. Sasaki: Über die Vererbung des Seralogischen Ausscheidungstypus. Z. Immunitätsf., **77:** 129, 1932.

771. Schiff, F., and H. Sasaki: Den Ausscheidungstypus ein auf Seralogischem Wege Nachweisbares Mendelndes Merkmal. Klin. Wschr., **11:** 1426, 1932.

772. Schmidt, G.: Die Hämagglutination, im Besonderen Menschlicher B-Blutzellen durch Extrakte aus Samen von Euonymus Vulgaris (Pfaffenhütchen). Z. Immunitätsf., **111:** 432, 1954.

773. Schneider, H. A.: Nutritional and Genetic Factors in Natural Resistance of Mice to Salmonella Infections. Ann. N.Y. Acad. Sci., **66:** 337, 1956.

774. Schneider, R. G., and M. E. Haggard: A New Hemoglobin Variant Exhibiting Anomalous Electrophoretic Behaviour. Nature (Lond.), **180:** 1486, 1957.

775. Schneider, R. G., and M. E. Haggard: Hemoglobin *P* (The "Galveston" Type). Nature (Lond.), **182:** 322, 1958.

776. Schnek, A. G., R. T. Jones, and W. A. Schroeder: Unpublished work.

777. Schnek, A. G., and W. A. Schroeder: The Relation between the Minor Components of Whole Normal Human Adult Hemoglobin as Isolated by Chromatography and Starch Block Electrophoresis. J. Amer. Chem. Soc., **83:** 1472, 1961.

778. Schramm, V. G., J. W. Schneider, and A. Anderer: Zur Bestimmung der Amino-endgruppen Verschiedener Hämoglobine und des Tabakmosaikvirus mit Phenylisothiocyanat. Z. Naturforsch., **11:** 12, 1956.

779. Schroeder, W. A.: The Chemical Structure of the Normal Human Hemoglobins. Fortschritte der Chemie Organischer Naturstoffe, **17:** 322, 1959.

780. Schroeder, W. A.: Separation of Peptides, in Ion Exchangers in Organic and Biochemistry. C. Calmon and T. R. E. Kressman, eds. Interscience Publishers, Inc., New York, 761, 1957.

781. Schroeder, W. A., L. M. Kay, and I. C. Wells: Amino-acid Composition of Hemoglobins of Normal Negroes and Sickle-cell Anemics. J. Biol. Chem., **187:** 221, 1950.

782. Schroeder, W. A., and G. Matsuda: N-Terminal Residues of Human Fetal Hemoglobin. J. Amer. Chem. Soc., **80:** 1521, 1958.

783. Schull, W. J.: Empirical Risks in Consanguineous Marriages: Sex Ratio, Malformation, and Viability. Amer. J. Hum. Genet., **10:** 294, 1958.

784. Schull, W. J.: The Problem of Inadequate Counseling in Heredity Counseling. H. G. Hammons, ed., Paul B. Hoeber, Inc., New York, 102, 1959.

785. Schull, W. J., and J. V. Neel: Radiation and the Sex Ratio in Man. Science, **128:** 343, 1958.

786. Schultz, J., and P. St. Lawrence: A Cytological Basis for a Map of the Nucleolar Chromosome in Man. J. Hered., **40:** 31, 1949.
787. Schwartz, H. A., T. H. Spaet, W. W. Zuelzer, J. V. Neel, A. R. Robinson, and S. F. Kaufman: Combinations of Hemoglobin *G*, Hemoglobin *S*, and Thalassemia Occuring in One Family. Blood, **12:** 238, 1957.
788. Schwentker, F. F.: The Epidemiology of Rheumatic Fever. Rheumatic Fever, A Symposium. L. Thomas, ed., U. of Minn. Press, Minneapolis, 17, 1952.
789. Scott, D. B. McN., A. M. Pakaskey, and K. K. Sanford: Analysis of Enzymatic Activities of Clones Derived from Variant Cell Lines Transformed to Malignant Cells in Tissue Culture. J. Nat. Cancer Inst., **25:** 1365, 1960.
790. Scott, E. M., and I. V. Griffith: The Enzymatic Defect of Hereditary Methemoglobinemia: Diaphorase. Biochim. Biophys. Acta, **34:** 584, 1959.
791. Scottish Council for Research in Education: Trend of Scottish Intelligence. University of London Press, 151, 1948.
792. Scottish Council for Research in Education: Social Implications of the 1947 Scottish Mental Survey. University of London Press, 356, 1953.
793. Shatin, H. S., and W. C. Boyd: The Use of Adjuvants in the Production of *Rh* Antisera in Animals. J. Immunol., **62:** 237, 1949.
794. Shaw, M. W., H. F. Falls, and J. V. Neel: Congenital Aniridia. Amer. J. Hum. Genet., **12:** 389, 1960.
795. Shaw, R. F., and J. D. Mohler: The Selective Significance of the Sex Ratio. Amer. Nat., **87:** 337, 1953.
796. Sheldon, J. M., R. G. Lovell, and K. P. Mathews: A Manual of Clinical Allergy. W. B. Saunders Publishers, Philadelphia, 56, 1953.
797. Shelton, J. R., and W. A. Schroeder: Further N-Terminal Sequences in Human Hemoglobins *A*, *S*, and *F* by Edman's Phenylthiohydantoin Method. J. Amer. Chem. Soc., **82:** 3342, 1960.
798. Shooter, E.: Private communication.
799. Siemens, H. W.: Über Einen in der Menschlichen Pathologie noch nicht Beobachteten Vererbungsmodus: Dominant-Geschlechts-Gebundene Vererbung. Arch. f. Rass. Ges. Biol., **17:** 47, 1925.
800. Simms, H. S., and M. S. Parshley: The Effect of Proteins and Amino Acids on the Growth of Adult Tissue *in vitro* in Protein and Amino-acid Requirements of Mammals. A. A. Albanese, ed., Academic Press, Inc., New York and London, p. 155, 1959.
801. Simpson, H. R.: The Estimation of Linkage on an Electronic Computer. Ann. Hum. Genet., **22:** 356, 1958.
802. Singer, S. J., and H. A. Itano: On the Asymmetrical Dissociation of Human Hemoglobin. Proc. Nat. Acad. Sci. (Wash.), **45:** 174, 1959.
803. Sjögren, T., and T. Larsson: Microphthalmos and Anophthalmos with or without Coincident Oligophrenia; Clinical and Genetic-Statistical Study. Acta Psychiat. et Neurol., **56** Suppl.: 103, 1949.
804. Slatis, H. M.: Comments on the Inheritance of Deaf Mutism in Northern Ireland. Ann. Hum. Genet., **22:** 153, 1958.
805. Slatis, H. M.: Comments on the Rate of Mutation to Chondrodystrophy in Man. Amer. J. Hum. Genet., **7:** 76, 1955.
806. Slatis, H. M., R. H. Reis, and R. E. Hoene: Consanguineous Marriages in the Chicago Region. Amer. J. Hum. Genet., **10:** 446, 1958.
807. Smith, C. A.: The Detection of Linkage in Human Genetics. J. Roy. Stat. Soc., **15:** 153, 1953.
808. Smith, C. A.: Some Comments on the Statistical Methods Used in Linkage Investigations. Amer. J. Hum. Genet., **11:** 289, 1959.
809. Smith, C. A.: A Note on the Effects of Method Ascertainment on Segregation Ratios. Ann. Hum. Genet., **23:** 311, 1959.
810. Smith, C. A.: Counting Methods in Genetical Statistics. Ann. Hum. Genet., **21:** 254, 1957.
811. Smith, C. E., R. R. Beard, E. G. Whiting, and H. G. Rosenberger: Varieties of Coccidioidal Infections in Relation to the Epidemiology and Control of the Disease. Amer. J. Publ. Hlth., **36:** 1394, 1946.
812. Smith, E. L., and R. D. Greene: The Isoleucine Content of Seed Globulins and β-Lacto-globulin. J. Biol. Chem., **172:** 111, 1948.
813. Smith, E. L., R. D. Greene, and E. Bartner: Amino-acid Composition of Seed Globulins. J. Biol. Chem., **164:** 159, 1946.

814. Smith, E. L., and R. D. Greene: Further Studies on the Amino-acid Composition of Seed Globulins. J. Biol. Chem., **169**: 833, 1947.

815. Smith, E. W., and J. V. Torbert: Study of Two Abnormal Hemoglobins with Evidence for a New Genetic Locus for Hemoglobin Formation. Johns Hopkins Hosp. Bull., **102**: 38, 1958.

816. Smith, I.: Color Reactions on Paper Chromatograms by a Dipping Technique. Nature, **171**: 43, 1953.

817. Smith, S. M., and L. S. Penrose: Monozygotic and Dizygotic Twin Diagnosis. Ann. Hum. Genet., **19**: 273, 1954.

818. Smith, S. M., and A. Sorsby: Retinoblastoma: Some Genetic Aspects. Ann. Hum. Genet. (Lond.), **23**: 50, 1958.

819. Smithies, O., and G. E. Connell: Biochemical Aspects of the Inherited Variations in Human Serum: Haptoglobins and Transferrins. Ciba Foundation Symposium on Human Biochem. Genetics. G. E. W. Wolstenholme and C. M. O'Connor, eds., Churchill, London, 178, 1959.

820. Smyth, C. J., C. W. Cotterman, and R. H. Freyberg: The Genetics of Gout and Hyperuricemia: An Analysis of Nineteen Families. J. Clin. Invest., **27**: 749, 1948.

821. Snell, G. D.: The Induction by X Rays of Hereditary Changes in Mice. Genetics, **20**: 545, 1935.

822. Snyder, L. H., and P. R. David: Clinical Genetics. A. Sorsby, ed., C. V. Mosby Co., St. Louis, p. 580, 1953.

823. Soulie, P., Y. Bouvrain, J. DiMatteo, and C. Rey: Pulmonary Tuberculosis in Congenital Heart Disease. Arch. Mal. Coeur., **46**: 1057, 1953.

824. Spuhler, J. N.: Estimation of Mutation Rates in Man. In: Clinical Orthopedics. J. B. Lippincott Co., Philadelphia, 34, 1956.

825. Spuhler, J. N.: Physical Anthropology and Demography. The Study of Population. Univ. of Chicago Press, 1959.

826. Stecher, R. M., A. H. Hersh, and W. M. Solomon: The Heredity of Gout and Its Relationship to Familial Hyperuricemia. Ann. Intern. Med., **31**: 595, 1949.

827. Stein, W. H., H. G. Kunkel, R. D. Cole, D. H. Spackman, and S. Moore: Observations on the Amino-acid Composition of Human Hemoglobins. Biochim. Biophys. Acta, **24**: 640, 1957.

828. Steinberg, A. G.: Methodology in Human Genetics. J. Med. Educ., **34**: 315, 1959.

829. Steinberg, A. G.: The Genetics of Diabetes. Ann. N.Y. Acad. Sci., **82**: 197, 1959.

830. Steinberg, A. G., B. D. Giles, and R. Stauffer: A Gm-like Factor Present in Negroes and Rare or Absent in Whites: Its Relation to *Gma* and *Gmx*. Amer. J. Hum. Genet., **12**: 44, 1960.

831. Steinberg, A. G., and N. E. Morton: Sequential Test for Linkage between Cystic Fibrosis of the Pancreas and *MNS* Locus. Amer. J. Hum. Genet., **8**: 177, 1956.

832. Steinberg, A. G., and R. M. Wilder: A Study of the Genetics of Diabetes Mellitus. Amer. J. Hum. Genet., **4**: 113, 1952.

833. Stern, C.: The Chromosomes of Man. Amer. J. Hum. Genet., **11**: 301, 1959.

834. Stern, C.: Color-blindness in Klinefelter's Syndrome. Nature (Lond.), **183**: 1452, 1959.

835. Stern, C.: Use of the Term "Superfemale." Lancet, **2**: 1088, 1959.

836. Stern, C., G. Carson, M. Kinst, E. Novitski, and D. Uphoff: The Viability of Heterozygotes for Lethals. Genetics, **37**: 413, 1952.

837. Stern, C., and E. Novitski: The Viability of Individuals Heterozygous for Recessive Lethals. Science, **108**: 538, 1948.

838. Stern, C., and G. L. Wallis: The Cunier Pedigree of "Color Blindness." Amer. J. Hum. Genet., **9**: 249, 1957.

839. Stern, K.: The Ratio of Monozygotic to Dizygotic Affected Twins and the Frequencies of Affected Twins in Unselected Data. Acta Genet. Med. et Gemell., **7**: 313, 1958.

840. Stevenson, A. C.: Comparisons of Mutation Rates at Single Loci in Man, Effect of Radiation on Human Heredity. World Health Organization, Geneva, 125, 1956.

841. Stevenson, A. C.: Achondroplasia: An Account of the Condition in Northern Ireland. Amer. J. Hum. Genet., **9**: 81, 1957.

842. Stevenson, A. C., and E. A. Cheeseman: Hereditary Deaf Mutism, with Particular Reference to Northern Ireland. Ann. Hum. Genet., **20**: 177, 1956.

843. Stewart, J. S. S.: The Chromosomes in Man. Lancet, **1**: 833, 1959 (letter to the editor).

844. Stokes, J. H., and A. D. King: Acne Vulgaria: Heredity in the Etiologic Background. A. M. A. Arch. of Derm. and Syph., **26**: 456, 1932.

845. Stratton, F., and P. H. Renton: Practical Blood Grouping. Thomas Publ., Springfield, Ill., 1958.
846. Stretton, A. O. W., and V. M. Ingram: An Amino-acid Difference between Human Hemoglobins A and A_2. Fed. Proc., **19**: 343, 1960.
847. Strobel, V. D., and F. Vogel: Ein Statistischer Gesichtspunkt für das Planen von Untersuchungen über Änderungen der Mutationsrate beim Menschen. Acta Genet. (Basel), **8**: 274, 1958.
848. Sturgeon, P., H. A. Itano, and W. R. Bergren: Clinical Manifestations of Inherited Agnormal Hemoglobins. I. The Interaction of Hemoglobin-S with Hemoglobin-D. Blood, **10**: 389, 1955.
849. Sturtz, G. S., and E. C. Burke: Syndrome of Hereditary Hematuria, Nephropathy, and Deafness. Proc. of Mayo Clin., **33**: 289, 1958.
850. Sutphin, A., F. Albright, and D. J. McCune: Five Cases (Three in Siblings) of Idiopathic Hypoparathyroidism Associated with Moniliasis. J. Clin. Endocr., **3**: 625, 1943.
851. Sutter, J., and L. Tabah: Effets de la Consanguinité et de l'Endogamie. Population, **7**: 249, 1952.
852. Sutter, J., and L. Tabah: Fréquence et Nature des Anomalies dans les Familles Consanguines. Population, **9**: 425, 1954.
853. Sutter, J., and L. Tabah: Structure de la Mortalité dans les Familles Consanguines, Population, **8**: 511, 1953.
854. Sutton, H. E.: β-Aminoisobutyricaciduria. In the Metabolic Bases of Inherited Diseases. J. B. Stanbury, J. B. Wyngaarden, and D. S. Fredrickson, eds., McGraw-Hill, New York, 792, 1960.
855. Sutton, H. E.: Unpublished results.
856. Sutton, H. E., and J. H. Read: Abnormal Amino-acid Metabolism in a Case Suggesting Autism. A. M. A. J. Dis. Child., **96**: 23, 1958.
857. Sutton, H. E., and R. E. Tashian: Inherited Variations in Aromatic Metabolism. Symposium on Hereditary Metabolic Diseases. Metabolism, **9**: 284, 1960.
858. Swenson, R. T., and R. L. Hill: Unpublished experiments.
859. Swim, H. E.: Microbiological Aspects of Tissue Culture. Ann. Rev. Microbiol., **13**: 141, 1959.
860. Sydenstricker, E.: Reporting of Notifiable Diseases in Typical Small City. Hagerstown Morbidity Studies. Publ. Hlth. Rep. (Wash.), **41**: 2186, 1926.
861. Sydenstricker, E., and A. W. Hedrich: Completeness of Reporting of Measles, Whooping Cough, and Chicken Pox at Different Ages. Hagerstown Morbidity Studies. Publ. Hlth. Rep., **44**: 1537, 1929.
862. Symposium on Genetic Approaches to Somatic Cell Variation. J. Cell Comp. Physiol., **52** Suppl., 1: 410, 1958.
863. Szybalski, W.: Genetics of Human Cell Lines. II. Method for Determination of Mutation Rates to Drug Resistance. Exp. Cell Res., **18**: 588, 1959.
864. Takahara, S.: Progressive Oral Gangrene Probably Due to Lack of Catalase in the Blood (Acatalasemia). Lancet, **2**: 1101, 1952.
865. Takahara, S., H. B. Hamilton, J. V. Neel, T. Y. Kobara, Y. Ogura, and E. T. Nishimura: Hypocatalasemia: A New Genetic Carrier State. J. Clin. Invest., **39**: 610, 1960.
866. Tanaka, K., and K. Ohkura: Evidence of Genetic Effects of Radiation in Offspring of Radiological Technicians. Jap. J. Hum. Genet., **3**: 135, 1958.
867. Tashian, R. E.: Unpublished results.
868. Taylor, E. S.: Chronic Ulcer of the Leg Associated with Congenital Hemolytic Jaundice. J. Amer. Med. Assn., **112**: 1574, 1939.
869. Taylor, I., and J. Knowelden: Principles of Epidemiology. Little, Brown & Co., Boston, 191, 1957.
870. Taylor, J. H.: Asynchronous Duplication of Chromosomes in Cultured Cells of Chinese Hamster. J. Biophys. and Biochem. Cytol., **7**: 455, 1960.
871. Thannhauser, S. J.: Lipoidoses, Diseases of the Cellular Lipid Metabolism. Oxford Univ. Press, New York, 456, 1950.
872. Thoday, J. M.: Components of Fitness. Symp. Soc. Exp. Biol. Med., **7**: 96, 1953.
873. Thomsen, O., V. Friedenreich, and E. Worsaae: Die Wahrscheinliche Existenz eines neuen, mit den drei Bekannten Blutgruppengenen (O, A, B) Allelomorphen, A'benannten Gens mit den daraus folgenden Zwei neuen Blutgruppen A' und $A'B$. Klin. Woch., **9**: 67, 1930.
874. Thomsen, O., V. Friedenreich, and E. Worsaae: Über die Möglichkeit den Existenz Zweien Neuer Blutgruppen; auch ein Beitrag zun Beleuchtung Sogenannter Untergruppen. Acta Path. Microbiol. Scand., **7**: 157, 1930.

875. Tiselius, A.: Electrophoresis of Serum Globulin. II. Electrophoretic Analysis of Normal and Immune Sera. Biochem. J., **31:** 1464, 1937.
876. Tiselius, A., and F. L. Horsfall, Jr.: Mixed Molecules of Hemocyanins from Two Different Species. J. Exp. Med., **69:** 83, 1939.
877. Tjio, J. H., and A. Levan: Chromosome Analysis of Three Hyperdiploid Ascites Tumors of the Mouse. K. Fysiogr. Sällsk. Lund. Förh., **65:** 1, 1954.
878. Tjio, J. H., and A. Levan: The Chromosome Number of Man. Hereditas (Lund.), **42:** 1, 1956.
879. Tjio, J. H., and G. Östergren: The Chromosomes of Primary Mammary Carcinomas in Milk Virus Strains of the Mouse. Hereditas (Lund.), **44:** 451, 1958.
880. Tjio, J. H., and T. T. Puck: Genetics of Somatic Mammalian Cells. II. Chromosomal Constitution of Cells in Tissue Culture. J. Exp. Med., **108:** 259, 1958.
881. Tjio, J. H., and T. T. Puck: The Somatic Chromosome of Man. Proc. Nat. Acad. Sci. (Wash.), **44:** 1229, 1958.
882. Tjio, J. H., T. T. Puck, and A. Robinson: The Human Chromosomal Satellites in Normal Persons and in Two Patients with Marfan's Syndrome. Proc. Nat. Acad. Sci., **46:** 532, 1960.
883. Tjio, J. H., T. T. Puck, and A. Robinson: The Somatic Chromosomal Constitution of Some Human Subjects with Genetic Defects. Proc. Nat. Acad. Sci. (Wash.), **45:** 1008, 1959.
884. Tolmach, L. J. and P. I. Marcus: Development of X-ray induced Giant HeLa Cells. Exp. Cell. Res., **20:** 350, 1960.
885. Trager, W.: The Nutrition of an Intracellular Parasite; Avian Malaria. Acta Trop (Basel), **14:** 289, 1957.
886. Trowell, O. A.: The Culture of Mature Organs in a Synthetic Medium. Exp. Cell Res., **16:** 118, 1959.
887. Tucker, D. P., A. G. Steinberg, and D. G. Cogan: Frequency of Genetic Transmission of Sporadic Retinoblastoma. A. M. A. Arch. Ophthal., **57:** 532, 1957.
888. Turpin, R., J. Lejeune, and M. O. Rethore: Étude de la Descendance de Sujets Traités par Radiothérapie Pelvienne. Acta Genet. (Basel), **6:** 204, 1956.
889. Turpin, R., J. Lejeune, J. LaFourcade, and M. Gautier: Aberrations Chromosomiques et Maladies Humaines; La Polydysspondylie à 45 Chromosomes. C. R. Acad. Sci. (Paris), **248:** 3636, 1959.
890. United Nations Report of the United Nations Scientific Committee on the Effects of Atomic Radiation. Official Records: Thirteenth Session, 17 Suppl. (A/3838) New York, 172, 1958.
891. United States Bureau of the Census, Statistical Abstract of the United States (seventy-fifth edition), Wash., D.C., p. 1040, 1950.
892. United States National Health Survey. Origin and Program of the U.S. National Health Survey. Wash., U.S. Department of Health, Education, and Welfare, Div. of Publ. Hlth. Met., 1958.
893. Verschuer, O. Von, and H. C. Ebbing: The Mutation Rate of Man, Studies on Its Determination. Z. Menschl. Vererb. Konst., **35:** 93, 1959.
894. Vickery, H. B., and A. White: Proportion of Cystine Yielded by Hemoglobins of the Horse, Dog, and Sheep. Proc. Soc. Exp. Biol. Med. (N.Y.), **31:** 6, 1933.
895. Videbaek, A.: Heredity in Human Leukemia. Nyt Noridsk Forlage, Copenhagen, p. 279, 1947.
896. Vinograd, J. R., W. D. Hutchinson, and W. A. Schroeder: C^{14}-Hybrids of Human Hemoglobins. II. The identification of the Aberrant Chain in Human Hemoglobin. S. J. Amer. Chem. Soc., **81:** 3168, 1959.
897. Vogel, F.: Neue Untersuchungen zur Genetik des Retinoblastoms (Glioma Retinae). Zschr. Menschl. Vererb.-u. Konstitutionslehre, **34:** 205, 1957.
898. Vogel, F.: Über Genetik und Mutationsrate des Retinoblastoms (Glioma Retinae). Zschr. Menschl. Vererb.-u. Konstitutionslehre, **32:** 308, 1954.
899. Vogel, F.: Über die Prüfung von Modellvorstellungen zur Spontanen Mutabilität an Menschlichem Material. Zschr. Menschl. Vererb.- u. Konstitutionslehre, **33:** 470, 1956.
900. Von Dungern, E., and L. Hirschfeld: Z. Immunitätsforschung und Experimente Therapie. Ztschr. f. Immunitätsforsch, **6:** 284, 1910.
901. Von Tavel, P., and R. Signer: Countercurrent Distribution in Protein Chemistry. Advanc. Protein Chem., **11:** 237, 1956.
902. Waardenburg, P. J.: Das Menschliche Ange und seine Erbanlagen. M. Nijhoff, Hagg, 1932.
903. Waardenburg, P. J.: A New Syndrome Combining Developmental Anomalies of the Eyelids

Eyebrows, and Nose Root with Pigmentary Defects of the Iris and Head Hair and with Congenital Deafness. Amer. J. Hum. Genet., **3:** 195, 1951.

904. Wald, A.: Sequential Analysis. John Wiley & Sons, New York, p. 212, 1947.

905. Wald, A., and J. Wolfowitz: Optimum Character of the Sequential Probability Ratio Test. Ann. Math. Stat., **19:** 326, 1948.

906. Walker, G. C., and L. B. Ellis: Congenital Heart Disease in Families. Proc. New England Heart Assn., 26, 1940.

907. Wallace, B.: The Role of Heterozygosity in *Drosophila* Populations. Proc. X Int. Congr. Genetics, **1:** 408, 1959.

908. Wallace, H. W., K. Moldave, and A. Meister: Studies on Conversion of Phenylalanine to Tyrosine in Phenylpyruvic Oligophrenia. Proc. Soc. Exp. Biol. (N.Y.), **94:** 632, 1957.

909. Walsh, R. J., and C. M. Montgomery: New Human Iso-agglutinin Subdividing *MN* Blood Groups. Nature (Lond.), **169:** 504, 1947.

910. Waterhouse, J. A. H., and L. Hogben: Incompatibility of Mother and Fetus with Respect to the Iso-agglutinogen A and Its Antibody. Brit. J. Soc. Med., **1:** 1, 1947.

911. Watkins, W. M., and W. T. Morgan: Inhibitions by Simple Sugars of Enzymes which Decompose the Blood-group Substances. Nature (Lond.), **175:** 676, 1955.

912. Watkins, W. M., and W. T. Morgan: Specific Inhibition Studies Relating to the Lewis Blood-group System. Nature (Lond.), **180:** 1038, 1957.

913. Watkins, W. M., and W. T. Morgan: Possible Genetical Pathways for the Biosynthesis of Blood-group Mucopolysaccharides. Vox Sang., **4:** 97, 1959.

914. Watkins, W. M., and W. T. Morgan: Vox Sang., **5:** 1, 1955.

915. Waymouth, C.: Rapid Proliferation of Sublines of NCTC Clone 929 (Strain L) Mouse Cells in a Simple Chemically-defined Medium (MB 752/1). J. Nat. Cancer Inst., **22:** 1003, 1959.

916. Weinberg, W.: Beitrage zur Physiologie und Pathologie der Mehrlingsgeburten Beim Menschen. Arch. Ges. Physiol., **88:** 346, 1901.

917. Weliky, N., R. T. Jones, and L. Pauling: unpublished work.

918. Welshone, W. J., and L. B. Russell: The Y-Chromosome as the Bearer of Male Determining Factors in the Mouse. Proc. Nat. Acad. Sci. (Wash.), **45:** 560, 1959.

919. Wermer, P.: Genetic Aspects of Adenomatosis of Endocrine Glands. Amer. J. Med., **16:** 363, 1954.

920. Westfall, B. B., E. V. Peppers, V. J. Evans, K. K. Sanford, N. M. Hawkins, M. C. Fioramonti, H. A. Kerr, G. L. Hobbs, and W. R. Earle: The Arginase and Rhodanese Activities of Certain Cell Strains after Long Cultivation *in vitro*. J. Biophys. and Biochem. Cytol., **4:** 567, 1958.

921. Westphal, O.: Die Struktur der Antigene und das Wesen der Immunologischen Spezifität. Naturwissenschaften, **46:** 50, 1959.

922. White, P.: In Treatment of Diabetes Mellitus, Tenth edition. Joslin, Root, White, and Marble, eds., Lea and Febiger, Philadelphia, 48, 1959.

923. White, P. R.: The Cultivation of Animal and Plant Cells. The Ronald Press Co., New York, p. 239, 1954.

924. Wiener, A. S.: Origin of Naturally Occurring Hemagglutinins and Hemolysins. J. Immunol., **66:** 287, 1951.

925. Wiener, A. S.: *Rh-Hr* Blood Types, Applications in Clinical and Legal Medicine and Anthropology. Grune and Stratton, New York, p. 763, 1954.

926. Wiener, A. S., and H. R. Peters: Hemolytic Reactions Following Transfusions of Blood of the Homologous Group with Three Cases in which the Same Agglutinogen Was Responsible. Ann. Intern. Med., **13:** 2306, 1940.

927. Wildy, P., and M. Stoker: Multiplication of Solitary HeLa Cells. Nature (Lond.), **181:** 1407, 1958.

928. Wilkins, L.: The Diagnosis and Treatment of Endocrine Disorders in Childhood and Adolescence. C. C. Thomas, Springfield, Ill., p. 384, 1950.

929. Williams, R. J., and others: Biochemical Institute Studies. IV. Individual Metabolic Patterns and Human Disease: An Exploratory Study Utilizing Predominantly Paper Chromatographic Methods. Univ. of Texas Publ. No. 5109, 1951.

930. Willmer, E. W.: Tissue Culture (The Growth and Differentiation of Normal Tissues in Artificial Media). Methuen and Co., Ltd., London, 1958.

931. Wilson, M. G.: Rheumatic Fever, New York Commonwealth Found., Oxford Univ. Press, London, p. 595, 1940.

932. Wilson, G. S., and A. A. Miles: Topley and Wilson's Principles of Bacteriology and Im-

munology. G. S. Wilson and A. A. Miles, eds., The Williams and Wilkins Company, Baltimore, Maryland, 1453, 1955.

933. Wilson, S., and D. B. Smith: Separation of the Valyl-leucyl- and Valyl-glutamyl-polypeptides Components of Horse Globin by Fractional Precipitation and Column Chromatography. Canad. J. Biochem. Physiol., **37**: 405, 1959.

934. Wintrobe, M. M.: Clinical Hematology. Lea and Febiger, Publishers, Philadelphia, p. 1186, 1961.

935. Woodrow, J. C.: Liverpool Medical Institution. A Symposium on the Role of Inheritance in Common Diseases. Lancet, **2**: 526, 1959.

936. Woods, M. W., K. K. Sanford, D. Burk, and W. R. Earle: Glycolytic Properties of High and Low Sarcoma-producing Lines and Clones of Mouse-tissue-culture Cells. J. Nat. Cancer Inst. **23**: 1079, 1959.

937. Woolf, C. M.: A Genetic Study of Carcinoma of the Large Intestine. Amer. J. Hum. Genet., **10**: 42, 1958.

938. Woolf, C. M.: Investigations on Genetic Aspects of Carcinoma of the Stomach and Breast. Univ. California Pub., Public Health, **2**: 265, 1955.

939. Woolf, C. M.: A Further Study of the Familial Aspects of Carcinoma of the Stomach. Amer. J. Hum. Genet., **8**: 102, 1956.

940. Woolf, C. M., F. E. Stephens, D. D. Mulaik, and R. E. Gilbert: An Investigation of the Frequency of Consanguineous Marriages among the Mormons and Their Relatives in the United States. Amer. J. Hum. Genet., **8**: 236, 1956.

941. Woolsey, T. D.: Sampling Methods for a Small Household Survey, Public Health Monograph No. 40. Publ. Hlth. Rep., **71**: 827, 1956.

942. Wright, S.: The Genetics of Quantitative Variability. Quantitative Inheritance. H. M. S. O., London, 5, 1952.

943. Wright, S.: Physiological Genetics, Ecology of Populations, and Natural Selection. Persp. Biol. Med., **3**: 107, 1959.

944. Wright, S.: Adaptation and Selection. Genetics, Paleontology, and Evolution. Princeton Univ. Press, 369, 1949.

945. Wright S.: On the Genetics of Silvering in the Guinea Pig with Especial Reference to Interaction and Linkage. Genetics, **44**: 387, 1959.

946. Wright, S.: The Analysis of Variance and the Correlations between Relatives with Respect to Deviations from an Optimum. J. Genet., **30**: 243, 1935.

947. Wu-Min: On Methods for Isolation of Clones Derived from Several Tumor Strains *in vitro*. Vopr. Onko., **5**: 5, 1959.

948. Wyngaarden, J. B.: Gout and Hyperuricemia. The Metabolic Bases of Inherited Disease. J. B. Stanbury, J. B. Wyngaarden, and D. S. Fredrickson, eds., McGraw-Hill, New York, p. 1477, 1960.

949. Wyngaarden, J. B., and D. M. Ashton: Feedback Control of Purine Biosynthesis by Purine Ribonucleotides. Nature (Lond.), **183**: 747, 1959.

950. Yates, F.: Sampling Methods for Censuses and Surveys, Second ed. Hafner, Hew York, p. 401, 1953.

951. Yates, R. A., and A. B. Pardee: Control of Pyrimidine Biosynthesis in *Escherichia coli* by a Feed-back Mechanism. J. Biol. Chem., **221**: 757, 1956.

952. Yerganian, G.: Cytologic Maps of Some Isolated Human Pachytene Chromosomes. Amer. J. Hum. Genet., **9**: 42, 1957.

953. Yerganian, G., R. Kato, M. J. Leonard, H. Gagnon, and L. A. Grodzins: Sex Chromosomes in Malignancy, Transplantability of Growths, and Aberrant Sex Determination. Cell Physiology of Neoplasia. The University of Texas, M. D. Anderson Hospital and Tumor Institute, 14th Annual Symposium of Fundamental Cancer Research, Austin, Texas. The University of Texas Press, 1961.

954. Yerganian, G., and L. A. Grodzins: unpublished.

955. Young, L. E.: Hereditary Spherocytosis. Amer. J. Med., **18**: 486, 1955.

956. Zamecnik, P. C., M. L. Stephenson, and L. I. Hecht: Intermediate Reactions in Amino Acid Incorporation. Proc. Nat. Acad. Sci. (Wash.), **44**: 73, 1958.

957. Zinkham, W. H., and R. E. Lenhard, Jr.: Metabolic Abnormalities of Erythrocytes from Patients with Congenital Nonspherocytic Hemolytic Anemia. J. Pediat., **55**: 319, 1959.

958. Zuckerkandl, E., R. T. Jones, and L. Pauling: A Comparison of Animal Hemoglobins by Tryptic Peptide Pattern Analysis. Proc. Nat. Acad. Sci. (Wash.), **46**: 1349, 1960.

959. Zuckerman, H. S., and L. R. Wurtzebach: Kartagener's Triad. Dis. of Chest, **19**: 92, 1951.

AUTHOR INDEX

SUBJECT INDEX

Error
  diagnostic, 24
  in determining mutation rate, 204
  of metabolism, 32
  sources of, in calculating mutation rate by
    the indirect approach, 206
  standard, 38
  type I, 10, 39, 42
  type II, 10, 14, 39
Erythroblastosis fetalis, 338
Erythrocyte receptors, 361
Erythrocytes, studies of biochemical potential
  in vitro, 299
Eskimos, 105
Estimate
  maximum likelihood, 52
  of gene frequency, 77
  of mutation rates, 210
Estimation
  iterative, 31
  of proportion of sporadic cases, 79
Etiologic heterogeneity, 149
Etiology, genetic, 18
Examination of concomitant variables, 17
Example of segregation analyses, 32, 33,
  34
Excess heterozygote, 34
Expressivity, 20
Extramarital conceptions, 23, 24, 34

Falciparum malaria, 198
False diagnosis, 23
Familial
  cases, 26
  neutropenia and impaired cellular response,
    162
  probands, 24, 32
  urinary tract infections, 173
Families selected, 26
Family
  multiplex, 24, 30
  relationships, records identifying, 101
  simplex, 24
Fanconi syndrome, 292
Feedback
  inhibition, 293
  mechanisms, 290
Fertility
  differences, 74
  differential, 64
  equal, 5
  selective, 74
Fetal
  deaths, 34, 63
  hemoglobin, 314, 327
Fetal and newborn tissues, induction of enzymatic systems in, 303
Fetuin, use in tissue culture, 282
File, size of, 100
"Fingerprinting," 300, 315

limitations of method, 334
of hemoglobin, 315
Fisher-Race notation for *Rh* genes, 339
Fitness, 53, 54, 69
  definition, 54
Fixed sample size, theory of, 38
  estimation, 43
Forssman antigens, 336
Fractions, recombination, 40
Fraternal twins, 18
Frequency
  chiasma, 40
  of affected offspring, 14
  of rare recessive genes, 76, 77
  of trait, 57, 80
  segregation, 26, 35, 37
Friedreich's ataxia, 179
Function
  mapping, 40
  monotonic, 40
Fungus infections, 163
Fy blood groups, 86, 87

Galactosemia, 12, 146, 265, 290, 297, 299
  empiric risk, 146
Gallstones, 177
γ-chain of hemoglobin, 314, 321, 330
Gastric carcinoma, 150
  related to blood groups, 142
Gastrointestinal system and susceptibility to
  infection, 177
Gaucher's disease, 175, 265
Gene
  action and enzymatic variation, 287
  autosomal, 20
    recessive, 32
  dominant, 21, 25
  frequency, 6, 8, 33, 76
    equilibria, 60
    rate of change, 59
  nonallelic recessive, 33
  number, estimate of, 77
  recessive, 5, 6, 20, 24, 32, 33
  segregation models, 28, 29
Genealogical Society, 112
Genetic
  abnormalities and susceptibility to infec-
    tious disease, 156
  alterations in drug metabolism, 180
  control of hemoglobin production in man,
    331
  control of virulence, 158
  information existing in records, 113
  etiology, 18
  factors predisposing to infection, 160
  load, 34, 53, 66, 67, 72
  map lengths, 140
  methods, 12, 17, 23
Genetically normal child, probability of an
  unaffected carrier, 25